THE ANDROMEDA GALAXY

ASTROPHYSICS AND SPACE SCIENCE LIBRARY

A SERIES OF BOOKS ON THE RECENT DEVELOPMENTS
OF SPACE SCIENCE AND OF GENERAL GEOPHYSICS AND ASTROPHYSICS
PUBLISHED IN CONNECTION WITH THE JOURNAL
SPACE SCIENCE REVIEWS

VOLUME 176

THE ANDROMEDA GALAXY

by

PAUL HODGE

Department of Astronomy,
University of Washington,
Seattle, U.S.A.

KLUWER ACADEMIC PUBLISHERS

DORDRECHT / BOSTON / LONDON

Library of Congress Cataloging-in-Publication Data

Hodge, Paul W.
The Andromeda Galaxy / by Paul Hodge.
p. cm. -- (Astrophysics and space science library ; v. 176)
Includes bibliographical references and index.

1. Andromeda Galaxy. I. Title. II. Series.
QB858.42.H63 1992
523.1'12--dc20 92-13215

ISBN 978-90-481-4139-5

Published by Kluwer Academic Publishers,
P.O. Box 17, 3300 AA Dordrecht, The Netherlands.

Kluwer Academic Publishers incorporates
the publishing programmes of
D. Reidel, Martinus Nijhoff, Dr W. Junk and MTP Press.

Sold and distributed in the U.S.A. and Canada
by Kluwer Academic Publishers,
101 Philip Drive, Norwell, MA 02061, U.S.A.

In all other countries, sold and distributed
by Kluwer Academic Publishers Group,
P.O. Box 322, 3300 AH Dordrecht, The Netherlands.

Printed on acid-free paper

TABLE OF CONTENTS

PREFACE

This book began about 10 years ago as a review paper. At that time the idea of a single article on M31 was not unthinkable. Soon, however, the situation changed. Perhaps it was just the general growth of astronomy, or perhaps the change was caused more directly by the increased numbers of large, well-equipped telescopes. Whatever the cause, it soon became the case that the amount of new material on the Andromeda Galaxy was growing faster than I was writing. It was gaining on me. Soon it overtook me and I realized that a single review paper would be far too limiting.

Fortunately, the good people at what was then Reidel (now Kluwer), especially N. M. Pols-v.d. Heijden were not averse to the idea of a monograph on M31. These same good people, especially Gerrit Kiers, have shown a remarkable amount of patience over the years in which I have tried to catch up with the advancing tide of material on M31. Although there was no sign that the tide might be changing, I decided at the end of 1991 that I would stop running with it. The result is on the following pages.

I have tried to include information on all major aspects of M31 research and citations to all major published papers through 1991. Experience tells me that I have not succeeded; surely some seminal papers, ones that I no doubt read and admired when they were published, became lost in the clutter of my desk and mind. I apologize to the authors of these papers and to the readers who will have to find them for themselves.

M31 represents a very special topic for study. It has many attractions (its proximity, its spiral arms, its kinematics) and provides certain difficulties (it is nearly edge-on); both have challenged our instruments and our ingenuity and, no doubt, will continue to for many years to come.

Table i, adapted from Hodge (1981) and van den Bergh (1991), summarizes some basic data on the Andromeda Galaxy, taken from the recent literature. Sources for the data and arguments for and against these choices will be found in the various chapters of the book.

TABLE i. Basic Properties of M31

Position (1950)	
Right Ascension	$00^h40^m00^s.3$
Declination	$+41°00'03''$
Angular diameter (optical)	$240'$
Apparent magnitude (V)	3.58
Color (B-V) and (U-B)	+0.91, +0.50
Distance modulus, apparent	24.5
Foreground reddening, E(B-V)	0.08
Distance modulus, true	24.3
Distance, kpc	725
Absolute magnitude, M_V	−21.0
Diameter, kpc	51
Angle of plane to line of sight	12°.5
Position angle of major axis	37°.7
Radial velocity (wrt sun), km/sec	−310
Mass (total, solar masses)	$> 4.1 \times 10^{11}$
Mass (neutral hydrogen, solar masses)	3.9×10^9

ACKNOWLEDGEMENTS

Many people helped with this volume. Scores of astronomers and observatories generously let me use published diagrams and images and I list these on the next page. Many astronomers sent me preprints and reprints. Three specialists in the study of M31 – Sidney van den Bergh, Elias Brinks and René Walterbos – read much of the manuscript in an attempt to uncover its many mistakes and shortcomings.

Local help was both generous and competent. Sandi Larsen typed many of the first versions of the manuscript and Karen Fisher produced the final version of the text, utilizing her mastery of the latest techniques in word processing. Jeffrey Goldader, Loren Kelley, Debbie Yun and Grace Lee spent several weeks chasing down figures and references; their help was vital.

CREDITS FOR ILLUSTRATIONS

This acknowledges the following persons and institutions, from whose publications and file illustrations and other material were obtained for this book:

Armandroff, T.
Arp, H.
Baade, W.
Babcock, H.
Baum, W.
Beck, R.
Beichman, C.
Berkhuijsen, E.
Brandt, J.
Brinks, E.
Buonanno, R.
Bystedt, J.
Cacciari, C.
Casoli, F.
Ciardullo, R.
Code, A.
Cohen, J.
Collura, A.
Combes, F.
Condon, J.
Cowley, A.
Crampton, D.
de Vaucouleurs, G.
Deharveng, J.
Dennefeld, M.
Dressler, A.
Durdin, J.
Effremov, Y.
Elson, R.
Fesen, R.
Ford, H.
Freedman, W.
Gottesman, S.
Hanes, D.
Harris, H.
Haud, U.
Hiromoto, N.
Hoessel, J.
Huchra, J.
Humphreys, R.
Hutchings, J.
Iye, M.
Kennicutt, R.
Kent, S.
Kitt Peak National Observatory
Koper, E.
Kutner, M.
Lada, C.
Lawrie, D.
Lee, M.
Light, E.
Lucke, P.
Lynga, G.
Makishima, K.
Massey, P.
Mcelroy, D.
Moffat, A.
Morton, D.
Nakai, N.
Newton, K.
Nikolov, N.
Nolthenius, R.
Odewahn, S.
Palomar Observatory
Pellet, A.
Pooley, G.
Pritchett, C.
Racine, R.
Reddish, V.
Rich, R.
Richer, H.
Roberts, M.
Rubin, V.
Ruiz, M.
Ryden, B.
Sandage, A.
Scientific American
Sharov, A.
Simien, F.
Sitko, M.
Sofue, Y.
Soifer, B.
Spinrad, H.
Stark, S.
Sullivan, W.
van de Hulst, H.
van den Bergh, S.
van Speybroeck, L.
Vetesnik, M.
Vogel, H.
Walterbos, R.
Welch, D.

CHAPTER 1

EARLY HISTORY

The Andromeda Galaxy is the nearest spiral galaxy. It is one of only three or four galaxies that can be seen without telescopic aid, and for that reason it has been mentioned in texts and marked on maps for over a thousand years. Although it is not easy to see from our modern, well-lit environments, from a dark site it is a conspicuous and anomalous feature in the sky and was surely noticed by curious men and women far back into human prehistory.

1.1 First Records

One of the first recorded instances of the recognition of the unusual nature of the faint smudge of light that is the visual Andromeda galaxy, is found in the *Book on the Constellations of the Fixed Stars* by the famous Islamic

Figure 1.1 Al Sufi's representation of the constellation Andromeda. The dots at the nostrils of the fish represent the Andromeda Nebula (from Kunitsch, 1987).

author Abu I-Husain al-Sufi (A.D. 903–986) (for details see papers by Coddington 1898, and Kunitzsch 1987, and references therein). Based originally on Ptolemy's *Almagest*, al-Sufi's book showed the 48 classical constellations and gave a commentary on each of them. In the chart of the constellation *Andromeda*, his chart shows the traditional figure of a woman, but with a large fish placed across the figure, representing the Arabic tradition for certain stars there. In the manuscript, the fish has a peculiar double nature, with one image superimposed on the other, larger fish. Both fish have a distinctly human-shaped nose, and at the nostrils of the larger fish is a scattering of dots, which are described as a "nebulous spot" (lathka sahabiya). This is the Andromeda Nebula.

Figure 1.2 A representation of the nebula as it appears in a later, Latin version of the Ptolemy star catalog (from Kunitsch, 1987).

In succeeding centuries, Al-Sufi's book of constellations, along with his charts and lists of star names, provided a thread of continuity from the Greek astronomers, through Ptolemy, to the European astronomers of the middle ages. In the European branch of the al-Sufi tradition, the "nebulous spot" at the fish's mouth was retained. For example, Gerard of Cremona, in his Latin version of Ptolemy's star catalogue, included al-Sufi's "nebulous spot". In the figure of Andromeda reproduced from the Latin manuscript by Kunitzsch (1987), the nebula is even more conspicuous than in al-Sufi's original, even though Kunitzsch believes that the authors could not have known what it meant, as the Arabic text was not available to them in Latin translation.

1.2 Early Telescopic Observations

The first telescopic view of the Andromeda galaxy was reported by Simon Marius in 1612, who observed it on 15 December of that year and reported that it had the appearance of a "candle shining through horn" (Coddington 1898). In those days, pieces of translucent horn were used to diffuse light in lanterns, and Simon Marius' description in these terms is quite graphic.

For several centuries very little was said about the Andromeda galaxy. Nebulous stars were considered mere curiosities and not something to be studied systematically. Galileo, having found the nebulous Milky Way to be made up of unresolved stars, apparently concluded that all nebulous objects among the fixed stars must similarly consist of unresolved stars, and most astronomers of the time seem to have agreed.

By the eighteenth century, astronomy was beginning to take on a more modern attitude and the first systematic catalogues of non-stellar objects were prepared. The best known of these is the Messier catalogue (Messier 1784), which listed over 100 nebulous objects, the 31st of which is the Andromeda galaxy. Astronomers still refer to it most often as M31. Messier was not so much interested in these objects themselves, but catalogued them in the process of his comet hunting so as not to mistake one for a comet. He did describe it, however, as resembling "two luminous cones or pyramids opposite at their bases."

Messier's description was at least clear. Some years before, Edmund Halley gave a description of it as being triangular in shape, with its apex on its southwest side. He went on (Halley 1715) to say, "the spot is nothing else but the light coming from an extraordinary great space in the ether, through which a lucid medium is diffused that shines with its own proper lustre." Although one must allow for the level of physical science at the time, this nevertheless seems particularly uninformative.

A really serious study of the various nebulae was not made until Sir William Herschel began his meticulous searches of the sky with his reflecting telescopes, which were the most powerful of his time. Herschel found hundreds of nebulae scattered about the sky and wrote about their characteristics. At first, he followed Galileo's idea that all consisted of unresolved stars, but he later considered it likely that some at least were made up of gaseous clouds. He said of the Andromeda nebula, "the stars which are scattered over it appear to be behind it, and seem to lose part of their lustre in the passage of light through the nebulosity" (Whitney 1971). It is somewhat ironic that he should be right about many nebulae, but about this one, his rejection of Galileo's hypothesis should be wrong.

Among the deductions that Herschel made about the nebulae that he observed is a curious one: he thought that the evidence indicated that nebulae evolve rapidly, changing drastically in just a few years. This idea was based on comparison of drawings made at different times, and sometimes by different people. Without a permanent photographic record, the astronomers of the time were forced to rely on sketches, and could not know how much the appearance of these faint nebulous structures depended on the telescope used, the condition of the sky, and the eye of the observer. This problem persisted into the nineteenth century, when photography showed that the changes were almost entirely illusory (a few cases of rapid change, as in Hubble's nebula and certain HH objects, of course, are real, but these objects are much closer to us than the Andromeda galaxy).

In succeeding years there continued to be speculation about changes in the Andromeda nebula, especially changes detected in its nearly stellar nucleus. Such reports continued right up to the end of the nineteenth century. See, for example, the disclaimer by Dreyer (1886), the defense of changes in the nucleus made by Hartwig (1886), the photographic evidence for variability advanced by Roberts (1891), its confirmation by Seraphimoff (1898), and its final disproof by photography published by Barnard (1898).

An important advance in the study of Andromeda was made when G. P. Bond turned the Great Harvard Refractor in its direction on 14 September, 1847. He was able to distinguish for the first time the two dark dust lanes that cut across the image on its northwest side. This is the first observation of what later became recognized as the spiral arms of the galaxy. Visually, these features are fairly difficult to see clearly, and previous observations always had shown them as straight lines (Coddington 1898).

1.3 S Andromeda

1885 was an important year in the history of M31, as it was in August of that year that a major discovery was made — a truly variable object was discovered in it; an object that produced both insight and confusion for decades afterwards.

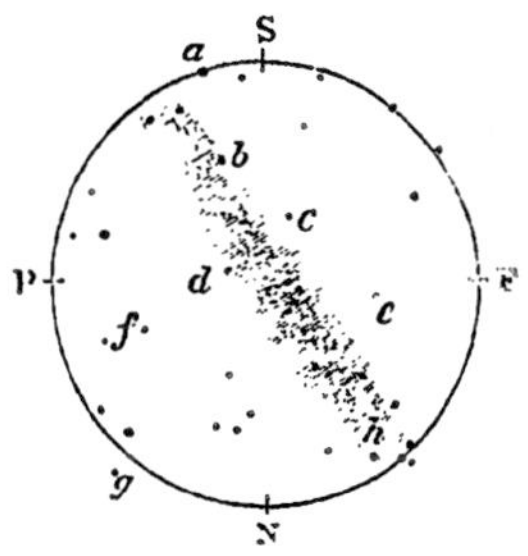

Figure 1.3 A sketch of the central 22 arcmin of M31, with field stars indicated, made in 1886 in Jamaica (Hall, 1886).

Apparently the sudden appearance of a "new star" near the nucleus of M31 (Figure 1.3) was discovered by several astronomers, including L. Gully on 17 August, 1885 and I. Ward (1885) on 19 August, 1885. Eventually called S Andromedae, the new star was extensively observed visually, photographically and spectroscopically before it eventually faded from view. At maximum reaching a visual magnitude of 6.7 (Schroder 1885), it apparently began its rise sometime before 17 August, 1885 (Wolf 1885). Although Gully (1886) reports having seen it on 17 August, others claim that it was not there on 16 August (Temple 1885), and perhaps even as late as 25 August (Wolf 1885). By the middle of September it had faded to about ninth magnitude, and was at eleventh magnitude in November (Engleman 1886). It continued to fade and most observers saw nothing in its place when it was searched for in 1886. A complete light curve (Figure 1.4) was assembled on its 100th birthday in an excellent commemorative article by de Vaucouleurs and Corwin (1985).

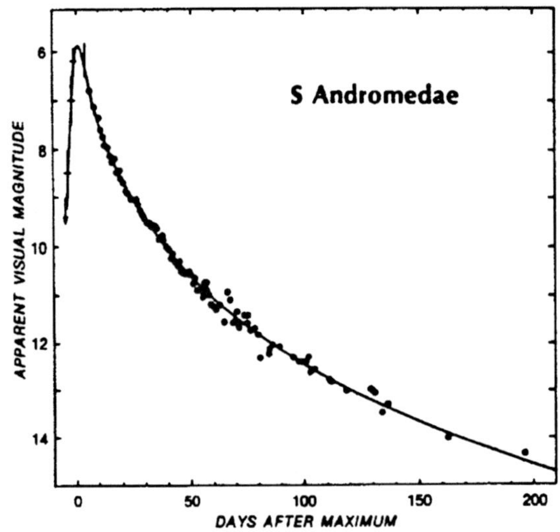

Figure 1.4 The visual light curve of the supernova S And, assembled on its 100th anniversary (from de Vaucouleurs and Corwin, 1985).

The "new star" was recognized as being like a Galactic nova, though there was some speculation about its possibly being a long-period variable star. This latter suggestion seems to have been supported by the fact that a bright star in the nebula is told about in folklore (Tromholt 1885). Although folklore is, of course, notoriously easy to misinterpret, it is nevertheless tempting to speculate about whether any of M31's previous supernovae might have been seen by some of our sharper-eyed antecedents.

Spectroscopy of stars was at the time a new and burgeoning field. The spectrum of S And was observed by several astronomers (*e.g.*, Gothard 1885, Huggins and Rosse 1885, Konkoly 1885, Ricco 1885, Seabroke 1885, and Vogel 1885). It was found to be a continuous spectrum, with some of the investigators reporting bright emission lines, including hydrogen lines.

The significance of the appearance of S And in M31 was not a matter of universal agreement. Some astronomers, according to de Vaucouleurs and Corwin (1985), insisted that its presence in the nebula was evidence that the nebula must be made up of stars and could, therefore, be a very distant object. Others, including Kammermann (1885) and Hasselberg (1886), argued that it could not be connected with the nebula, but must be a chance coincidence. Attempts to measure the parallax to S And failed (Hall 1886, Franz 1887), leading to the conclusion that, if S And is associated with the nebula, they both must be farther away than the bright stars in the field.

Eventually S And was responsible for a delayed recognition of the true nature of M31 in the minds of many people. Its characteristics were so much like those of ordinary novae that it was assumed to be one; since novae seemed to be associated with the Milky Way system, and since S And was not much fainter than a typical naked eye nova, it was concluded that M31 must also be within our Galaxy. Only in the twentieth century did the ordinary novae become distinguished from their far more luminous cousins, the supernovae, so that the true picture could emerge. We now know that S And reached an absolute magnitude of $M_V \approx -19$, ten magnitudes brighter than a Galactic nova.

As a matter of curiosity, it is interesting to note that a second "new star" in M31, similar in brightness to S And, was reported 13 years later (Espin 1898). This object, however, was not confirmed (*e.g.*, Brenner 1898, Möller 1898). As another matter of curiosity, nearly a century later, while this chapter was being rewritten, another spurious supernova was reported in M31, this one by Soviet astronomers, but it was also quickly found to be a false alarm (Marsden 1988).

1.4 Structure

Perhaps the most important nineteenth century discovery about the form of M31 was made by Roberts (1887), who photographed it with his 50-cm reflector. In an anniversary article 100 years later, de Vaucouleurs and Corwin (1985) reproduced some of Roberts' pioneering photographs and assays the importance of Roberts' studies. In his first report in 1887, Roberts pointed out the outer structure of the disk, which he likened to "rings" that appeared to circle the center of the nebula. This interpretation led him to conclude that M31 must be a Laplace nebula, in which planets will form out of the rings surrounding a newly forming star.

By 1899, however, he had obtained more photographs of M31 (as well as many other nebulae) and concluded that, instead of rings, the outer structure was spiral in nature. Thus, its spiralicity (a topic that still evokes controversy — see Hodge 1979) was at last recognized.

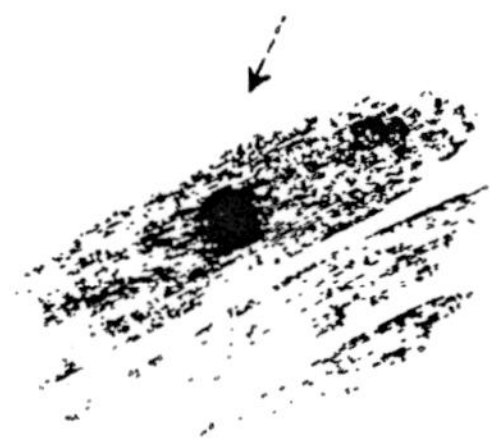

Figure 1.5 A sketch of the central part of Andromeda made in 1871 by Lord Rosse with the 6-foot reflector at Birr Castle (from Rosse, 1885).

Further progress in the study of M31's structure developed as photography became more advanced. By 1907, for example, Götz (1907) measured the positions and magnitudes of over 1,000 stars in (and superimposed on) the nebula and analyzed them to produce a comparison of the structure of the unresolved portion of the nebula with the positions of the resolved stars.

A remarkable study by Reynolds (1913) provides an interesting lesson, which proves that more complete and accurate data does not necessarily lead to more accurate understanding. Reynolds did a rather modern-sounding analysis of the structure of Andromeda by measuring the plate density of an image as a function of distance from the center. He found that the resulting curves showed that the image was made up of a central bulge and wings, to which a curve could be fit with the shape,

$$\frac{1}{I} = (x+1)^2, \qquad (eq.1.1)$$

where I is the brightness at a position that has a distance x from the center. This is nearly the same formula as derived several years later by Hubble for elliptical galaxies. Reynolds, however, used this piece of evidence, remarkably advanced for its time, to support the hypothesis that M31 was a reflection nebula, with the nucleus being the central star.

Several years later Reynolds (1926) published a discussion of the structure of M31 that recognized its true nature. Here, he derived the inclination angle from the ellipticity of its image (obtaining a value of 70°, rather large compared to present values)* and measured the arm's

* This book uses the convention of quoting the inclination angle as the angle between the plane of the galaxy and the plane of the sky and the tilt as the angle between the plane of the galaxy and the line of sight.

structure. He also noted the much higher density of resolved stars in the outer parts of the image compared to the inner, a feature that has been rediscovered many times since.

1.5 Spectra of M31

Most early observations of the spectrum of the nebula referred, because of its brightness, to the nucleus. For example, Gothard (1885) described the spectrum of the nuclear object as continuous with no lines visible. Köveslizethy (1886) claimed to have detected variations in the spectrum. Scheiner (1898) obtained a photographic spectrum in which he detected solar-type lines and claimed that this indicated that M31 must be made up of solar-type stars. On the other hand, Hussey (1899), at about the same time, found only a continuous spectrum. Fath (1909), using the Lick Observatory Crossley reflector with a nebular spectrograph, found a solar-type spectrum and identified 14 absorption lines. He puzzled over the apparently obvious conclusion that all of the stars in M31's nucleus should be of the same spectral type. Less puzzling, apparently, was his conclusion that, based on Bohlin's parallax (see below), all of these stars must be about the size of asteroids!

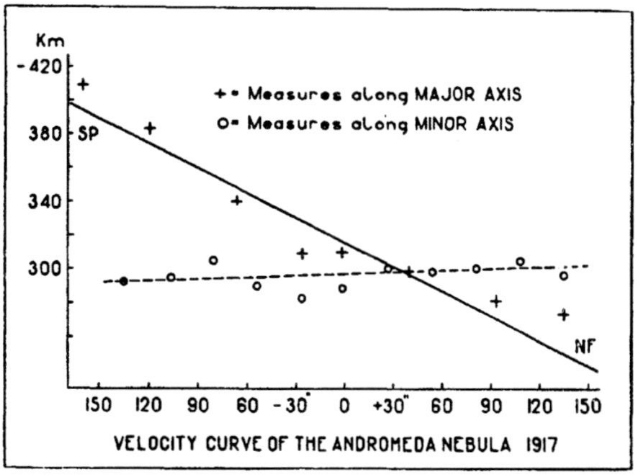

Figure 1.6 Demonstration of the rotation of M31 (from Pease, 1918).

Five years later, when V. M. Slipher (1914) obtained a spectrum of M31 at the Lowell Observatory, it could be seen that the positions of the lines were displaced to the blue. Slipher determined a large negative radial velocity, which suggested to him that a proper motion study should be made, as was done (see below), and that a natural explanation of S And was suggested. He put forward the idea that the star flared up when the nebula, rushing towards us at this high velocity, collided with a "dark star."

Slipher, one of the pioneers in using photographic spectrographs for nebulae, found that the mean radial velocity of M31 was −300 km/sec, almost exactly the velocity measured today, and detected an inclination of the lines, indicating rotation. At a distance of 20 arcsec from nucleus, the radial velocity on the major axis was 100 km/sec different from that of the nucleus.

A much better spectrum was obtained by Pease (1918), who found a systemic velocity of −316 km/sec and a rotation that extended out to five times as far as Slipher's limit.

1.6 Dust

Following Bond's visual discovery of the dark lanes that parallel the major axis and are conspicuous near the minor axis, Ritchey (1904) published photographs that showed many additional sharply defined rifts and dark holes. There was at the time uncertainty about whether such "dark nebulae" represent obscured regions or holes where there are no stars (see a discussion in Curtis 1918).

A more quantitative approach was taken by Kapteyn (1909), who was making a general study of absorption of light in space. He quoted results of an experiment carried out by H. D. Babcock at Mt. Wilson, who compared photographs of M31, one taken on blue-sensitive emulsion and one with a red filter. He found that M31 is about one magnitude redder than a star of similar spectral type, indicating the presence of a considerable amount of absorbing dust.

1.7 The Distance

In the pre-Hubble era probably the most important and most poorly determined fact about M31 was its distance. Its parallax was measured in several attempts, starting with the report by Hall (1886) that no parallax was detected for S And. Franz (1887) also attempted to measure a parallax for S And, finding a negative parallax with respect to three neighboring stars, and concluding (correctly) that the nebula must be behind these stars. Hale (1899) attempted to measure the parallax of the nucleus of M31, also finding a negative value.

Figure 1.7 The dome of the Mt. Wilson 100-inch telescope, where the first intensive study of M31 was carried out.

While these early attempts to measure the distance by detecting a parallax failed, they at least were consistent with the facts as we now know them. That cannot be said for a later study, in which Bohlin (1907) used 15 photographic plates of M31 and determined a positive parallax of 0.17 arcsec, implying a distance of only 6 parsecs! This represents an error of a factor of 100,000. A rather better measurement was that carried out at Mt. Wilson by van Maanen (1918), who found a value of 0.004 ± 0.005 arcsec, indicating that it was undetectably small.

Other attempts to gauge the distance to M31 included measurements of its proper motion. While not giving a definite determination of a distance, a proper motion, if large, would at least indicate that it must be a relatively nearby object. For example, Curtis (1915) used Keeler's original photographs together with new plates to obtain a mean proper motion for the nebulae observed, which included M31, of 0.033 arcsec/year. Using Slipher's velocities and assuming random motions, this implied average distances for the nebulae of about 3 kpc. On the other hand, Barnard (1917), using Mt. Wilson plates with a seven-year time span, found no detectable proper motion.

The argument continued, however. Lee (1926) reviewed all the proper motion studies, including his own, and concluded that M31 has a proper motion of 0.0184 arcsec.

Of course, S And had been used in the nineteenth century to measure the distance to M31, giving an incorrect answer because it was first thought to be an ordinary nova. By 1917, however, true novae were being discovered, primarily at Mt. Wilson (see, for example, Ritchey 1917*a*, 1917*b*, Shapley 1917, Duncan 1918, and Sanford 1918). By 1923 enough data were available on M31's ordinary novae to allow Lundmark (1923) to derive a distance. Equating the luminosity of the M31 novae to that of Nova Sagittarii, he calculated a distance to M31 of a little over 10^6 parsecs (50% larger than the presently accepted value).

One year earlier, Öpik (1922) found a comparable distance for M31 by a novel argument. He assumed that the ellipsoidal shape of the inner parts was due to rotation and then calculated, from the velocities, that its distance must be 450 kpc, a value only 50% smaller than the current value.

These various attempts to measure the distance to M31 were a prelude to the eventual determination of that quantity by Hubble, which is described in Chapter 2.

1.8 Summary

For a thousand years the Andromeda Nebula remained a largely unexplored and poorly understood curiosity. By the nineteenth century, which was notable for the appearance of S Andromeda, a "new star", interest in its nature began to grow. Most scientists thought that it was a nearby cloud of gas, possibly a solar system in formation.

CHAPTER 2

DISCOVERY OF ITS GALAXIAN NATURE

The modern history of M31 is dominated by the pioneering work of Edwin Hubble, using the Mt. Wilson 100-inch telescope during the first few years of its operation. This chapter is largely a summary of Hubble's accomplishments in pulling together a wide variety of observations that showed beyond a doubt that M31 is an external galaxy and that it contains many of the features and contents that we had already recognized in our Galaxy. Hubble published a comprehensive paper on the galaxy in 1929 (Hubble 1929).

2.1 Pre-Hubble Results

Although there was a lack of consensus about the nature of M31 in the nineteenth century, by the beginning of the twentieth century clues as to its true nature were beginning to be found. Probably the best turned up in 1899, when Scheiner concluded that its solar-like spectrum could only mean that the object is made up of ordinary stars, rather than being a diffuse, gaseous nebula. Subsequent measurements by a number of observers showed that M31's radial velocity is on the order of −300 km/sec (see Chapter 1). Furthermore, the radial velocity showed a gradient that suggested rotation. These velocities were too large to be easily explained in terms of a small, nearby nebula, and thus further indicated the probable large distance to M31.

Another kind of evidence was the discovery of stellar-like images that were variable. There was at the time no clear way to establish how intrinsically bright the apparent stars of M31 might be, as there was no clue about the absolute magnitude for a non-variable star from the kind of information then available. However, a star that showed variations in brightness could provide the much needed information on absolute magnitude. Some astronomers (*e.g.*, Shapley and van Maanen) had interpreted the "nova" of 1885, S And, as being an ordinary nova, which would indicate a rather small distance to M31, leaving it within the Milky Way Galaxy. However, using the Mt. Wilson 60-inch telescope, and later the 100-inch, Ritchie and others found much fainter objects that behaved like novae (Ritchie 1917). By 1929 85 such objects had been recorded. As discussed below, Hubble's detailed analysis of these objects eventually established the fact that they, not S And, were novae like those in our Galaxy. But the most convincing evidence came from Hubble's discovery and detailed study of the Cepheid variables.

2.2 The Cepheids

In 1923 Hubble found two faint variables in M31 that were clearly not novae. They showed a periodicity and Hubble soon determined that they had periods (20 and 31 days) and light curves like Cepheid variables. They were very faint and thus, if correctly identified as Cepheids, they proved the great distance to the object. Hubble immediately began an intensive search for more Cepheids so that he could apply the period-luminosity law and obtain a precise distance, like that which Shapley (1925) had obtained for the Magellanic Clouds.

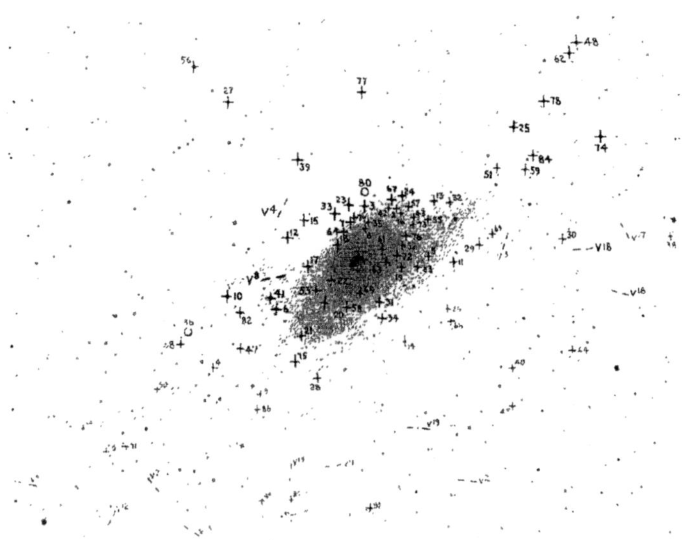

Figure 2.1 Some of the variable stars discovered by Hubble in M31. Novae are marked by crosses, variables by the prefix V (from Hubble, 1929).

Hubble surveyed about 40% of the nebula with the 100-inch in his search for variables, concentrating on four regions. By 1929 he had available 270 plates, covering 18 years. The result was a list of 50 variables and he felt that his search was complete to about 19th (blue) magnitude. Subsequent searches, *e.g.*, by Baade and others, have shown that the search was not very complete, but that Hubble had found most of the large-amplitude Cepheids of that brightness in the regions surveyed.

Using visual step scale methods, incorporating local sequences and extrapolating down from preliminary magnitudes determined in Selected Areas by Sears, he measured magnitudes for the variables by eye estimates and determined light curves. Forty of them turned out to be Cepheid variables, with periods ranging from 10 to 48 days. Magnitudes at maximum ranged from 17.9 to 19.3 on his scale (which has subsequently been found to be in error by about 1 magnitude at the faint end). He found that the Cepheids were primarily limited in position to the spiral arms, there being none in the nuclear bulge region, but he detected no difference in the period-magnitude relation with position in the galaxy, an important issue at the time because it seemed to confirm, at least to some extent, the universality of the period-luminosity law.

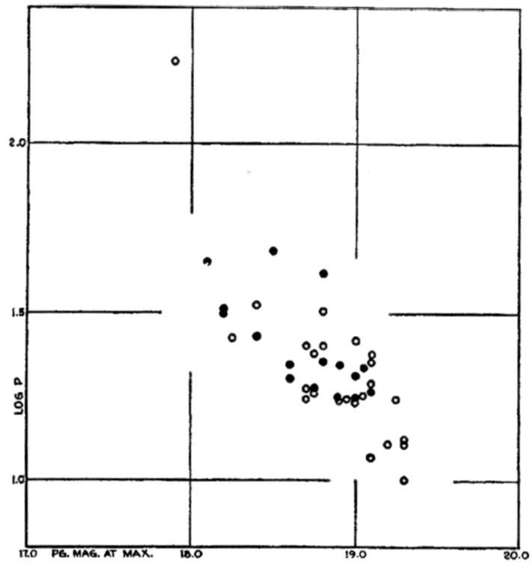

Figure 2.2 Hubble's Cepheid period-luminosity diagram (from Hubble, 1929). Open and closed circles are for different areas in the galaxy.

As had been found for Cepheids in the Magellanic Clouds, the number of Cepheids increased with decreasing period, down at least to his effective limit of about 17 days. One variable stood out; Hubble 42 was the brightest one and its period of 175 days was more than three times as long as the next longest one.

The period-luminosity diagram that Hubble plotted is shown in Figure 2.2. It is similar to that obtained for the Small Magellanic Cloud (Shapley 1925), and so Hubble felt justified in assuming that he was dealing with the same kinds of stars with the same intrinsic properties. A direct comparison led him to conclude that the difference in apparent magnitude for the two galaxies is 4.7 ± 0.1 magnitudes.

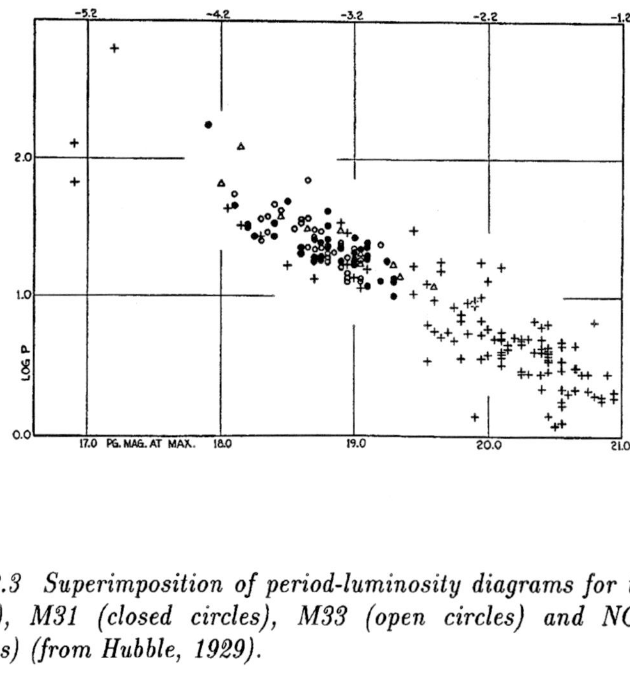

Figure 2.3 Superimposition of period-luminosity diagrams for the SMC (crosses), M31 (closed circles), M33 (open circles) and NGC 6822 (triangles) (from Hubble, 1929).

Because the number of Cepheids was small and because he felt that the data that he had for M33 were so similar, Hubble decided to combine the data for the two galaxies in making his final determination of the distance to M31, after adjusting for a 0.1 magnitude difference between them. The resulting fit for 75 variables, using Shapley's data for 106 SMC Cepheids, offset by 4.65 magnitudes, is shown in Figure 2.3. Adopting the then-current distance for the SMC (which was grossly in error because of the misidentification of cluster-type Cepheids as classical Cepheids) led Hubble to derive a distance to M31 of 275 kpc, less than half of the presently derived distance. It is, perhaps, instructive to note that he estimated the accuracy of the distance to M31 to be on the order of 2.5%. Many astronomers, of course, have continued in the tradition of greatly underestimating the errors in extragalactic distance estimates. In Hubble's defense it should be mentioned that he considered the result to be internally reliable, but externally uncertain, saying "there still remains the possibility of a considerable correction when more data on galactic Cepheids becomes available."

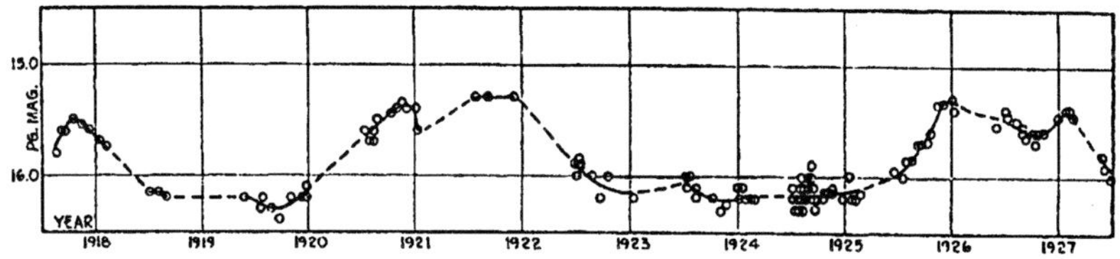

Figure 2.4 Light curve of Hubble's V19, a very luminous blue irregular variable (from Hubble, 1929).

2.3 The Non-Cepheids

Ten of the variable stars discovered by Hubble were not obviously Cepheids, although four were considered possible candidates for which insufficient coverage was available. Six were found to be irregular, with long-term variability. Hubble 19, for instance, was brighter than any of the other variables, ranging in blue magnitude from 15.3 to 16.3 (Figure 2.4). It, as well as two of the other blue variables in the list, is one of the class of objects now called Hubble-Sandage variables, which are among the most luminous stars in galaxies (see Chapter 13). Hubble suggested that they might be useful distance indicators, which they were for several decades, although they are now rather inferior to other methods because of the intrinsic spread in their absolute magnitudes.

2.4 The Novae

By 1923 nearly 20 ordinary novae had been found in M31. The list of discoverers is almost a who's who of Mt. Wilson astronomers: Ritchie, Shapley, Duncan and Humason. In that year Hubble began his systematic study of the galaxy, and by 1929 he had increased the number of novae to 85.

That these objects were normal novae similar to Galactic examples was shown by both their light curves (Figure 2.5) and by a spectrum of one of them obtained by Humason, who found that the continuum extended deeply into the ultraviolet and that there were Balmer lines in emission, as in Galactic novae. To some extent, Hubble's study had the opposite of the usual effect; he found that he could generalize better about some aspects of novae behavior than could astronomers from Galactic samples, as the M31 novae were systematically observed and all were at the same known distance.

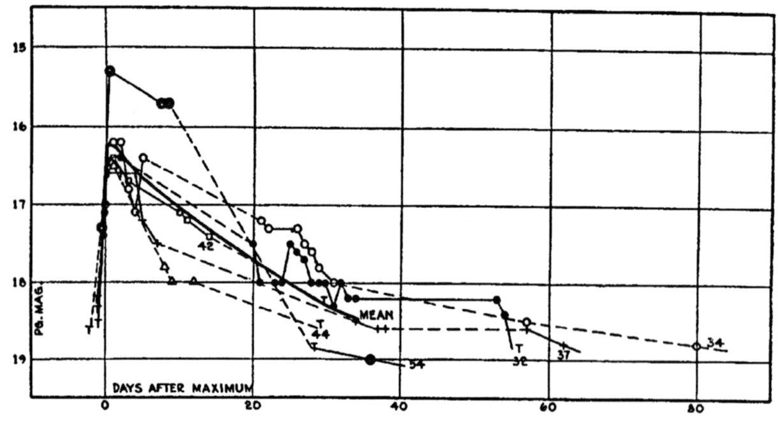

Figure 2.5 Light curves for six novae in M31 (from Hubble, 1929).

Hubble found that the mean maximum (adjusted in various ways to take into account certain differences in light curve coverage) was at magnitude 16.4 and that, after a 10-day-long steeper drop, the light curves then decreased in brightness linearly at a rate of 0.05 mag per day. Not all novae had the same maximum brightness (Figure 2.6), but the distribution of maximum magnitudes showed a dispersion of only about 0.5 mag. With enough novae in the sample, he was confident of knowing the mean rather well, and so could try finding a distance that was independent of the period-luminosity distance. The absolute magnitude of the mean maximum for novae implied by using the Cepheid distance to M31 would be −5.7.

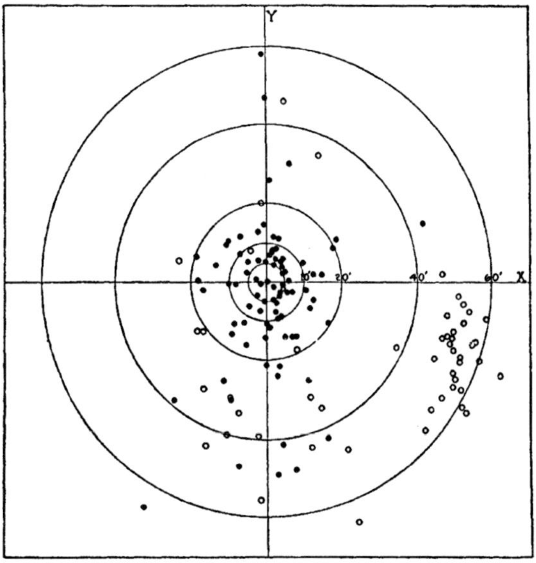

Figure 2.6 Distribution in space, corrected to face-on, of M31's novae (solid dots) and Cepheids (open circles), the blank area on the left was not searched (from Hubble, 1929).

Unfortunately, only two Galactic novae had accurately known distances at that time — Nova Persei 1901, with a maximum magnitude of −5.0, and Nova Aquilae 1918, with a maximum at −9.2. Hubble chose to use Lundmark's (1927) value of −6.1 to conclude that the novae gave essentially the same distance as that given by the Cepheids. This agreement, however, was fortuitous; had he used a better value for the true nova absolute magnitude, he might have seen that the Cepheid distance was too small, a fact that was not discovered until 23 years later.

Figure 2.6 shows Hubble's plot of nova positions, corrected to the plane of M31. It clearly shows a strong concentration toward the center, which is quite different from the distribution of Cepheids, which avoid the center. Hubble noted that this amorphous central region (now referred to as the central "bulge") was notable for (1) the lack of dust clouds, (2) the lack of extremely luminous stars, (3) the presence of novae, and (4) the presence of a solar-type spectrum. He concluded, with characteristic prescience, that this part of M31 is different in basic character from the outer parts (the arms), with the key difference being the lack of very luminous stars (giants and Cepheids). The novae he associated with the fainter stars that dominate the bulge. In a sense he was planting the seeds at this time for the concept of stellar populations that followed some 20 years later, when Baade continued the study of M31 and other Local Group galaxies.

It is a matter of interest, which should not detract from the enormous accomplishments that characterize Hubble's studies of M31, that he appears to have been persuaded by Jeans' (1928) view that the central parts of M31 must be gaseous. Hubble rationalized the presence of novae there in a rather cavalier way by suggesting that they may be "due to penetration into the gaseous region of stars from the outer regions."

2.5 NGC 206

Although he did not study the bright, resolved non-variable stars in general, Hubble did analyze in a wonderfully modern way the stellar content of NGC 206, which he identified as a star cluster. Now, based on its immense size ($\sim 1,000$ pc) and its loose structure, it is considered a star cloud, even larger than an OB association. Using newly developed yellow-sensitive plates, he was able to measure magnitudes and colors for nearly 100 stars in the object. These were displayed as a color-magnitude diagram (CMD) (Figure 2.7), making it the very first CMD to be obtained for such an extragalactic object. When compared with modern data for NGC 206 (Chapter 10), it looks remarkably good, though Hubble turned it 90° with respect to modern convention.

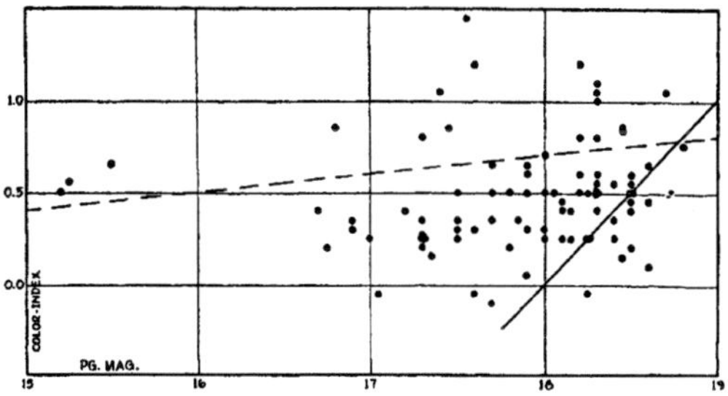

Figure 2.7 Color-magnitude diagram for NGC 206 (from Hubble, 1929).

From the CMD for NGC 206, supplemented by inspection of photovisual plates of other areas, Hubble concluded that the brightest stellar members of the arms of M31 are bluer than the Cepheids (which he concluded to be of approximately solar spectral type). He further noted that the brightness of the brightest stars was fainter than in more open type "nebulae" (he did not use the term "galaxy" at that time) like M33, and suggested that there might be a systematic difference in this quantity with galaxy type, an idea that was reconfirmed several times in succeeding decades by astronomers working on the distance scale.

2.6 Integrated Properties

Since M31 appeared to be the best example of a massive distant galaxy, Hubble was interested in attempting to determine its integrated properties: its volume, mass and total luminosity. The volume was estimated from his distance determination and from its dimensions, assuming (based on edge-on galaxies) that its minor/major axis ratio is 1/8. The total volume calculated was 7×10^{10} pc^3.

To obtain a mass, he applied a Keplerian model to the measured rotational velocity, which was traced out to about 2.5 arcmin from the nucleus, where he used a velocity of 72 km/sec. This gave a mass of 2.4×10^8 solar masses for this inner portion of the nebula, which he then, somewhat arbitrarily, extrapolated to a total mass of 3.5×10^9 solar masses, which is far too small by today's estimates, which gives a mass that is at least two orders of magnitude larger (Chapter 8). Nevertheless, this calculation showed that M31 was a large galaxy, though much smaller in mass than the Galactic system (he quoted Eddington's value of 2.7×10^{11} solar masses for the latter, a remarkably good value for the time). We now know, of course, that his mass estimate was wrong for a number of reasons, including the incorrectness of the period-luminosity relation zero point and the incompleteness of the velocity curve.

Hubble's pioneering work on M31 had a profound effect on extragalactic astronomy. When combined with his equally thorough studies of the smaller galaxies M33 and NGC 6822, it showed astronomers the true nature of the non-gaseous nebulae and demonstrated how to study these objects, paving the road for moving outward towards more and more distant reaches of space.

2.7 Summary

Following several years of intensive study with the world's largest telescope, Hubble concluded that M31 is an external galaxy with many characteristics (size, mass, stellar content) similar to those of our Milky Way Galaxy.

CHAPTER 3

MODERN HISTORY

In order to place the many recent diverse studies of M31 in perspective, this chapter gives a brief summary of the more important recent discoveries and studies. Following chapters then take up the different components individually in more depth. For simplicity, references are not given here, but will be found in later chapters as appropriate.

3.1 Baade's Work

Until the 1940s virtually all of the modern history of the study of M31 was the work of Edwin Hubble. However, as other astronomers began to turn the large Mt. Wilson reflectors toward nearby galaxies, new developments came about. Walter Baade, an astronomer from Germany whose early work on a nova in M33 had attracted Hubble's attention, moved to Mt. Wilson and began a long career that centered on M31 and its population of stars and interstellar matter.

Shortly after joining the Mt. Wilson staff, Baade began a systematic search of the Andromeda Galaxy with the 100-inch telescope. This led to two major discoveries that revolutionized our views of stars and our scale of distances. The first breakthrough came in 1944. Baade had, in the previous years, almost exclusive use of the telescope for galaxy work, and he experimented freely with new photographic emulsions and filters of different colors. During this period Los Angeles was occasionally subjected to defensive night-time blackouts, at which times the usually brightly lit city completely vanished from view from Mt. Wilson. On these nights the sky was unusually black and Baade was able to push his photography of M31 deeper than would otherwise have been possible.

The problem that Baade was trying to solve involved the puzzling fact, noted years before by Hubble, that while M31 was well resolved into individual stars in its outer parts, the central, bright amorphous region defied resolution on even Hubble's best photographic plates. The light there could be inferred to be coming from millions of stars, but why could none of them be seen? To compound the mystery, four small companion galaxies (NGC 147, NGC 185, NGC 205 and M32) were also unresolved into stars.

Finally, he tried the right combination — red-sensitive emulsion, a red filter, perfect atmospheric conditions, a blacked-out Los Angeles, and long exposures. The resulting photographs not only resolved the stars in M31's central bulge and in its companions, but also led Baade to the idea of two stellar populations, an idea that fostered the later successful unravelling of the whole scheme of stellar and galaxy evolution. Baade found that the stars that finally showed up were red giants, too faint in the blue to be resolved by the blue-sensitive emulsions previously used to hunt for them. They belonged to a newly recognized class of stars, called by Baade Population II. Also found in the small elliptical galaxy companions, these kinds of stars were the same, Baade argued, as the stars in globular clusters of our Galaxy.

Baade's second major discovery was also largely dependent on M31. In our Galaxy the globular clusters contain a few Cepheid variables. Previously, Shapley had used this apparently fortunate fact to calibrate the Cepheid period-luminosity relation. Baade noticed, however, that there was a discrepancy when the globular clusters were compared with M31. In the clusters the brightest giants had about the same brightness as the Cepheids that had periods of about 30–40 days, while in Andromeda the Population II giant stars were much fainter than Cepheids of these periods. It was unclear at first what might be wrong: was the period-luminosity relation different in different galaxies, were Population II giants different from place to place, or were the globular cluster Cepheids basically different from those in M31?

The answer was announced at the 1952 meeting of the International Astronomical Union in Rome. Baade showed that the Cepheids in the globular clusters were the problem. They were shown to be a different kind of Cepheid, found only among Population II stars, and not in M31's spiral arms or in our Galaxy's main disk of Population I. Baade found that the Population II Cepheids are on the average four times fainter, intrinsically, than normal Cepheids of the same period. Thus, the normal Cepheids studied by Hubble in M31 and other nearby galaxies must be twice as far away as had been thought. Since all galaxies' distances depended on this measurement, Baade's discovery doubled the current estimate of the size of the universe.

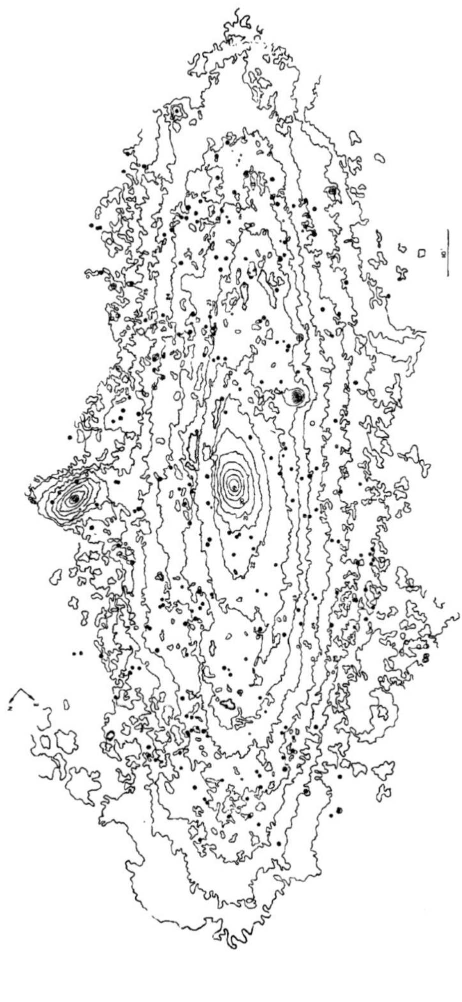

Figure 3.1 Isophotal diagram in B for M31 (from Hodge and Kennicutt, 1982).

Baade's interest in stellar populations led him to investigate as well as he could the optically visible gas in the Andromeda Galaxy. He had found that gas and dust seemed to be associated almost exclusively with Population I stars, a fact that came to be recognized as circumstantial evidence that these were areas where new stars were formed. To try to discover both how star formation proceeds and how a spiral galaxy arranges its positions of active star formation, Baade examined the distribution of H II regions. Using red filters to isolate Hα, he and H. C. Arp were able to map out the locations of 688 H II regions, finding that they were concentrated in the spiral arms, being most conspicuous in the intermediate-distance arms, about 10 kpc from M31's center. That is where most bright blue stars exist, also, and it is in that region that Andromeda is now producing most of its new stellar members.

3.2 Radio Observations

With the advent of radio astronomy, new ways became available for looking at this problem. Most important for M31 was the radio emission of H I at 21 cm wavelength, detection of which allowed early radio astronomers to measure how much neutral hydrogen exists in M31. But the astronomical world had to wait many years before the exact locations of this gas could be discovered, as the first small radio telescopes could not resolve the Andromeda Galaxy very well at such long wavelengths.

An example of a more complete study came in 1966 when Morton Roberts used the 300-foot radio telescope at Greenbank, West Virginia to get a good high resolution (10-arcminute) look at the galaxy. Instead of a disk of gas, Roberts found the gas to be arranged as a giant ring, with a hole in the center and a maximum density about 10 kpc from the nucleus. Thus, in this part of the galaxy the neutral gas was found to mimic the gas clouds found by Baade. However, beyond the thickest part of the ring the two components differed. The bright gas clouds tapered off at about 15 kpc, while the neutral hydrogen could be detected out to about twice that distance. This behavior has now been found to be fairly common for galaxies of the same type as M31 and is generally understood to be due to a mixture of causes. In the inner regions, star formation was completed long ago and little gas is left to form stars. In the middle regions where gas is still abundant, strong shocks due to passing waves of density occur, leading to repeated triggering of star condensation. In the outer regions, however, the gas density is too low for stars to condense out of it. These and other effects now explain M31's ring of hydrogen.

Even more detail can be gained with the use of multiple-antenna radio interferometers. The first studies to be completed with this new generation of telescope included that of Emerson and his colleagues at Cambridge University, who used the Cambridge Half-Mile telescope to achieve a resolution of about 300 pc. The increased detail allowed Emerson to show that the "hole" in Roberts' ring extends out to about 4 kpc from the center. He demonstrated a close agreement from that point outward between the hydrogen gas and the dust and found hydrogen extending out to 30 kpc along the southwest major axis, farther than any optical counterpart of the disk. Even more detailed studies have been made by Brinks and collaborators working with the Westerbork Aperture Synthesis Telescope and by Braun with the VLA.

The radio continuum emission has also been the subject of analysis. In 1969 Pooley at Cambridge showed that most of M31's radio continuum emission came from the nucleus and from the bright parts of the spiral arms. Van der Kruit in Holland and Terzian, Balick and their co-workers at Cornell confirmed Pooley's results at other wavelengths and demonstrated that a large percentage of the radio noise must come from non-thermal sources. Ordinary H II regions could not be the primary source. Instead, the radio spectra indicated that much of the radiation must come from some source, such as supernova remnants and the high energy electrons emitted by supernovae into the interstellar gas, that emits synchrotron radiation.

Subsequently, the 100-meter telescope near Bonn allowed Elly Berkhuijsen and her collaborators to map out the various sources of continuum radio emission. Berkhuijsen's maps showed that the thermal continuum radiation mostly comes from the area of the "bright ring" of hydrogen, about 10 kpc from the center. It is therefore probably due to the numerous H II regions concentrated there. The non-thermal radiation, on the other hand, is more broadly based. Observed supernova remnants tend to have a similar distribution in our Galaxy, and probably also in M31. Thus, the continuum radiation may be largely explained as the combined emission from the remnants of numerous past supernovae in Andromeda.

3.3 Planetary Nebulae

Planetary nebulae are another kind of object that has been recognized in M31. Baade had found five of them many years ago but they were only discovered in large numbers in the 1970s. Ford and Jacoby used image intensifiers and narrow band filters to penetrate down to faintness levels that would have been impossible for Baade to achieve. Using the Lick Observatory's 3-meter telescope, they found 315 planetary nebulae and calculated that the total number detectable in M31 is probably roughly 10,000. About five stars die in this way every 100 years in Andromeda's nuclear bulge, according to these data. Over the last few billion years the gas lost in this process should accumulate in a gaseous disk that rotates around the nucleus. Astronomers have found a central disk of gas that might be part of that ejected material (Chapter 7).

3.4 H II Regions

Baade's searches for H II regions led to the first comprehensive catalogue of these objects, published by Baade and Arp (1964). A deeper survey was completed in 1978 at Marseilles by a group of French astronomers using the 2-meter telescope at the Haute Provence Observatory. They catalogued almost 1,000 objects and penetrated deeply into the complex interconnecting system of gas clouds that populate M31's fragmented arms. Even deeper surveys are in progress (Chapter 14).

3.5 Globular Clusters

Hubble had, in 1932, published a list of 140 objects that he tentatively identified as globular clusters. Subsequently, more candidates, plus confirming color and structural data have been published. For example, Sargent, Hartwick, van den Bergh and Kowal carried out a search for distant globular clusters, using photographic plates taken with the 4-meter telescope of the Kitt Peak National Observatory. They catalogued 355 probable globular clusters, about doubling the number that had previously been discovered by others. Subsequent surveys have added even more. This collection of objects is two to three times larger than the number of globular clusters known in our Galaxy, in itself evidence that the mass of M31 is probably substantially larger than ours.

Van den Bergh's study of the spectra and colors of the brighter globular clusters showed that the early history of M31 may have been somewhat different from that of our own Galaxy. Whereas in our Galaxy the globulars in the outer regions are very poor in heavy elements and the inner ones tend to be somewhat more heavy element-rich, van den Bergh found that the globulars of M31 show a large range of heavy element abundances, and less correlation with location. This is, however, still a controversial point.

3.6 Young Open Clusters and Dust

The OB associations of M31 have been studied in some detail, following van den Bergh's pioneering paper published in 1964. Large star clouds can be recognized, as well as about 200 OB associations roughly similar to those in the solar neighborhood. They are similar (in their ages, star distributions, and luminosity functions, for example) to the local OB associations, but their sizes and their mean surface density, normalized to the galaxy luminosity is anomalously small.

Until recently very few open clusters had been found in M31, except for a few objects that Hubble had identified, but recent studies have identified 403 such clusters. Most are small, only about 20 pc across, and similar in age and population to the richer young clusters in our Galaxy, such as h and χ Persei. They allow the tracing of the recent history of star formation in a crude way and show that the star formation rate has recently been anomalously high, compared to the rest of M31, in the region 10 kpc out, where the neutral hydrogen density and bright stars are concentrated. The open clusters are good spiral arm tracers.

Figure 3.2 NGC 206, a giant star cloud in M31 (Palomar Observatory photo).

Dust clouds are common in most spiral galaxies and the dust of M31, because of its high angle of tilt (its plane is only about 12°.5 from the line-of-sight), is particularly conspicuous. A search in 1980 led to a map of 730 discrete dust clouds. These complexes of obscuring matter range from only about 3 pc up to about 10^3 pc in major dimension. They appear concentrated towards the nuclear region and show no clear spiral pattern.

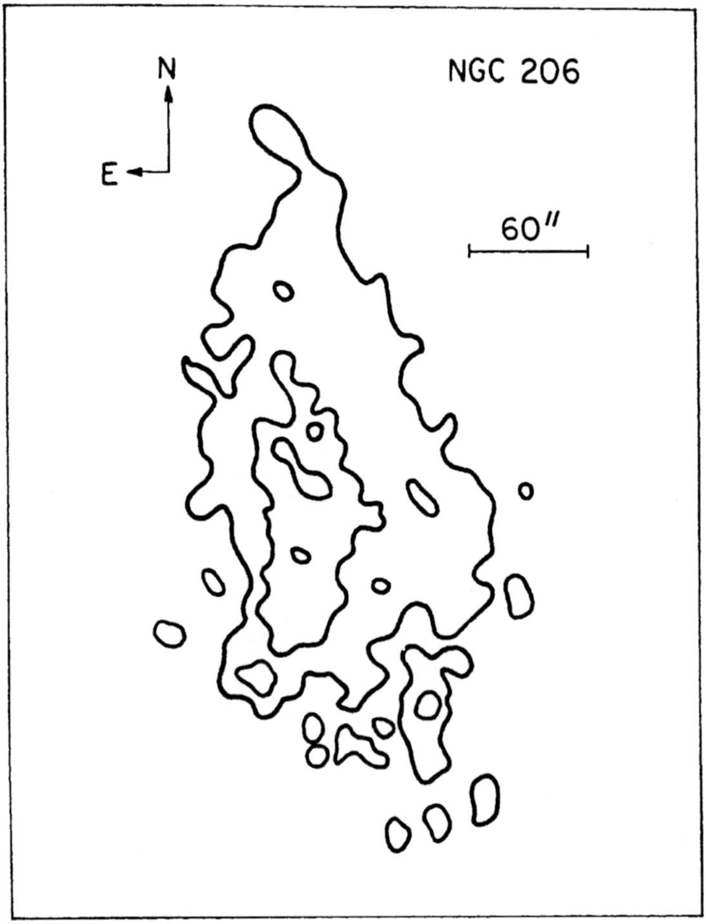

Figure 3.3 NGC 206's structure from B isophotes (from Hodge and Kennicutt, 1982).

3.7 Spiral Structure

The first good attempt to disentangle the spiral arms of the Andromeda Galaxy was that of Arp in 1964, who used the gas clouds to attempt to delineate the arm pattern. Arp found that a two-armed spiral with the arms trailing seemed to fit fairly well, but suggested that the imperfections of the fit might be due to gravitational disturbance of the disk by the mass of M32, its close companion galaxy. Computer modeling by Byrd subsequently seemed to confirm this early suggestion and seemed also to be able to explain the warping of the inner parts of the hydrogen disk seen at radio wavelengths.

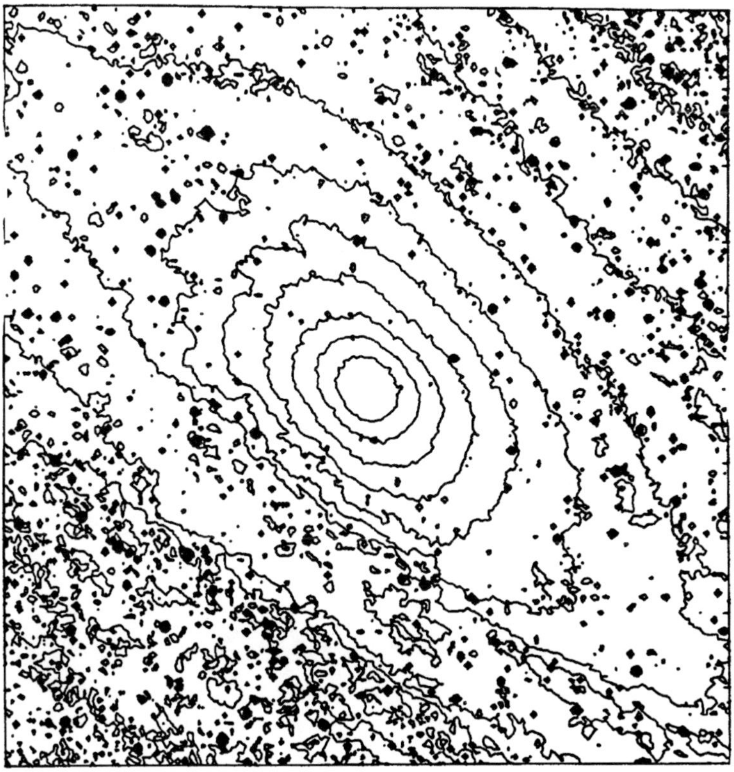

Figure 3.4 Isophotes in B of the smooth central bulge in M31 (from Hodge and Kennicutt, 1982).

However, other ideas were then put forward that were radically different. Employing new analytical techniques, Kalnajs (1974) restudied the evidence and concluded that there is only *one* spiral arm and that it emerges in the opposite direction to that plotted by Arp. It would be a leading spiral arm. Kalnajs suggested that the presence of M32 probably generates a resonance between the two galaxies that caused the one-armed structure. Subsequently, Simien and collaborators analyzed a large collection of M31's Population I objects and attempted to deduce its spiral structure using a purely morphological approach. They also concluded that the best fit was with a leading one-armed spiral pattern.

The controversy was resolved convincingly in 1991 when Braun (1991) showed that the H I data strongly favor a two-armed trailing spiral pattern. The dust clouds, on the other hand, don't fit any spiral pattern very well, and serve to point out how imperfect most galaxy spiral patterns are.

3.8 Summary

Baade's many years of study of M31 with the Mt. Wilson and Palomar telescopes erected a framework for understanding it, based on the foundations laid by Hubble. He corrected the distance and explored its spiral arms, their stars and nebulae. Subsequent studies have refined our knowledge in detail, sharpening our understanding of the differences found between it and our Galaxy.

CHAPTER 4

OPTICAL STRUCTURE

The optical structure of M31 provided one of the earliest puzzles about this galaxy and it has continued to present us with intriguing questions. It took good photographs to show that the object has a spiral shape and it has taken a number of more sophisticated techniques to discover the more subtle aspects of its structure. This chapter begins, as a way of introducing the development of the field, with a brief survey of the various techniques used.

4.1 Techniques

Microphotometry

Among the earliest quantitative studies of M31 were those that used photographic plates analyzed through microphotometric scans. The first results from such a technique were those of Reynolds (1913) and Shapley (1934), who analyzed long-exposure small-scale plates. Redman and Shirley (1937) did the most complete early investigation, using plates similar to Shapley's and working with calibrations that showed that they were reaching to surface brightnesses of about 26.7 mag per arcsec^{-2} in the blue. A large-scale, exhaustive analysis was carried out by Lyngå (1959), who used large-scale plates in two colors, who performed detailed calibrations, and who published an unusually complete presentation of his data.

Isophotometry

An alternative to the method of hand-reducing a series of microphotometric scans was the use of a mechanically-driven isophotometer (Williams and Hiltner 1941). Historically, this technique provided the first quantitative information about M31's shape.

Photoelectric Scanning

With the advent of the more precise photometry provided by photoelectric photometry, astronomers tried reproducing the kind of image-building based on scans, previously photographically obtained, but now with direct photoelectric measurements. Stebbins and Whitford (1934), pioneers in this technique, provided the first photoelectric study of M31. In a more complete survey, de Vaucouleurs (1958) used this technique with a small telescope, letting the image drift past the stationary telescope at the diurnal rate. Sharov and Lyutyi (1980) used an offset photometer, with positions carefully controlled with respect to field stars and the nucleus, to build up a photometric map of the central parts of M31. Hoessel and Melnick (1980), on the other hand, used a photometer that was driven at a chosen rate across the galaxy.

Digital Analysis

With the introduction of high-speed digital analysis of images, originally photographic ones, it became possible to do more detailed isophotometry. This technique has been used by Hodge and Kennicutt (1982) to examine the structure of M31 and by Walterbos and Kennicutt (1987, 1988) to look at both the structure and the color distribution.

Infrared Scanning

Near and far infrared imaging has, until recently, involved scanning across the image. Sandage *et al.* (1969) used the Palomar 5-meter telescope to examine the optical and near-infrared structure of the inner 60 arcsec of M31. More recently, Hiromoto *et al.* (1983) extended this work to greater distances from the nucleus. Near-infrared profiles were obtained by Battaner *et al.* (1986). Far infrared images, obtained from IRAS, have also been analyzed (Walterbos and Schwering 1987).

Star Counts

Although it led to early results of interest in the nineteenth century (see Chapter 1), the technique of using star counts to study M31's structure was only recently extensively exploited. Reddish (1962) used this method, as did van den Bergh (1966) and Berkhuijsen and Humphreys (1988).

CCD Photometry

The use of digital array detectors, particularly CCDs, has meant another step of reliability and convenience for the study of galaxy structure. The small size of current CCDs and the large size of M31, however, has meant that only a limited number of applications have been carried out. Kent (1983) and Kormendy (1987) both used this technique to examine the structure of M31's central regions, and Kent (1987) has used it for most of the galaxy by employing very small telescopes for the outer parts.

Table 4.1
Some Optical Photometric Studies of M31
(Adapted from Walterbos and Kennicutt, 1988)

Author	Year	Color	Range in intensity (mag arcsec^{-2})	Range in Position: Major axis (′)	Range in Position: Minor axis (′)
Lyngå	(1959)	P	18.0–23.4	−40, +40	−17, +17
		V	17.0–22.7	−32, +40	−17, +17
Fricke	(1954)	u	20.4–24.5	−34, +50	−9, +14
		b	18.6–23.9	−40, +56	−17, +17
		r	18.2–22.5	−37, +56	−17, +17
Sharov and Lyutyj	(1980)	U	19.0–22.8	−5, +5	−5, +5
		B	18.4–22.2	−5, +5	−5, +5
		V	17.4–21.0	−5, +5	−5, +5
Burgess	(1976)	U	18.9–24.0	−60, +50	−20, +15
		B	18.0–24.5	−60, +50	−20, +15
		V	17.1–24.0	−60, +50	−20, +15
		I	15.6–20.2	−10, +10	−13, +13
de Vaucouleurs	(1958)	U	22.1–24.9	–	–
		B	21.5–26.8	−120, 120	−58, +38
		V	20.8–24.6	–	–
Hoessel and Melnick	(1980)	v	17.9.–22.9	−44, −1	−8, +10
		g	17.0–22.3	−44, −1	−8, +10
		r	17.4–21.8	−44, −1	−8, +10
Hodge and Kennicutt	(1982)	B	18.90–25.50	−100, +100	−40, +40
Walterbos and Kennicutt	(1988)	U	20.11–26.07	−124, +124	−40, +20
		B	19.53–25.91	−124, +124	−50, +36
		V	18.56–25.10	−124, +124	−46, +28
		R	17.76–24.61	−124, +124	−46, +28

Table 4.1 is adapted from Walterbos and Kennicutt (1988), who compared the extent and the other properties of many of the various photometric studies of M31 published previous to theirs.

4.2 Dimensions

The size of the Andromeda Galaxy was not appreciated at first because of the low surface brightness of its disk. Before modern isophotometry, its major axis was thought to have an extent of only about 160 arcmin, although Hubble (1932) found evidence from faint stellar clusters (now recognised as star clouds) that the disk might have a diameter of about 200 arcmin, deprojected. A much larger size was derived by Stebbins and Whitford (1934), who found dimensions measured at a level of 27 mag $arcsec^{-2}$ of 110 arcmin by 450 arcmin, extrapolated from a profile measured in RA. This very large extent of the major axis was corroborated by the isophotometry of Williams and Hiltner (1941).

More recently, results of photometry of the outer parts of M31 have been limited to distances from the nucleus of about 120 arcmin radius (de Vaucouleurs 1958, Hodge and Kennicutt 1982, Walterbos and Kennicutt 1987), corresponding to a surface brightness of about B = 26. Accurate, modern photometry beyond this distance, though difficult, should be possible. It would be of interest to know more about the outer limits of the disk and halo and photometry in the region between 120 arcmin and Stebbins and Whitford's supposed limit of 225 arcmin are needed. Perhaps these parts of M31 will be better explored by the study of the distribution of resolved stars (*e.g.*, Crotts, 1986).

The ratio of the major to the minor axis depends, of course, on the radial position in which they are measured. Hubble found a value of about 4, which represents a good value for the brighter parts of the disk. The true inclination angle was measured by an ingenious technique by Simien *et al.* (1978), who found a best value of 12.5 degrees, corresponding to a ratio closer to 5. The relationship between the observed axial ratio and the inclination angle depends upon the thickness of the disk, of course, and so a comparison of these figures can lead to conclusions about the disk parameters. From his 1958 photometry, de Vaucouleurs calculated that the true axial ratio for the flat disk is $q = 0.042$.

For comparison with these photometric measures of M31's dimensions, it can be pointed out that the most distant open cluster so far detected is 110 arcmin from the nucleus (Hodge 1979), while the H I profile can be traced out to at least 163 arcmin (Emerson 1974).

4.3 Profiles

Most photometry of M31 has concentrated on finding the shape of the luminosity profile. Redman and Shirley (1937) were among the first to look at this problem quantitatively, and they characterized their results for the inner parts of the galaxy (now recognized as being dominated by the bulge) by a formula similar to that found for elliptical galaxies (see Chapter 1 for a discussion of Reynolds' similar and much earlier conclusion). The first detailed deconvolution of the profile into its bulge and disk components was made by de Vaucouleurs (1958) (Figure 4.1). He concluded that the inner part of M31 showed an elliptical galaxy-type profile (following his $r^{1/4}$ law), while the outer parts followed an exponential law, as do the disks of other spiral galaxies.

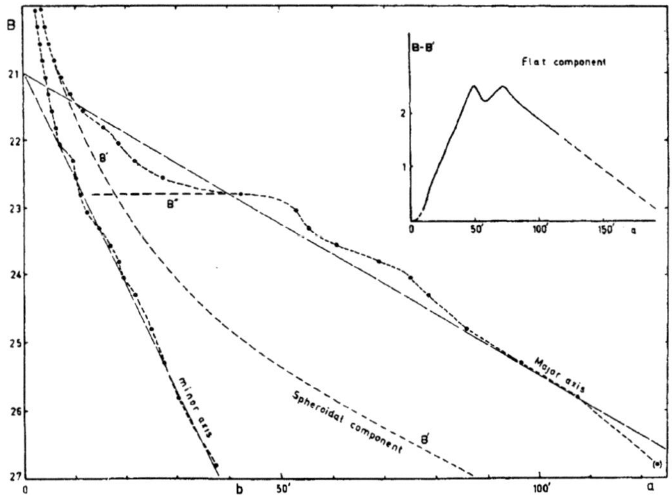

Figure 4.1 Profiles of M31 derived by de Vaucouleurs (1958). Observed major and minor axes are compared and the decomposition into a spheroidal component and flat component are shown (from de Vaucouleurs, 1958).

More recent studies of the photometric profile have served the purpose primarily of confirming de Vaucouleurs' general conclusions and of extending the data to other wavelengths and positions. The differences found, for example, between the four different axes are illustrated in Figure 4.2. The minor axis profiles from four different papers are compared in Figure 4.3, showing fairly good agreement, with a spread averaging about 0.4 mag. Figure 4.4 similarly compares three photometric studies of the major axis. Table 4.2 is taken from the paper by Walterbos and Kennicutt (1988), and can be taken as the most up-to-date determination of the mean optical profile for M31. Table 4.3 comes from Walterbos and Kennicutt (1988), who deconvolve their four-color data and provide the most complete set of parameters describing the bulge and disk now available.

Table 4.2
Average Radial Light Profile of M31 in U, B, V and R
(from Walterbos and Kennicutt, 1988)

Position	B	U–B	B–V	B–R	Position	B	U–B	B–V	B–R
4'	19.53	0.58	0.97	1.77					
5	19.80	0.56	0.98	1.78					
6	20.04	0.54	0.99	1.79	58'	23.24	0.26	0.90	1.60
7	20.26	0.52	1.01	1.79	61	23.34	0.26	0.89	1.58
8	20.45	0.51	1.01	1.79	64	23.43	0.26	0.88	1.57
8	20.61	0.51	1.00	1.79	67	23.50	0.25	0.87	1.56
10	20.76	0.51	1.01	1.79	70	23.61	0.23	0.87	1.54
12	21.03	0.52	1.01	1.79	73	23.77	0.27	0.85	1.53
14	21.23	0.51	1.00	1.78	76	23.94	0.26	0.86	1.53
16	21.41	0.52	1.00	1.78	79	24.07	0.26	0.86	1.53
18	21.58	0.51	1.00	1.79	82	24.19	0.26	0.84	1.51
20	21.76	0.52	1.00	1.79	85	24.35	0.24	0.84	1.51
22	21.96	0.51	1.01	1.80	88	24.47	0.23	0.82	1.50
24	22.12	0.51	1.02	1.80	91	24.59	0.22	0.81	1.50
26	22.22	0.50	1.01	1.78	94	24.73	0.20	0.80	1.50
28	22.28	0.48	1.00	1.75	97	24.84	0.20	0.80	1.47
30	22.34	0.47	0.98	1.73	100	24.94	0.20	0.81	1.44
32	22.42	0.47	0.98	1.72	103	25.10	0.22	0.80	1.45
34	22.52	0.44	0.98	1.72	106	25.24	0.20	0.80	1.43
36	22.63	0.43	0.99	1.72	109	25.36	0.16	0.77	1.43
38	22.71	0.39	0.99	1.72	112	25.49	0.17	0.80	1.42
40	22.73	0.35	0.97	1.69	115	25.62	0.11	0.81	1.41
43	22.82	0.33	0.96	1.68	118	25.70	0.19	0.81	1.38
46	22.82	0.29	0.92	1.63	121	25.87	0.13	0.83	1.38
49	22.89	0.26	0.90	1.62	124	25.91	0.16	0.81	1.30
52	23.00	0.25	0.90	1.61					
55	23.13	0.26	0.91	1.62					

Using a separation by color, Walterbos and Kennicutt (1988) calculated that the spheroidal component comprises 39% of the B light of M31, with the remainder in the exponential disk. Kent (1987) has compared his CCD-derived profile with anisotropic dynamical models and Figure 4.5 shows the quite good fit he was able to obtain.

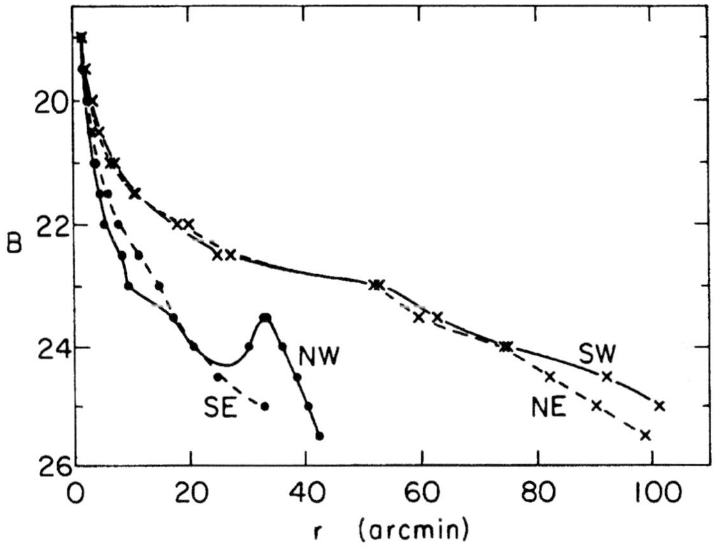

Figure 4.2 Comparison of the profiles on either side of the nucleus for the major and minor axes (from Hodge and Kennicutt, 1982). The inner part of the NW axis is depressed by dust obscuration and the bump on that axis is NGC 205.

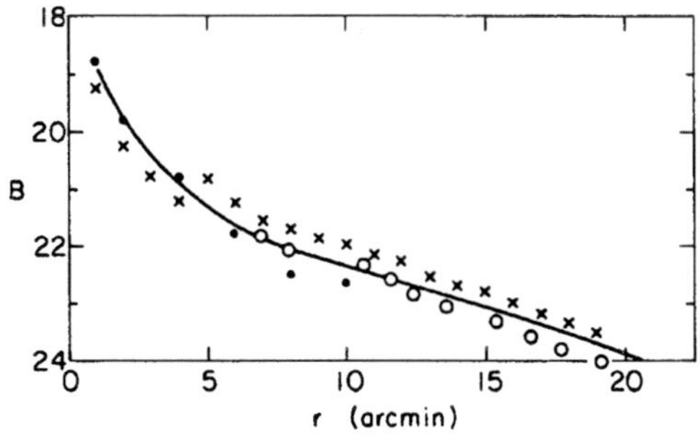

Figure 4.3 Minor axis photometry compared (with NGC 205 removed). Circles are data of de Vaucouleurs (1958), dots are from Hoessel and Melnick (1980), crosses are from Lyngå (1959), and the line is from Hodge and Kennicutt (1982).

Table 4.3
Bulge and Disk Parameters for M31
not corrected for extinction
(adapted from Walterbos and Kennicutt, 1988)

Parameter	Value
BULGE	
effective radius, r_e	2.1 kpc
true axial ratio	0.63
B intensity at r_e	22.2 mag arcsec^{-2}
U–B	0.50 ± 0.03
B–V	1.05 ± 0.03
B–R	1.75 ± 0.03
integrated magnitude in B	5.39
DISK	
central surface brightness	21.60 B mag arcsec^{-2} ±0.10
scalelength	
U	7.2 ± 0.4 kpc
B	6.1 ± 0.3 kpc
V	5.6 ± 0.3 kpc
R	5.5 ± 0.3 kpc
K	4.1 kpc
color in center	
U–B	0.55
B–V	1.05
B–R	1.80
color at 20 kpc	
U–B	0.00
B–V	0.70
B–R	1.37
color gradient	
U–B	0.03 ± 0.01 mag/kpc
B–V	0.018 ± 0 mag/kpc
B–R	0.019 ± 0.008 mag/kpc
integrated magnitude	
U	5.22
B	4.93
V	4.12
R	3.40
bulge to disk ratios, $L_b/(L_b + L_d)$	
U	0.35
B	0.39
V	0.45
R	0.45

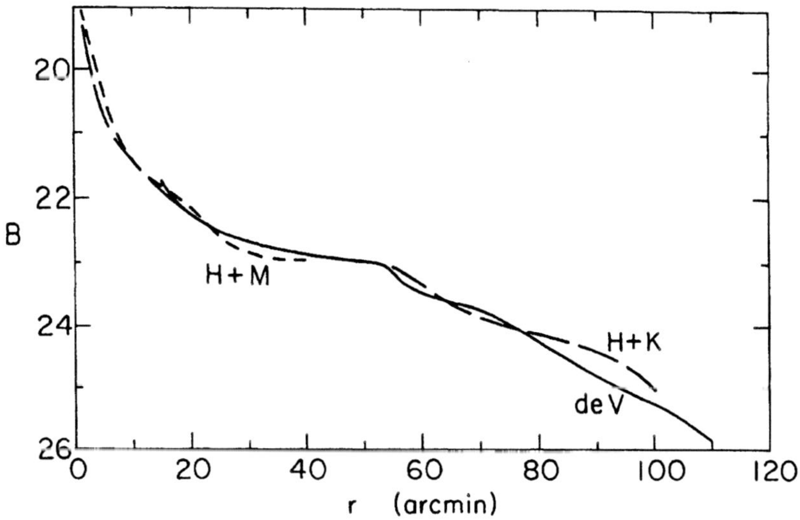

Figure 4.4 The major axis profile of M31, from Hoessel and Melnick (H & M, 1980), de Vaucouleurs (de V, 1958) and Hodge and Kennicutt (H & K, 1982).

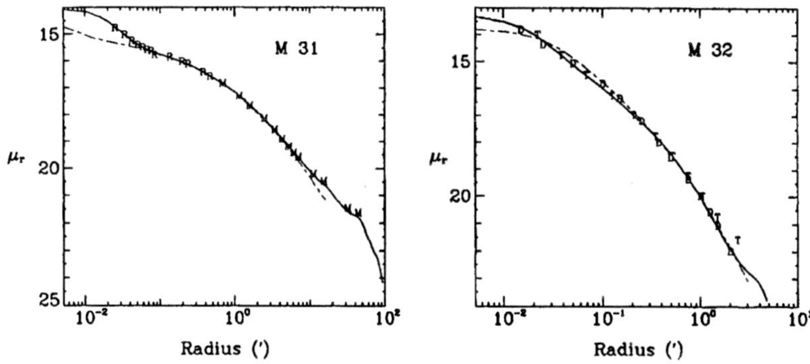

Figure 4.5 Logarithmic radial profile, comparing Kent's (1987) line with values from Redman and Shirley (R, 1937) and Hoessel and Melnick (M, 1980) (from Kent, 1987).

Near infrared profiles (*e.g.*, Iijima *et al.* 1976, Matsumoto *et al.* 1977 and Hiromoto *et al.* 1983) show a similar composite nature (Figure 4.6), but a longer scale length.

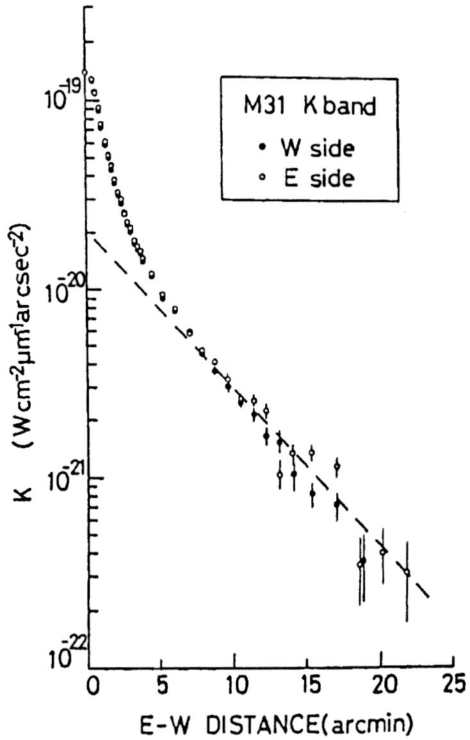

Figure 4.6 Infrared profile along the east-west axis of M31 (from Hiromoto et al. 1983).

4.4 Shape

Ellipticity

The optical shape of M31 has generally been described, to a first approximation, in terms of the position angle (PA) and ellipticity (ϵ) of proscribed ellipses fit at different distances from the center. (The position angle is the angle measured from north to the position of the major axis of the ellipse. The ellipticity is defined by $\epsilon = 1 -$ b/a, where a is the major axis and b is the minor axis of the ellipse.) After the first publication of isophotes by Williams and Hiltner (1941), there was a more quantitative determination of the shape parameters in the de Vaucouleurs (1958) study.

Figure 4.7 illustrates the behavior of the ellipticity curve with distance along the major axis, from the photometry of de Vaucouleurs (1958) and Hodge and Kennicutt (1982). The shape is like that typical for spiral galaxies, with a small value of ϵ at the center, increasing outwards to a maximum value, which for M31 is at a distance of about 50 arcmin. There is a slow and marginally significant decrease in ellipticity outwards from that position. Following the usual interpretation of this kind of curve, the central area has a small value because it is nearly spherical in three-dimensional shape, the maximum at 50 arcmin represents the value for the flat, thin disk, tilted at the angle of 12.5°, and the decrease in ellipticity outwards results from the increasing influence of a spherical halo on the measured luminosity density.

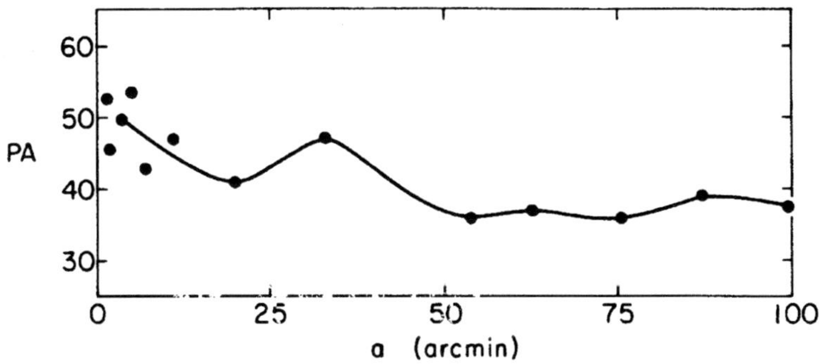

Figure 4.7 The variation of values for the position angle of ellipses fit to M31's isophotes (from Hodge and Kennicutt, 1982).

Position Angle

The variation of PA with distance from the center is also of interest. The mean value found by de Vaucouleurs (1958) was 37.7° ± 0.2°. As several astronomers have noted, the position angle is not constant (Figures 4.8 and 4.9. See also Peterson *et al.* 1977 and Feibelman 1979). The inner twists have been attributed to the probable presence of a weak bar or triaxiality in the central region (Lindblad 1956, Lyngå 1959, Stark 1977, Sharov and Lyutyi 1980), while the outer parts of the disk show a twist that is related to an apparent warp, discussed below. Furthermore, Kent (1987) noted that the shape of the isophotes of the bulge is somewhat "box-shaped" (Figure 4.9).

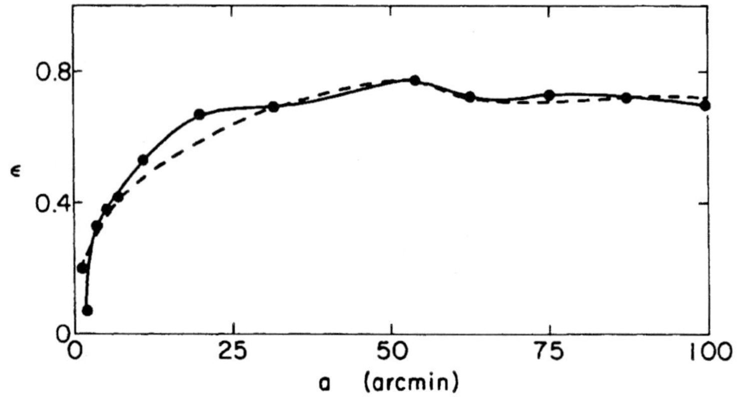

Figure 4.8 The ellipticity of the isophotes of M31 as a function of major axis distance. The dashed line is from de Vaucouleurs (1958) and the solid line and dots from Hodge and Kennicutt (1982).

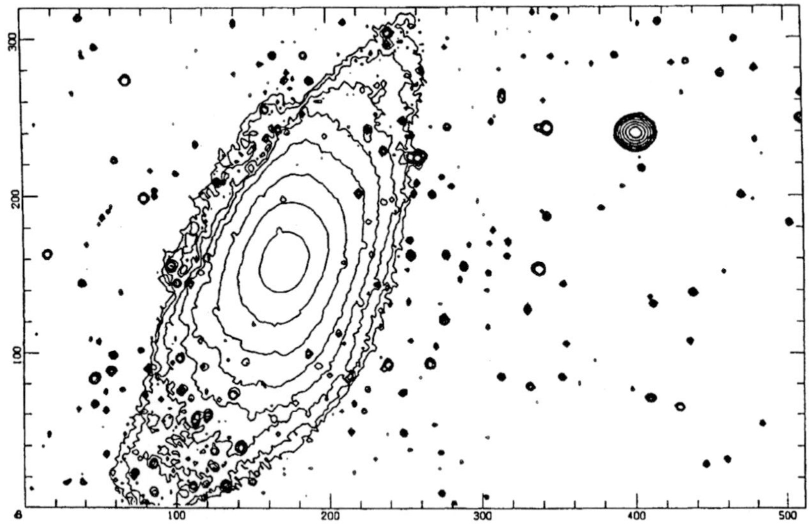

Figure 4.9 CCD photometry from Kent (1987) of the inner isophotes of M31. The image of M32 is at the upper right.

Starting with the work of Lindblad (1956), several astronomers have attempted to use photometric measures to derive the true three-dimensional shape of the bulge-bar area. Ruiz (1976) assumed a spheroidal bulge and demonstrated that the symmetry axes of the bulge and disk cannot then be parallel. On the other hand, Stark (1977) assumed the axes to be parallel, but allowed the bulge to be triaxial. Figure 4.10 shows one of his solutions that fit the observed variations in position angle. Matsumoto *et al.* (1977), from near-infrared surface photometry, at 7000–9000 Å and 2.2μm, respectively, find that the PA and ϵ values at the longer wavelengths change more markedly than at the shorter, and conclude that the bar-like structure is caused primarily by the arrangement of giant stars.

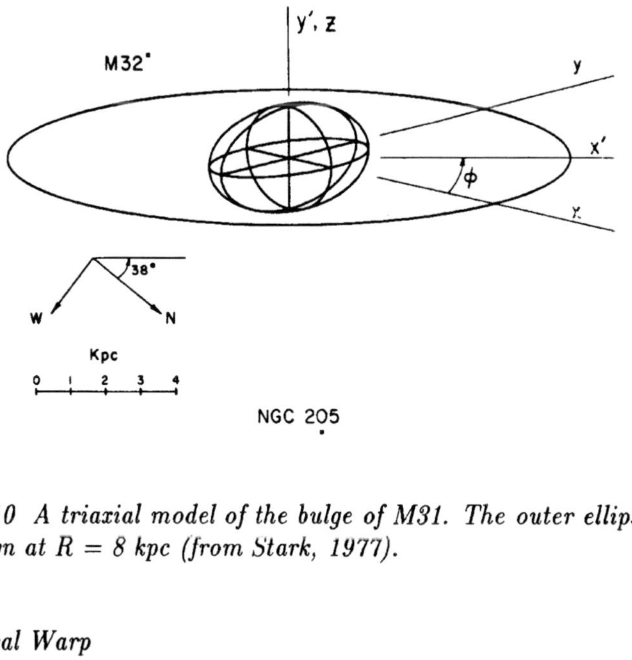

Figure 4.10 A triaxial model of the bulge of M31. The outer ellipse is the disk, drawn at R = 8 kpc (from Stark, 1977).

The Optical Warp

Baade (1963) found that, when he tried to trace the yellow giant stars of M31's disk out along the major axis, they departed from the straight line axis in a dramatic, antisymmetrical way, with the northeast axis turning towards the north and the southwest axis turning to the south. Arp (1964) also noted the effects of a possibly related asymmetry on the distribution of H II regions. This phenomenon was further explored by Innanen *et al.* (1981), who detected a turning of the major axis at a distance of about 1.7°. The effect was confirmed by Walterbos and Kennicutt (1987, 1988). The H I maps (*e.g.*, Newton and Emerson 1978) also show an antisymmetric warp, but not conspicuously until a distance from the center of 2.2° is reached. The cause of the warp has been explored by various geometrical and dynamical modeling, such as the models of Henderson (1979) and Brinks and Burton (1984). It is further discussed in relation to H I observations in Chapter 5.

4.5 Color Distribution

Several of the early measurements of M31's structure included determinations of the distribution of color in its image. For example, Redman and Shirley (1937) found that the central bulge is redder than the outer, spiral disk. A color study in the U, B, V system was included in de Vaucouleurs' (1958) analysis, which showed that the colors, particularly U−B, were correlated with the presence of dust lanes as well as with the dominance of different populations of stars. Figure 4.11 shows his data for the mean of two east-west cuts through the galaxy at positions 30 arcmin from the nucleus. The color of the disk is bluest where the measures cut across the arms and reddest where dust lanes are encountered.

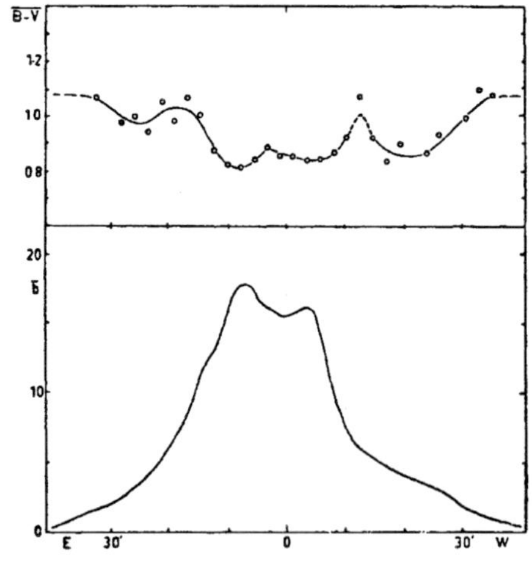

Figure 4.11 The color (B−V) and luminosity distribution across the major axis 30 arcmin north and south of the nucleus, averaged (from de Vaucouleurs, 1958).

Lyngå's (1959) photometry also covered the color distribution, though in the older m_{pg}, m_{pv} color system. His color curves resembled those of de Vaucouleurs to the extent that both showed a reddening in the areas of deep dust lanes. However, they did not agree on the extent of the reddening of the image at large distances from the center, where the faint light levels probably account for any disagreements.

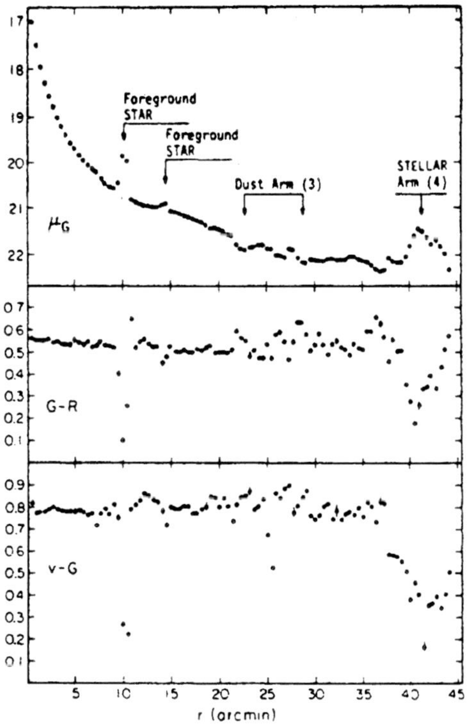

Figure 4.12 Color curves for the southwest half of the major axis (from Hoessel and Melnick, 1980).

More recent color work, in still a third color system, was published by Hoessel and Melnick (1980). Using the Thuan-Gunn system in order to keep sky brightness variation effects to a minimum, they found v−G and G−R color profiles from the center out to distances of about 45 arcmin along the major axis and to 8 arcmin on the minor axis. They detected a shallow slope outwards within the bulge from G−R of 0.57 at the center to 0.51 at the edge of the bulge, next to the inner dust arm. The latter showed a measurable reddening of about 0.1 mag in that color on the major axis and about 0.3 mag on the south preceding portion of the minor axis. The north following (NW) part of the minor axis showed reddening only over one small area at a distance from the nucleus of 7.8 arcmin. At this position lies the dust lane designated as D507 (Hodge 1980). Sample data are shown in Figure 4.12.

Kent (1987) determined a color profile for the central parts of M31 from his CCD frames, publishing a table of g−r colors that covers the inner profile in 6 arcsec intervals and the outer profile in larger intervals, out to a distance of 14.6 arcmin. In this color system the color is nearly constant at $g - r = 0.55$, except for a slightly redder value (0.60) at the nucleus.

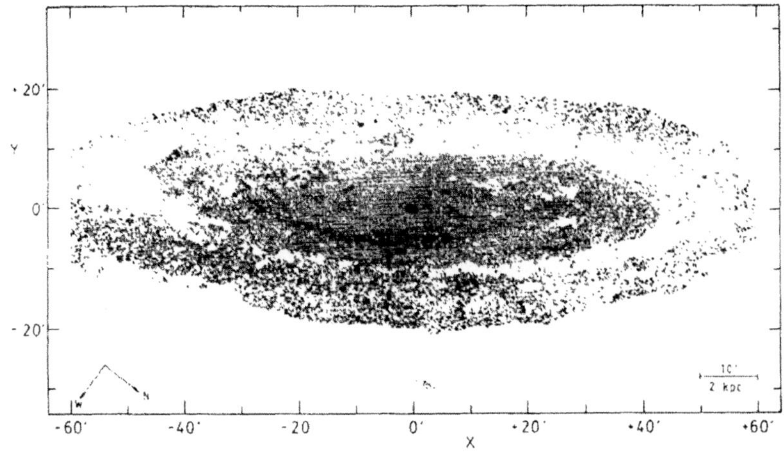

Figure 4.13 Color image of M31, from U and V images obtained from digitized photographs (from Walterbos and Kennicutt, 1987).

Walterbos and Kennicutt (1987) used modern digital techniques to analyze Schmidt plates of M31 in four colors, U, B, V, and R, for the whole detectable image. With resolutions ranging between 8 arcsec and 16 arcsec, which corresponds to 25 pc to 50 pc, they reduced the data in the form of color maps, which extend out to about 14 kpc and that have accuracies of about 0.1 to 0.2 mag. Table 4.2 is taken from their paper and gives what is probably the best and most complete set of global color data for M31. The color maps (see Figure 4.13 for a sample) show the redder color of the major dust lanes and the relatively bluer color of the bright 10 kpc ring. They find, furthermore, a continuing trend for bluer colors for distances beyond 10 kpc, out to their effective limits at radial distances of about 20 kpc. This trend is probably caused by a changing dust-to-gas ratio, an age gradient and a metallicity gradient for the giant stars. The bulge was found to have a nearly constant color at $B-V = 0.95$, $U - B = 0.50$, indicating the dominance of an old population of stars.

Infrared photometry has also been used to explore the color differences in different regions of M31. For example, Hiromoto *et al.* (1983) found that the difference $I' - K$ is nearly constant in the bulge, but increases somewhat in the outer parts (they carried out photometry to distances of 17 arcmin along the east-west axis). Their interpretation of the curve is that there is a nearly constant extinction of $A_V = 0.5$ in the dust lanes; it rises to values of ~ 1.4 in the 4–8 kpc region, and then increases further to as much as 2.4 mag in the outer parts of the disk (see Chapter 12 for a comparison with other data on extinction and for a discussion of results from IRAS measurements).

4.6 The Nucleus

M31 has a nearly unresolved, distinct nucleus (see Lord Rosse's sketch in Figure 1.5). This feature has been noted by most astronomers who have done photometry at high resolution. One of the first modern studies of its structure was published by Johnson (1961), who obtained photographic profiles along the major and minor axes using plates from the McDonald Observatory 82-inch reflector. He found a steep profile inside a distance from the center of about 4 arcsec. Subsequently, Kinman (1965) used photoelectric techniques and careful deconvolution to derive a detailed, calibrated profile (Table 4.4). He used the luminosity profile to examine the possible models of the nucleus that might be consistent with the remarkably rapid rotation found in the nucleus by Lallemand *et al.* (1960), as described further in Chapter 7.

Early looks at the color of the nucleus seemed to indicate that there was only a small (~ 0.05 mag) difference between it and the surrounding bulge. Sandage *et al.* (1969) carried out multicolor photometry (Figure 4.14) to determine whether there was any evidence for population differences between the nucleus and its surroundings, and to make direct comparisons in the infrared with infrared profiles of our Galaxy. They found evidence for a variation in the U−B color that did not show up for longer wavelengths, but were unable to decide from their data whether this variation was due to differences in the stellar luminosity function or in the metal abundances. A third possibility, that the effect may be due to dust was rejected (possibly prematurely; see Chapter 12).

Table 4.4
Central Surface Brightness along Major Axis
from Photoelectric Observations
(Kinman 1965)

Position along Major Axis	B per square Sec of Arc	Position along Major Axis	B per square Sec of Arc
26".98 N	18.01	6".74 S	17.27
20.23 N	17.78	13.49 S	17.55
13.49 N	17.54	20.23 S	17.82
6.74 N	17.23	26.98 S	17.98
0.00	16.35		

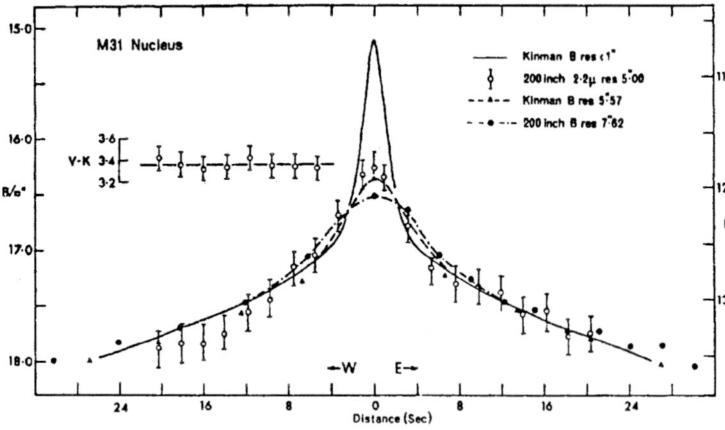

Figure 4.14 Structure of the nucleus of M31 in B and K (from Sandage et al. 1969).

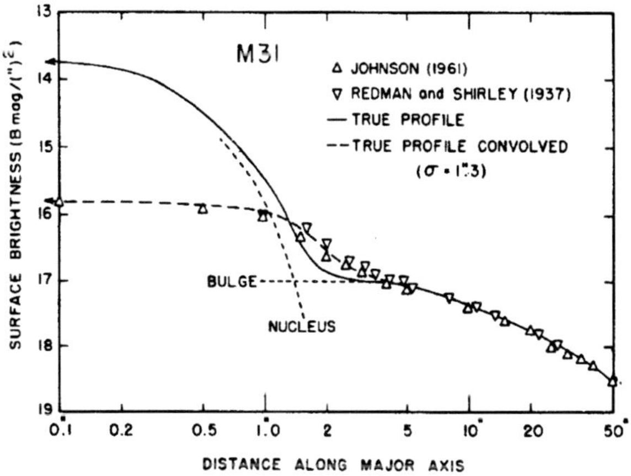

Figure 4.15 Profile of the nucleus from Project Stratoscope balloon observations. Outer observations are from earlier photographic studies (from Light et al. 1974).

Because of the small angular size of the nucleus, further progress in exploring it did not come until better resolution was available. Light *et al.* (1974) used Project Stratoscope, a balloon-borne telescope, to register the image of the nucleus at a resolution of 0.2 arcsec. They found the nucleus to be elliptical with half-intensity levels at 1.0 by 1.6 arcsec. The position angle of the nucleus is different from that of the bulge by nearly 30°; they quote a value of 63 ± 5°. It is clearly resolved as a distinct feature from the bulge. Its scale height is 0.5 pc and it has a mass, based on velocity dispersion measurements (see Chapter 8), of $\sim 10^8$ solar masses. Infrared speckle observations (Spillar *et al.* 1990) confirm the small size of the nucleus, which at K band wavelengths has a best-fitting FWHM size of 0.1 arcsec.

It is displaced from the center of M31 by 0.5 arcsec, a fact even observed in the infrared (Mould *et al.* 1989) and a source of continuing puzzlement. As described in Chapter 8, there is new evidence that the nucleus may not be spheroidal, but disk-shaped and that the mass may include a significant amount of matter in the form of a black hole.

4.7 Summary

The optical structure of M31 has been explored from its subarcsecond nucleus out to its faint outer envelope, 30 kpc from the nucleus. It can be divided into four components: the nucleus, the ellipsoidal central bulge, the disk, and the outer halo.

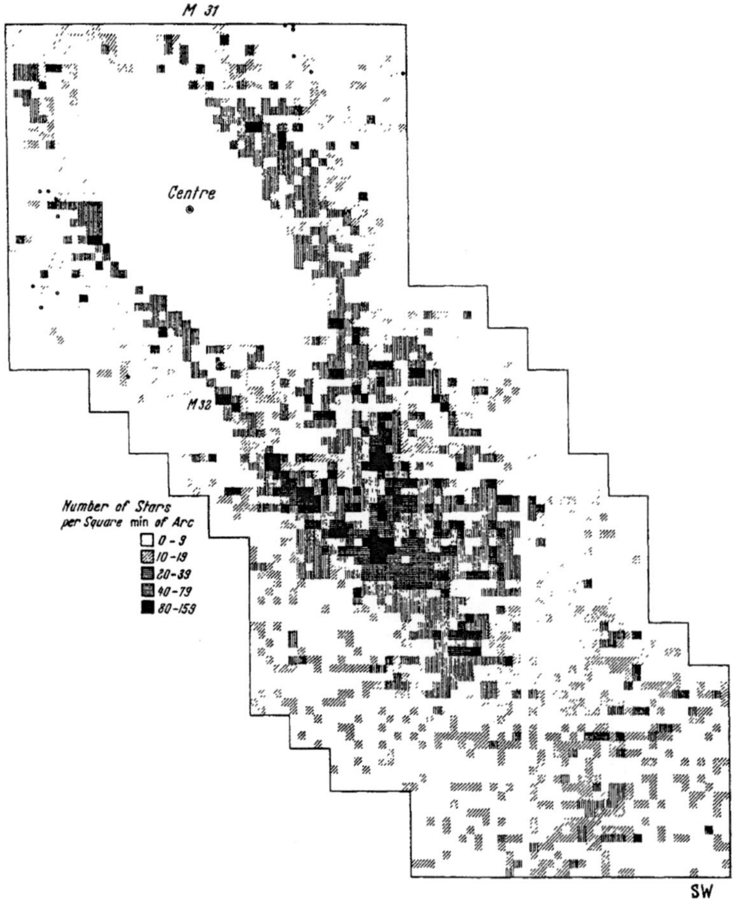

Figure 4.16 The distribution of stars in the southwest arms, from star counts on the Palomar Atlas (from Reddish, 1962).

CHAPTER 5

THE NEUTRAL HYDROGEN CONTENT

The study of the content, distribution, and kinematics of the H I in M31 has led to a great deal of important understanding about the galaxy. This chapter concentrates on the question of the location and morphology of the hydrogen gas, while the vast amount of information that comes from the measurements of the velocities of the gas are discussed in Chapter 7.

5.1 First Results

The first extensive survey of 21-cm emission from M31 was that published by Leiden astronomers, based on observations taken with the 25-meter radio telescope at Dwingeloo (van de Hulst *et al.* 1957). With a beam-width of a little over 0.5° and a velocity resolution of 29 km/sec, this survey was the first important step in exploring the distribution of gas in the galaxy and in interpreting its velocities in terms of a rotation curve. Line profiles were obtained at 20 locations along the major axis (Figure 5.1). The primary result regarding the distribution of gas was the discovery that the H I showed a maximum brightness at two points along the major axis about 1° away from the center on either side; the central region showed a gas density minimum that was about 0.25 the maximum value.

A somewhat more complete survey was subsequently published by Burke *et al.* (1963), who used the NRAO 300-foot radio telescope, which provided a spatial resolution of 10 arcmin. They explored both the major and minor axes and confirmed the Dwingeloo results on the distribution of gas.

As the 1960s saw an increase in the number and power of radio telescopes, more work on M31 became available with increasing resolution and sensitivity. Table 5.1 is a listing of the surveys published between the years 1957 and 1990, when the last giant survey, carried out at the VLA, was published.

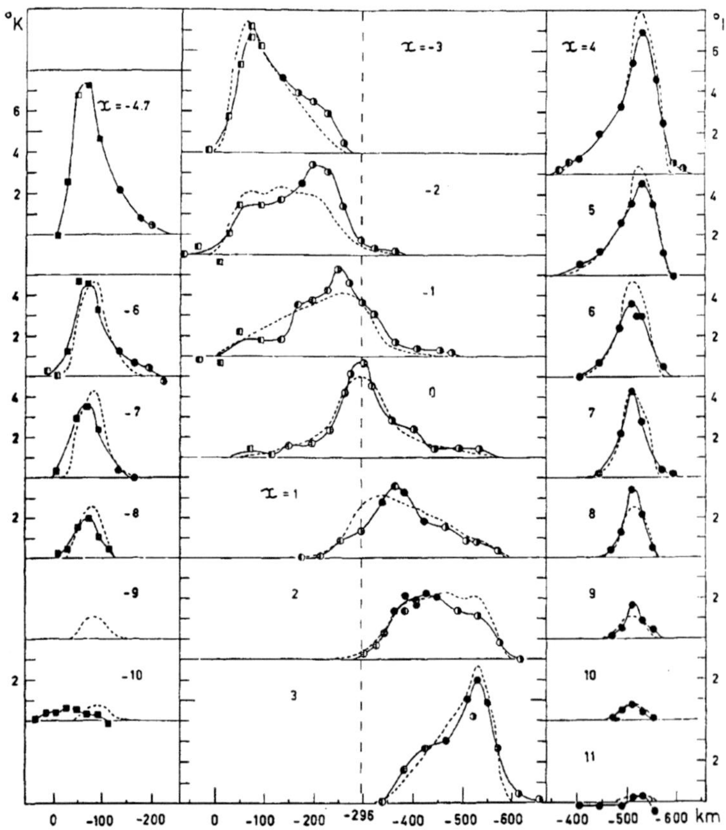

Figure 5.1 H I velocity profiles for various positions along the major axis of M31 (from van de Hulst et al. 1957).

5.2 The Ring

The main feature of the distribution of H I in M31, hinted at by the Dwingeloo survey and eventually confirmed by the studies that mapped the entire galaxy (Table 5.1), is the giant ring of gas, centered on the nucleus and with a maximum at an apparent distance along the major axis of a little under 1°. One of the best of these early maps of the ring is that obtained by Roberts (1966), reproduced in Figure 5.2. The features of this ring were further refined by additional observations. Some were based on single-dish observations (filled-aperture data), for example with the Jodrell Bank 250-foot telescope by Davies and Gottesman (1970) and the 100-meter Effelsberg telescope by Cram *et al.* (1980) (Figure 5.3).

Other studies of the ring, particularly of the details within it, were based on interferometry (see, for example, Bajaja and Shane 1982 and Unwin 1980a, b) or on combined observations (Brinks and Shane 1984 and Braun 1990).

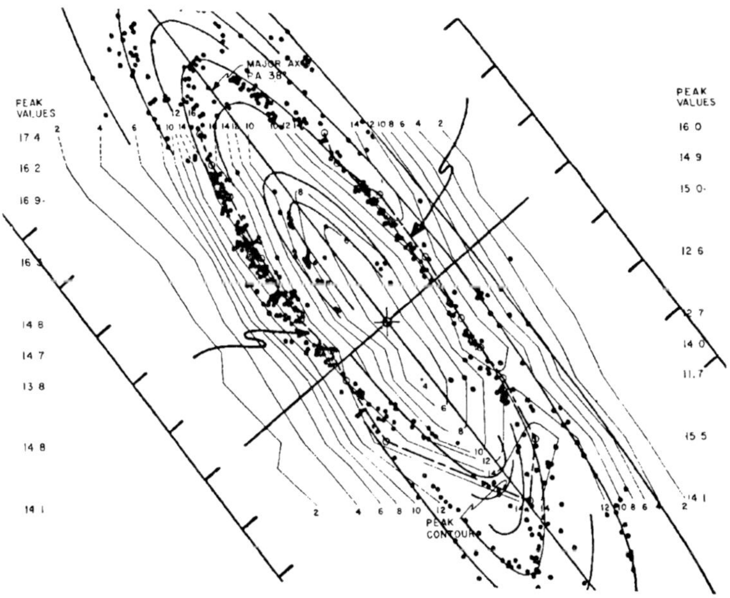

Figure 5.2 Comparison of 21-cm emission contours with the distribution of H II regions (from Roberts, 1966). North is at the top and east to the left.

One of the best ways to define the features of the ring is to transform the data to face-on, so as to have the entire structure available for analysis. The transformation from the inclined image to a face-on image, however, is not simple, because of the large inclination angle, irregularities in the velocities, the warp, and the thickness of the gas disk. Sofue and Kato (1981) used the Effelsberg data to combine the velocity curves into a velocity field diagram, which then was used to obtain a velocity-to-space conversion. The result is a face-on map of the H I gas density in the plane of M31 (Figure 5.5). Note that Sofue and Kato did not consider the blend

Table 5.1
HI observations of M31 (adapted from Brinks and Shane, 1984)

Authors	Telescope	HPBW	Velocity Resolution (km s^{-1}	Sensitivity (K)	Notes
Whitehurst and Roberts, 1972	Green Bank, 140 ft	$21'$	5.5	-	(1)
Emerson, 1974	Cambridge Half Mile	$1'.5 \times 2'.2$	39	1.5	(2)
		$6' \times 9.'4$	39	0.2	(3)
Roberts and Whitehurst, 1975	Green Bank, 300 ft	$10'$	6.6	-	(4)
Newton and Emerson, 1977	Cambridge Half Mile	$3'.6 \times 5'.5$	16	0.3	(5)
Landecker, 1978	Penticton Synthesis Telescope	$2' \times 3'$	5	2.0	
Cram, Roberts and Whitehurst, 1980	Effelsberg, 100-m	$8'.8$	5.5	0.1	
Unwin, 1980	Cambridge Half Mile	$0'.8 \times 1'.2$	16	3.6	
Bajaja and Shane, 1982	WSRT, 1.5 km	24" × 36"	27	1.4	(6)
Brinks and Shane, 1984	WSRT, 1.5 km	24" × 36"	8.2	2.5	
Braun, 1990	VLA, B, C, and D configurations	$10''$	5	6.6	(7)

(1) Single profile on the nucleus.

(2) Northern and central part.

(3) Southern part.

(4) South West major axis.

(5) Two fields on the major axis, centered $150'$ on each side of the nucleus.

(6) Central region. Velocities undersampled.

(7) Northeast half.

of velocities of the disk and warp and therefore that their transformation is unreliable at large distances from the center (for $r \gtrsim 15$ kpc).

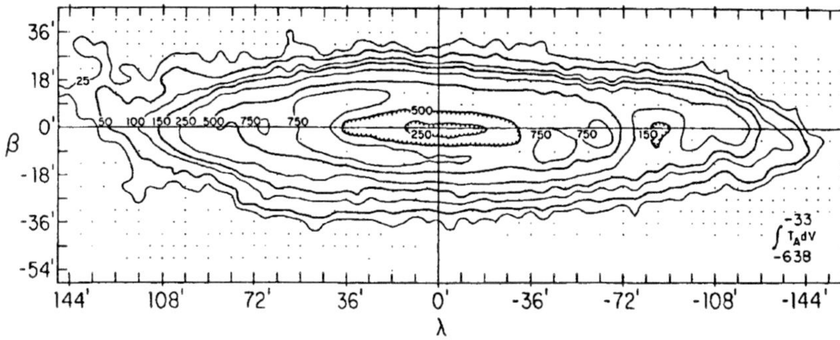

Figure 5.3 Integrated H I contours of M31 (from Roberts and Cram, 1980).

Figure 5.4 Grey-scale representation of the high resolution WSRT survey of M31 neutral hydrogen (from Brinks, 1984).

Sofue and Kato's map shows that the H I ring is a large-scale concentration of gas centered at a distance of about 12 kpc, with an extent from 8 to 15 kpc from the nucleus. The column density averages about 4×10^{20} (HI) atoms cm^{-2}, which corresponds to a number density about equal to that of the H I in the solar neighborhood, if the thickness is on the order of 100 pc. The interior of the ring has densities that fall to values of approximately 1×10^{20} (HI) atoms cm^{-2} at 2–3 kpc from the center.

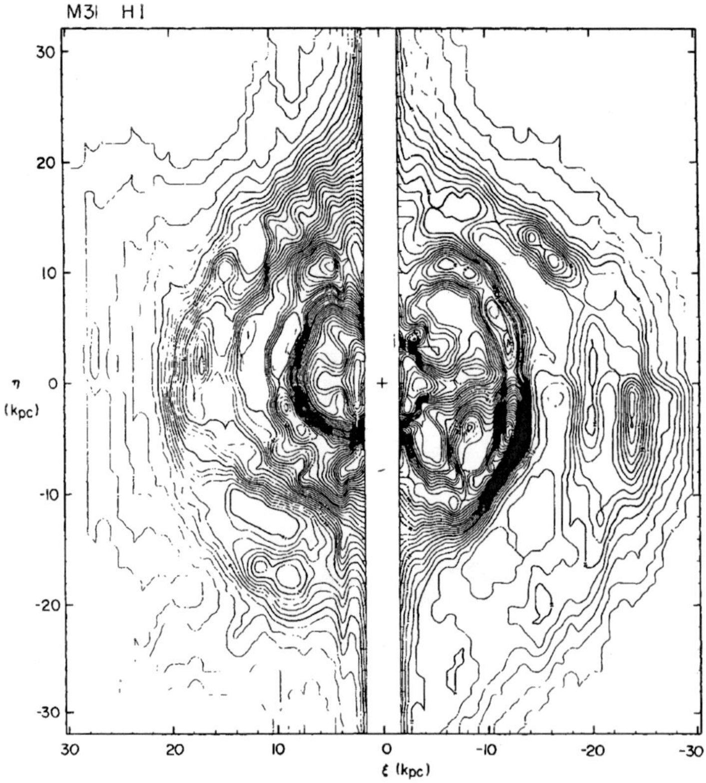

Figure 5.5 The face-on surface column density of H I in M31, reconstructed by Sofue and Kato (1981).

5.3 Spiral Arms

Although early surveys could show the gross features of the gas morphology, they did not at first have sufficient resolution to delineate any spiral structure (in this sense they paralleled the slow recognition of the spiral arms on the part of nineteenth century optical observers). With the superior data being obtained in the 1970s, however, a certain amount of fine structure within the ring was found to correspond to optically visible arm features. For example, Table 5.2, which is based on the study published by Davies and Gottesman (1970), gives details on their characterization of the H I in Baade's (1958) designated spiral arms. The neutral gas density in the arms is not uniform along them, but noticeably patchy in character. Figure 5.6 shows Sofue and Kato's (1981) plot of the H I ridges, obtained by subtracting a smoothed background gas disk, and it indicates the clear presence of arm-like features. These scientists describe a "giant patch" of H I, with a peak column density of 8×10^{20} atoms cm^{-2} and a size of 2×5 kpc located on the ridge of one of the arms (S-13, in their designation, which corresponds to Baade's arm S5).

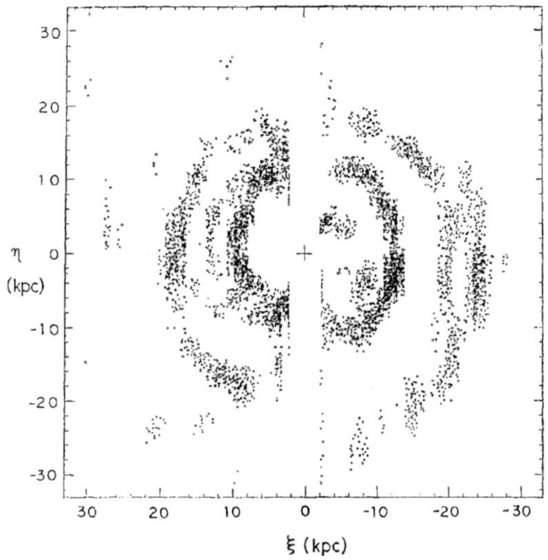

Figure 5.6 Distribution of H I ridges in M31, obtained by subtracting a smoothed global gas distribution (from Sofue and Kato, 1981).

There are some rather distant, outer arms in Sofue and Kato's maps, including one in the southwest, a trailing arm that starts at a distance of about 19 kpc, crosses the major axis at 22 kpc, and continues on outward to a radial distance of 30 kpc. It is connected by a bridge of gas to the main ring. Similarly, there is a "bright" outer arm in the NE, extending perhaps as far as 32 kpc from the center. In addition, numerous faint outer spur-like features are found, but only in the NE portion, which is less regular in its gas morphology than the SW portion of the outer disk.

The 21-cm data do not easily yield a spiral pattern, even when refined as in Figure 5.6.

5.4 The Radial Distribution of Gas

Moderate resolution mapping of the H I in M31 provided us with good information on the large-scale radial distribution of H I in the disk. The most obvious feature, of course, is the "gas ring" (Section 5.2), but there are other features of interest. Figure 5.7 is a series of plots derived by Sofue and Kato (1981) that show the radial distribution of H I density for the four quadrants of their face-on map, plus the mean curve for the entire galaxy. Gross properties of these curves are the "hole" in the center, the maximum at R = 12 kpc, and a "plateau" of gas that extends out to about 22 kpc, beyond which is a gradual, approximately exponential, decrease outwards to values near zero at distances of $\sim$ 35 kpc. In detail, the four quadrants show considerable difference: the W quadrant has a two-valued peak (at 7 and 13 kpc), as well as a secondary maximum at 21 kpc. The S quadrant is also double-peaked, and it has a very steep decrease in density beyond the peaks in the ring. The ring does not have a uniform width, being about 20% wider in the southwest than in the northeast. All of these differences can be correlated to details in the patterns that are delineated in the face-on maps.

There is an asymmetry in the radial distribution, with an elongation towards the north, where the outer gas extends approximately 3–5 kpc farther out than the mean, at a density level of 10^{19}cm^{-2}.

5.5 The Warp

While the apparent warp in the disk of M31 is not conspicuous in its optical images except in the most recent observations, a warp in the HI disk has been recognized for a long time, both because the ridge line of H I density turns away from the straight-line extension of the major axis (Roberts 1966) and because of the presence of double-peaked profiles in some areas of the galaxy (Emerson 1976; Shane 1975). Newton and Emerson (1977), based on their aperture synthesis observations of the outer parts of M31, found a warping of the plane that showed up as a displacement of the H I peak at distances on the order of 30 kpc from the nucleus, both in the SW and the NE (Figure 5.8). Several attempts to model the warp have been published (*e.g.*, Henderson 1979 and Sawa and Sofue 1982), and subsequent detailed studies of the galaxy have had to take it into account in order to make sense of the observations (*e.g.*, Bajaja and Shane 1982, who deconvolved the H I into two components, a warped outer disk and a flat inner disk).

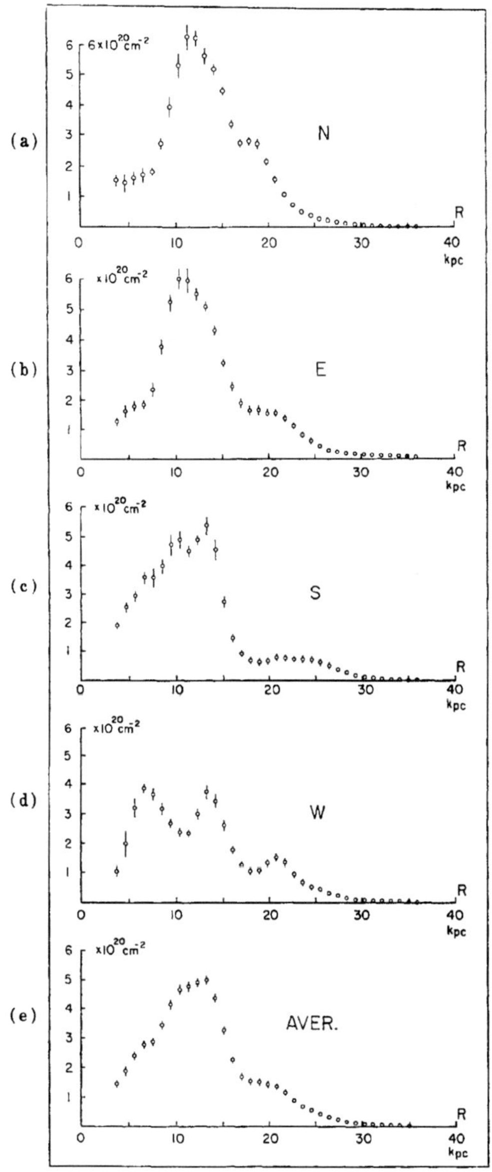

Figure 5.7 Comparisons of the H I radial distributions in four quadrants and for the whole galaxy (bottom panel) (from Sofue and Kato, 1981).

Table 5.2
The Gas Content of Baade's Spiral Arms in M31
(After Davies and Gottesman, 1970)

Arm	Distance from center to arm crossing	Main components	Surface density of H I	Comments
NI	3.4′	Dust	Low	Only small amounts of HI or HII
N2	8.0′	Dust	Low	
N3	25′	Dust and HII regions	Intermediate	Increasing gas density from the center
N4	50′	HII regions and supergiants	Maximum	Star formation at its peak (Baade), maximum hydrogen density
N5	70′	HII regions and supergiants	Maximum	
N6	91′	Blue supergiants, few HII regions	Intermediate	Sharply falling gas density, HII and blue supergiant density
N7	110′	Blue supergiants	Low	
SI	1.7′	Dust	Little	
S2	10.5′	Dust	Intermediate	Increasing amounts of gas in progressing out from center
S3	30′	Dust and HII regions	Intermediate	
S4	47′	HII regions and supergiants	Maximum	Star formation at its peak (Baade), maximum HI density
S5	66′	HII regions and supergiants	Maximum	
S6	95′	Blue supergiants, few HII regions	Intermediate	Still appreciable HI density, but decreasing HII and blue supergiant population
S7	116′	Blue supergiants	Intermediate	

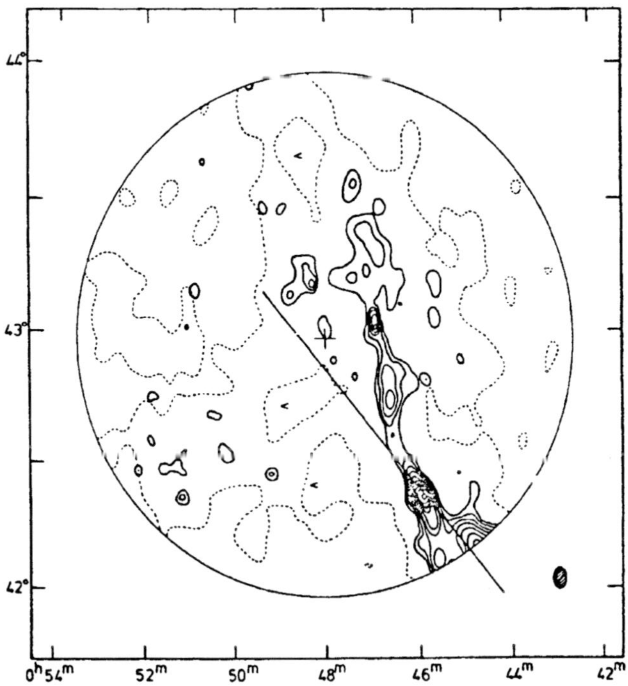

Figure 5.8 H I in the outer parts of the northeast portion of the disk. The line is the average major axis for the inner disk (from Newton and Emerson, 1977).

Sawa and Sofue's (1982) model is a simple composite, made up of a flat inner disk of radius 10 kpc and inclination of 13°. Beyond this distance the disk is warped to an extent that amounts to 1 kpc at a distance of $\sim$ 30 kpc. Even this regularly distorted model doesn't completely explain the data, however, and they find evidence for a peculiarly shaped arm at a distance of $\sim$ 35 kpc that extends as much as 8 kpc from the plane.

Brinks and Burton (1984), in a magnificent high-resolution analysis of the H I in M31 from the Westerbork array, compared two models of the H I distribution, one a simple model of a disk and a thin warp and one based on a more elaborate global flaring warp. The latter, which is able to reproduce position-velocity maps of M31 remarkably well, has the characteristics given in Table 5.3 and is shown graphically in Figure 5.9.

Table 5.3
Parameters of Brinks' (1984) Flared Warp Model of the HI in M31

Radius of the disk (kpc)	$R_d = 16.8$	
Extent of the warp (kpc)	$R_w = 15.2$	
Inclination of the disk	$i = 77°$	
Maximum warp angle	$\omega_m = 15°$	
Twist angle of warp line of nodes	$\gamma_o = 20°$	
Rotation curve (km s^{-1})	$V_c(r) = r \times 49.4$	r < 5.2 kpc
	$V_c(r) = 257$	r > 5.2 kpc
Scale height of gas layer (pc)	$h(r) = 120$	$r < R_d$
	$h(r) = 120 + (r-16) \times 108$	$r > R_d$
HI volume density at z=0 pc (at cm^{-3})	$n(r,0) = 0.6$	$r < R_d$
	var.	$r > R_d$
Temperature of the gas (K)	$T_g = 125$	
Velocity dispersion of the gas (km s^{-1})	$\sigma_g = 8$	

In an important paper published very recently, Braun (1991) analyzed the whole corpus of available data sets on M31's neutral hydrogen emission. He concluded that the inner disk, with r < 2 kpc, is tilted by 15° to the plane of the outer disk. His global gas distribution data are a good fit to a two-armed spiral with the arms trailing, resolving the difficulties mentioned in Chapter 3. An elliptical distortion of the spiral pattern in its inner portions was ascribed to the effects of the mass distribution in the triaxial bulge. It is proposed that the spiral pattern is caused by, and to some extent distorted by, the companion, M32. A linear relation between H I scale height and radial distance is found, with a value of 200 pc at 5 kpc radius and a value of 600 pc at 25 kpc radius.

5.6 Relation of H I to Other Features

Comparisons of the spatial distribution of H I in M31 with positions of other components of the galaxy have proven very informative. A pioneering study of this problem was published by Berkhuijsen (1977), who gathered material from optical and radio observations to show many close interrelationships between the components. In particular, the H I ring is found to be superimposed on a general ring-shaped distribution for such young stellar objects as H II regions and OB associations. Figure 5.10 provides an example, taken from the paper of Nakai and Sofue (1982), which plots the H II regions of Pellet *et al.* (1978) onto the H I density distribution, both corrected to face-on. It is seen that the H II regions are clumped in areas that are also especially dense in H I. The most luminous H II regions are found in the strongest H I concentrations. At a finer scale, it is seen that certain narrow linear arrangements of H II regions exist that seem to avoid the densest H I peaks.

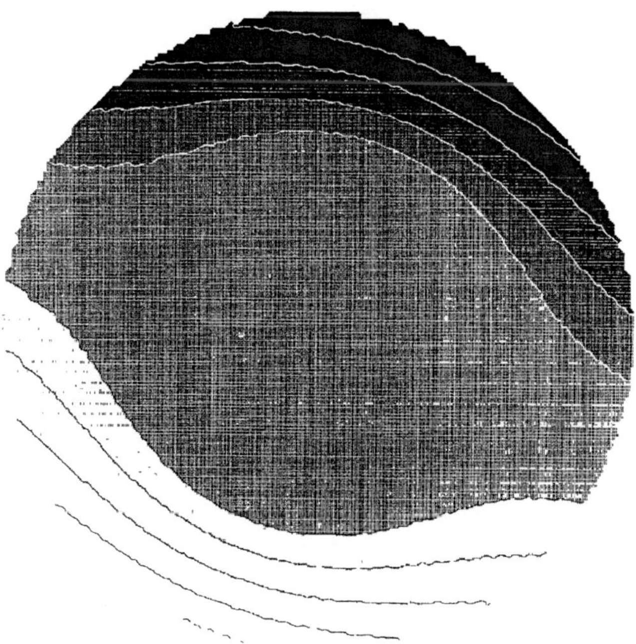

Figure 5.9 Brinks' model, shown face-on, of the warped H I plane of M31. Contour levels are ±3, ±2.25, ±1.5, and ±0.75 kpc (from Brinks, 1984).

Figure 5.11 compares the positions of OB associations with H I, showing a similar result to that for H II regions, except that the former show a more dispersed distribution, though they are largely confined to the H I-dense areas. There are several open clusters in the SW outer arm, where there are few H II regions (but see Lee and Hodge 1987), indicating that even in these very distant areas of relatively low H I density there is star formation going on, though at a muted pace.

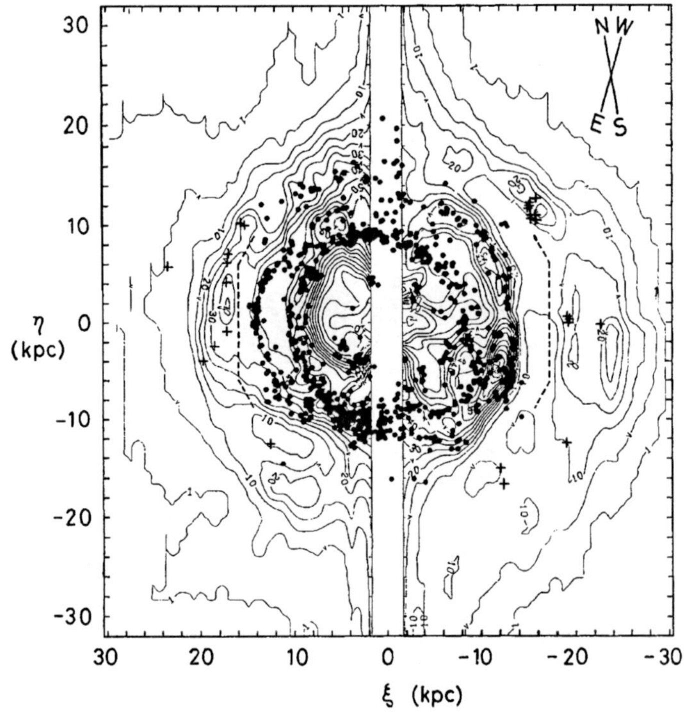

Figure 5.10 A comparison of the face-on H I contours with the H II regions catalogued by Pellet et al. (1978) (from Nakai and Sofue, 1982).

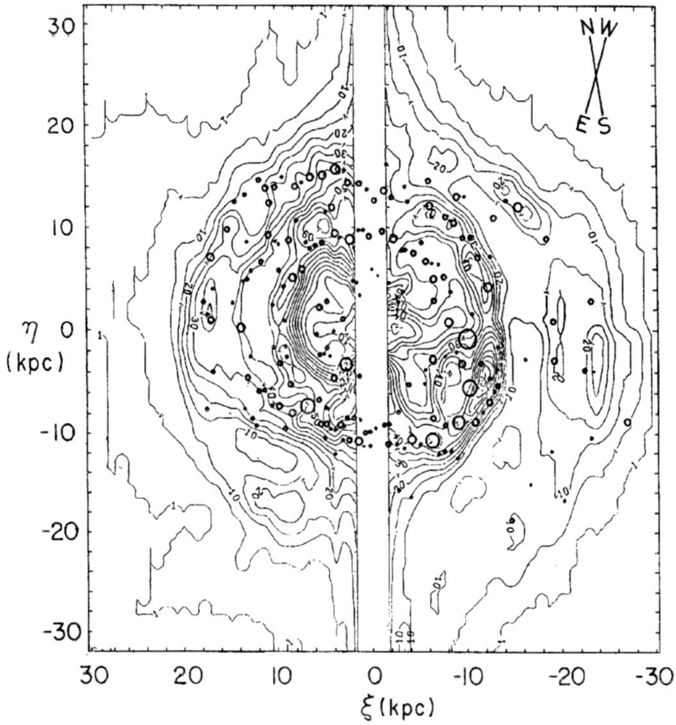

Figure 5.11 Comparing the face-on HI contours with van den Bergh's (1964) OB associations (from Nakai and Sofue, 1982).

A particularly nice summary of the radial distributions of different components of the galaxy was given by Nakai and Sofue (1982) in a series of graphs reproduced in Figure 5.12. Here we see that the open clusters most closely match the H I distribution, though the OB associations also are very similarly distributed. The dust clouds are more centrally concentrated and the H II regions more narrowly distributed.

Brinks and Bajaja (1986) discovered and explored a large number of H I "holes", areas with smaller than average H I column densities (Figure 5.13). They found 147 such holes and studied their properties, finding that they represented, on the average, a depletion of H I gas of about 10^4 solar masses (Figure 5.14). They probably represent areas where gas has been locally depleted by various mechanisms involved with recent star formation. A hole exists, for example, at the location of NGC 206, an immense star cloud, rich in recently formed, massive stars (Chapter 10).

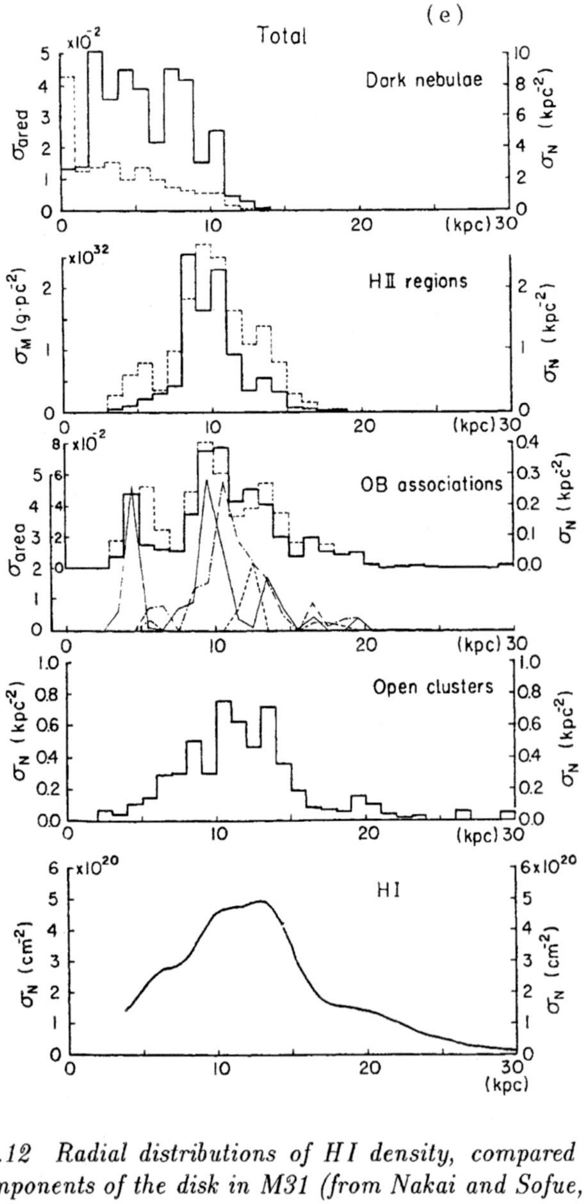

Figure 5.12 Radial distributions of HI density, compared to those for other components of the disk in M31 (from Nakai and Sofue, 1982).

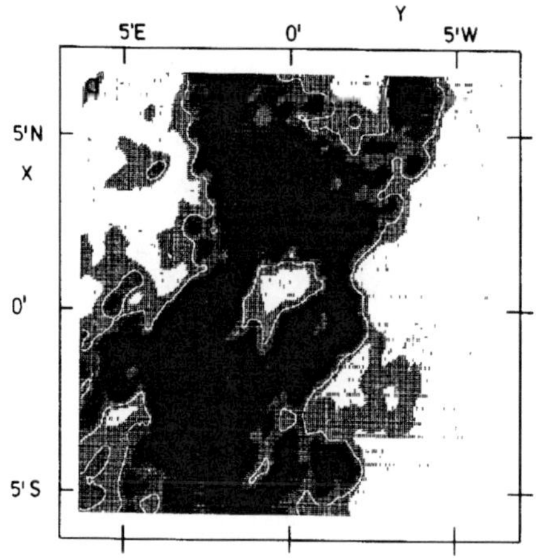

Figure 5.13 An example of a neutral hydrogen hole in M31 (from Brinks and Bajaja, 1986).

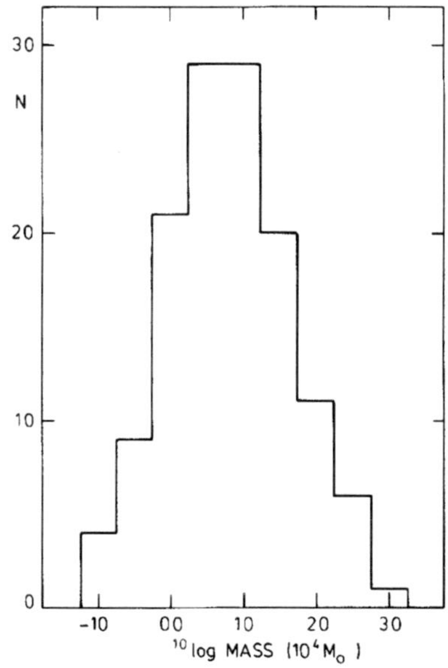

Figure 5.14 Mass distribution (the amount of HI mass missing) for Brinks' HI holes in M31 (from Brinks and Bajaja, 1986).

Braun's (1990, 1991) VLA survey of the northeast half of M31 showed considerable detail in the density distribution of H I; many high brightness temperature structures (140 to 180 K) were found, with characteristic sizes on the order of 100 pc. When compared to absorption measurements (see also Dickey and Brinks 1988), it is concluded that the properties of neutral hydrogen in both emission and absorption indicate a low volume filling factor of less than 30% and a two phase, semi-neutral interstellar medium, with a radial gradient showing temperatures increasing at large distances from the center.

5.7 Summary

The hydrogen gas in M31 is located in a thin, warped and flared disk, extending out almost 40 kpc from the center. The gas density shows a ringed structure, with the highest densities at a distance of about 10 kpc from the nucleus. It contains spiral arm segments, but shows holes that represent locations of enhanced star-formation activity.

CHAPTER 6

RADIO CONTINUUM RADIATION

Most galaxies are continuum radio sources and M31 is no exception. It is what radio astronomers classify as a normal galaxy, though, not a radio galaxy, as its radio luminosity is moderate. The spectrum of its radio radiation shows that the sources of this radiation are both thermal and non-thermal. In this sense, M31 is rather like our own galaxy, though the detailed structure and scale of the radio image is different.

6.1 Detection and Early Surveys

In 1944, Grote Reber, using his amateur's backyard radio telescope, was the first to attempt to detect M31 at radio wavelengths. He didn't succeed, but success came only a few years later, when Ryle, Smith and Elsmore (1950) and Hanbury Brown and Hazard (1951) reported its detection.

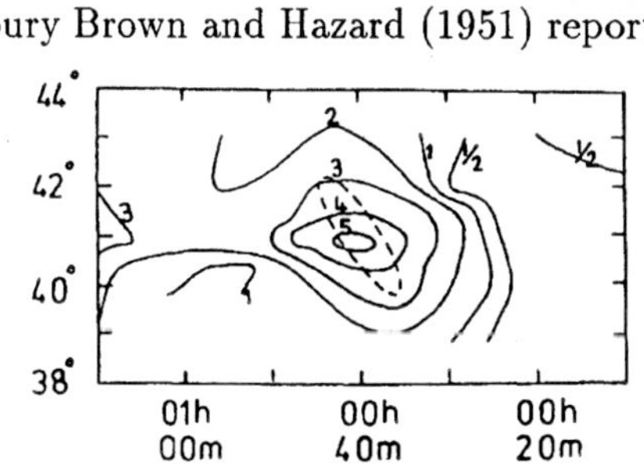

Figure 6.1 The first radio continuum map of M31, made by Hanbury Brown and Hazard (from Sullivan, 1984).

One story of the way by which this success was achieved has been given by Hanbury Brown in a compilation of historical recollections assembled by Sullivan (1984). It was the first object that Hanbury Brown and Hazard turned to after getting the Jodrell Bank 218-foot transit telescope in final form for observing. Figure 6.1 shows their result, based on observations in August and September 1950. They had been tempted to look at it as their first target, having calculated that, were our Galaxy placed at M31's distance, it should be detectable with their new antenna. It took them 90 nights, working during the radio-quiet time around midnight, to map out the radio radiation, as shown in the figure. M31 was the first object (after our Galaxy) to have its extended radio structure determined.

The most complete early survey was an extensive study made by Pooley (1969) with the Cambridge One-mile Synthesis Telescope. Using frequencies of 408 and 1407 MHz, Pooley was able to achieve good resolution compared to the earlier observations; his maps had resolutions of 80 × 120 arcsec and 23 × 35 arcsec, respectively. He found a total of 213 discrete sources above a limit of flux density equal to 12×10^{-29} W m^{-2} Hz^{-1}. The disappointing aspect of this achievement, however, is that only a few of these sources are related to M31, the number found being about equal to the number of background sources expected in the surveyed area of sky.

Table 6.1
Some Radiocontinuum Observations of M31
Showing Spiral Structure (from Berkhuijsen, 1978)

Telescope	λ (cm)	HPBW (')	Area (R in ')	Noise in T_b
Cambridge	200 TP	3 × 5	200	<20 K
	75 TP	4.0	80	1 K
Westerbork	49 TP	1.0 × 1.5	120	1 K
	POL	3.6 × 5.4	120	60 mK
	21 TP	0.4 × 0.6	12, 20	1 K
	21 TP	0.4 × 0.6	160	1 K
	POL			
	6 TP	0.1 × 0.2	–	–
Effelsberg	21 TP	9.2	130	20 mK
	POL		130	11 mK
	11 TP	4.8/4.5	125, 105	4.5 mK
	POL		90	2.5 mK
	6 TP	2.6	20	2.3 mK
	TP		90	

However, Pooley's search did show that there is extended radio emission from the region and that the one relatively bright source is found to coincide with M31's nucleus. Two components, the disk and the nucleus, showed up clearly in the maps (Figure 6.2). The disk portion of the source was found to correspond in position with the brightest parts of the optical spiral arms (at distances from the nucleus of $\sim$ 10 kpc). The nucleus broke into two parts, one almost unresolved, with a diameter of about 80 arcsec and one larger, extending out about 3 arcmin. Both contribute flux densities of about 2×10^{-27} W m^{-2} Hz^{-1}. Pooley calculated that the center of M31 was not as bright at radio wavelengths as Sgr A, the nuclear source at the center of our Galaxy.

All of the radio components were found to be largely non-thermal at 408 MHz, indicating that synchrotron radiation must play an important part in the mechanism responsible for much of the radiation. None of 15 or so possible discrete sources could be a supernova remnant like Cas A in our Galaxy, however, which would be below the detection limit of that M31 survey.

6.2 Radio Morphology

Central Region

In the years since Pooley's (1969) pioneering study, the morphology of the radio image of M31 has been explored with greater and greater resolution and more different frequencies. For example, Hjellming and Smarr (1982) used the VLA to explore the central 1° at a frequency of 1465 MHz. They found that the central source is nearly circular in shape with a diameter of 10 arcmin (2 kpc). The surface brightness decreases exponentially outward, with a scale length of about 1.1 arcmin. The maximum emission comes from a point located 20 arcsec from the optical nucleus at a position angle of 70°.

Berkhuijsen *et al.* (1983) observed the central region at 2695 and 4850 MHz, and Crane *et al.* (1981) found that, at a few arcsec resolution, the emission appears to be completely resolved. Figure 6.3 shows Berkhuijsen *et al.*'s (1983) 4850 MHz map.

Walterbos and Gräve (1985) used the WSRT and the Effelsberg 100-meter radio telescopes to examine it further, at frequencies of 610, 1412, and 10,700 MHz. A comparison of maps for all three frequencies is shown in Figure 6.4. From their analysis they separate the continuum emission into three components: an extended featureless component, unresolved sources, and fine-scale structure. The six unresolved sources at 12 arcsec resolution are, according to Crane *et al.*'s (1981) study, resolved at 2 arcsec. The fine-scale structure consists of features that appear to correspond to the spiral-like gas filaments detected optically (Chapter 14). There is one extended feature that emerges from the center towards the north and then bends to the northwest at a distance of 1.2 arcmin, with an almost opposing feature to the south and southeast. It is suggested that these may be filaments that lie in the nuclear bulge, out of the plane, similar to features detected near the nucleus of our Galaxy. It is possibly emitted by shock-ionized gas (see § 6.3).

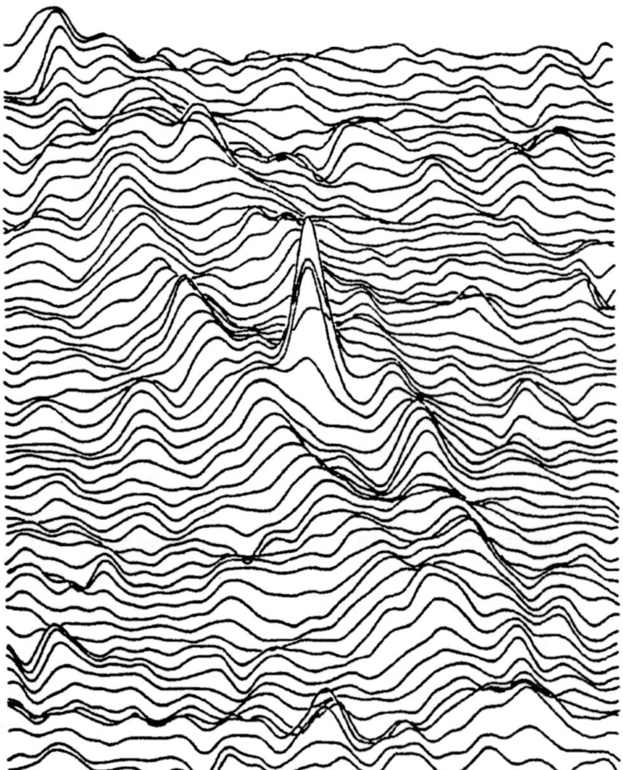

Figure 6.2 Radio continuum map of M31, showing the nucleus and disk emission at 408 MHz (from Pooley, 1969).

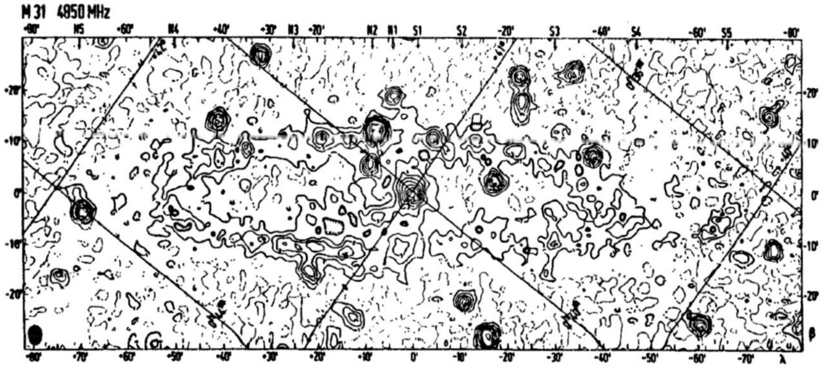

Figure 6.3 Contour map of M31 field at 4850 MHz, clearly showing the 10 kpc ring structure. Arrows indicate positions of Baade's arms N5 to S5 (from Berkhuijsen et al. 1983).

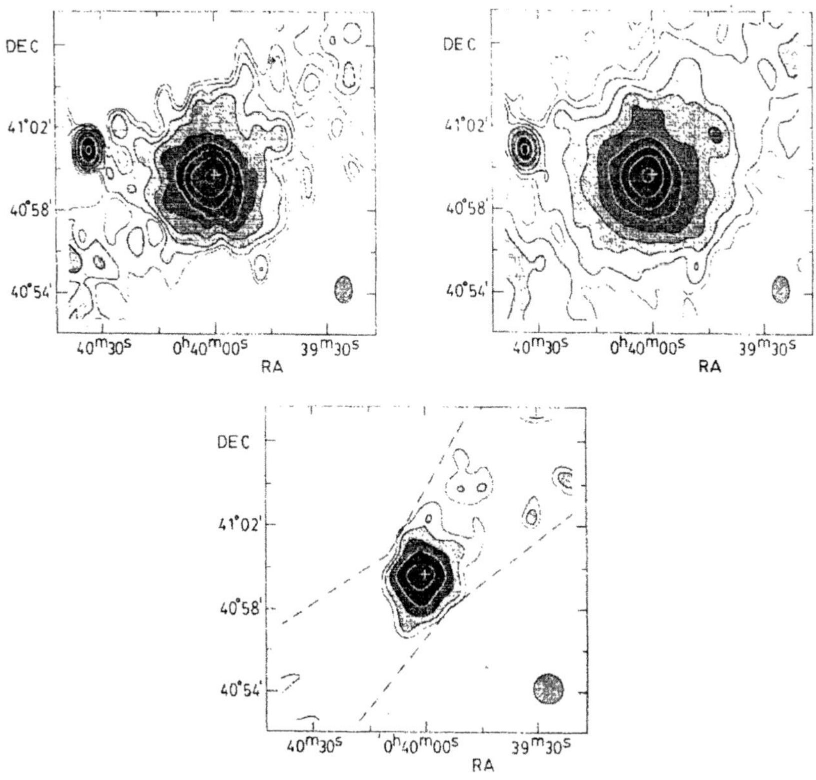

Figure 6.4 Three maps of the central source, taken at frequencies (from left to right) of 610, 1412, and 10,700 MHz (from Walterbos and Gräve, 1985). The dashed lines indicate the extent of the 10,700 MHz survey.

Table 6.2
Flux Densities of the Nuclear Region of M31
(from Walterbos and Gräve, 1985)

Frequency (GHz)	Flux (mJy)	Reference
0.408	520	Pooley (1969)
0.610	395	Walterbos and Gräve (1985)
1.412	215	Walterbos and Gräve (1985)
2.700	130	Berkhuijsen *et al.* (1983)
4.850	81	Berkhuijsen *et al.* (1983)
10.700	39	Walterbos and Gräve (1985)

Table 6.3
Physical Properties of the Nuclear Region of M31
(from Walterbos and Gräve, 1985)

Characteristic	Value	Remarks
Log of radio power (W/Hz) at 1415 MHz	19.13 $\pm$ .03	within 1 kpc radius
Radio luminosity (ergs/s)	1.8×10^{36} $\pm .1 \times 10^{36}$	
Radio spectral index	-0.75 $\pm$ .05	
Minimum energy (ergs)[1,2]	5.5×10^{53}	
Minimum magnetic field (microgauss)[1,2]	7	
Hα + [N II] luminosity (ergs)	1.8×10^{38}	inside 780 $\times$ 780 pc
Ratio of [N II] to Hα	2.2	for the inner 800 pc
Ionized gas mass (solar masses)	1500	
Filling factor of ionized gas	3×10^7	for a disk of 200 pc height
Neutral hydrogen mass (solar masses)	$< 1 \times 10^6$	

[1]Between 10^7 and 10^{11} Hz.

[2]Assuming spherical geometry with a radius of 1 kpc, a radio filling factor 1, and 100 times more energy in protons than in electrons.

A recent VLA map at 1.49 GHz by Condon (1987) confirms the morphology of the central region that had been discerned in earlier studies. Figure 6.5 shows this version, illustrating the squarish, resolved nature of the region.

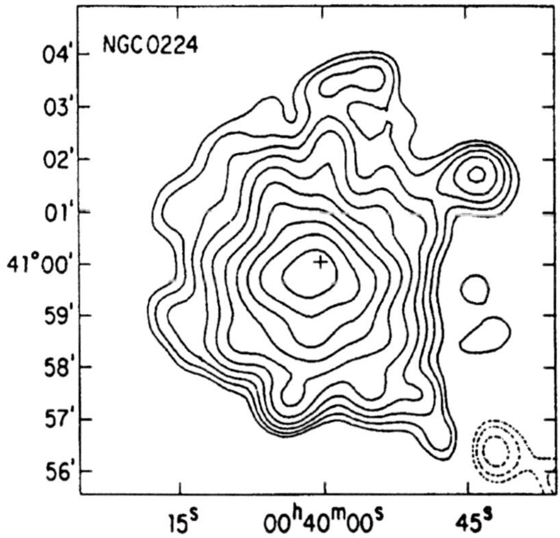

Figure 6.5 VLA map of the center source, taken at 1.49 GHz (from Condon, 1987).

A few arcmin northeast of the central sources is a bright triple source, called 5C3.116+5C3.119 in Pooley's (1969) catalogue and 37W158 in the Walterbos *et al.* (1985) catalogue (Figure 6.6). The central source of this triple system has an inverted spectrum, while the two side sources have steep spectra, suggesting that it is a normal, distant radio galaxy with radio lobes, seen through the Andromeda Galaxy.

Braun (1990) made a very detailed 20 cm study of the northeast half of M31, using the VLA. He cataloged 534 discrete sources. Possible identifications of these sources include 103 with H II regions, seven with known supernova remnents, 10 with X-ray sources, and 121 with previously identified radio sources.

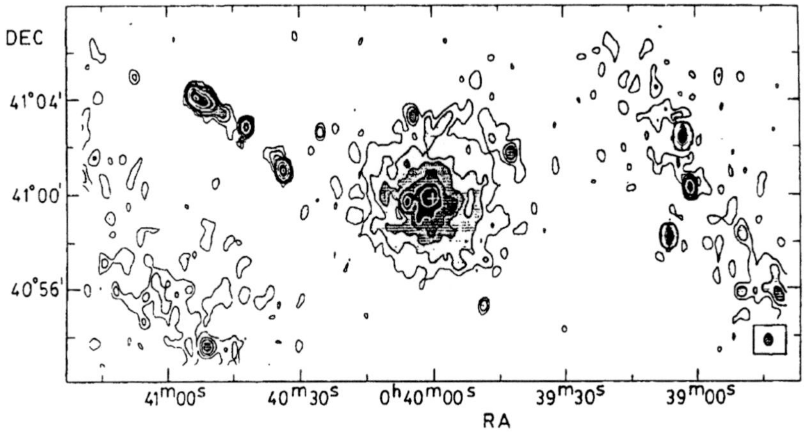

Figure 6.6 The central area, including the triple source 37W158, visible here to the northeast of the nuclear source (from Walterbos et al. 1985).

Spiral Arms

As discovered long before by Pooley (1969), more recent maps of M31 confirm the presence of an intense ring of continuum emission. Berkhuijsen and Wielebinski's (1974) survey at 2695 MHz, for example, showed that the extended emission, outside of the central source, was concentrated along Baade's (1958) optically identified arms 4, 5, and 7, with arm 4 being the most prominent. Arm widths are 1–2 kpc at this frequency and three spurs were identified, one appearing to form a bridge between arms 5 and 6.

Dennison *et al.* (1975) used the NRAO 300-foot radio telescope to map the galaxy at 1410 and 2695 MHz, with rather similar conclusions. The WSRT was subsequently used by Bystedt *et al.* (1984) to carry out a high resolution survey at 610 MHz. Resolution was 0.9 by 1.2 arcmin and 392 discrete radio sources were found in the M31 area, making up the 36W catalogue. Only a few of these are likely to be sources in M31, as there is only a slight excess in source counts compared to other, similar surveys in other parts of the sky. Subsequent, deeper surveys (Walterbos *et al.* 1985) suggest that even this slight excess is not real. The map shows clearly the 10 kpc ring of extended radio emission, as seen in Pooley's and other maps (Figure 6.7).

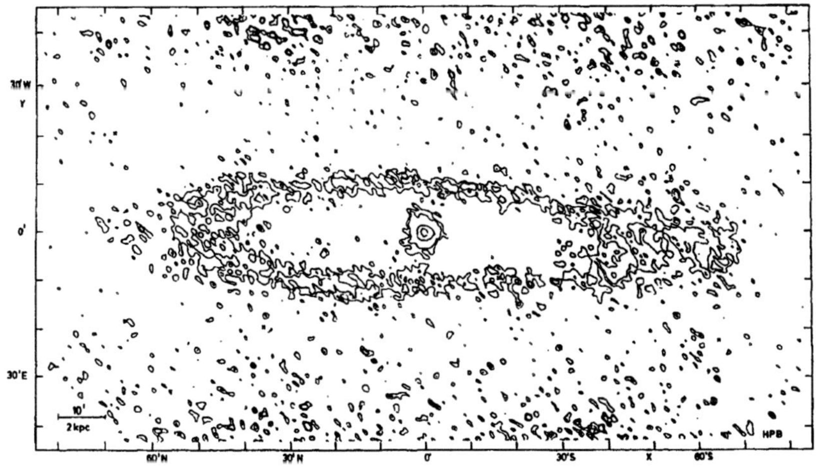

Figure 6.7 A WSRT map at 610 MHz, showing arcmin resolution in the arm regions (from Bystedt et al. 1984).

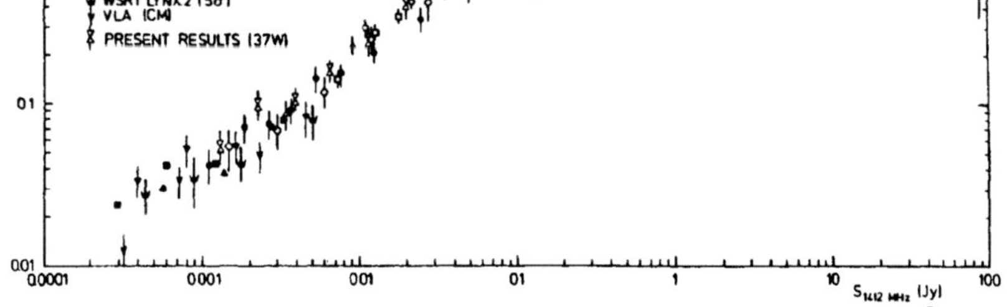

Figure 6.8 A comparison of source counts in the M31 region (the 37W catalogue) with those at other positions in the sky, showing no significant enhancement (from Walterbos et al. 1985).

Working at 21 cm (1412 MHz), Walterbos *et al.* (1985) provided an additional WSRT data set. The resolution was 23 × 35 arcsec and both the extended emission in the center and the 10 kpc ring emission was mapped. A list of 249 discrete sources with flux densities greater than 1 mJy is given, forming the 37W catalogue of radio sources. These do not, however, represent a significant increase over the normal background (Figure 6.8). The outer, 10 kpc ring is resolved into several individual maxima and it is found that these agree in position with areas of the most intense star formation in the arms (Figure 6.9).

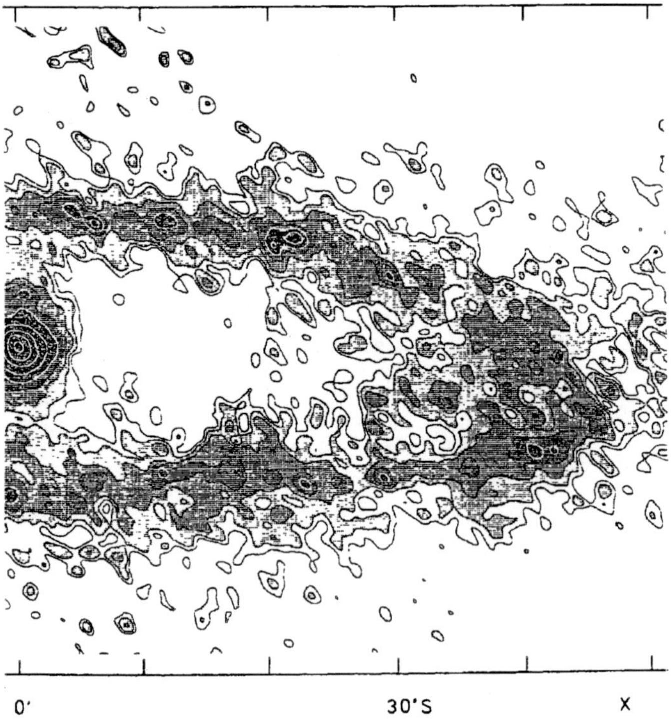

Figure 6.9 The southwest half of M31, from the WSRT 1412 MHz survey. Most (unrelated) discrete sources have been removed. Note the resolution of the arms into individual sources (from Walterbos et al. 1985).

Discrete Sources

The first discrete sources belonging to M31 to be detected at radio wavelengths were known supernovae remnants. Dickel *et al.* (1982), and Dickel and D'Odorico (1984) identified some with the VLA at 6 and 20 cm, and Walterbos *et al.* (1985) identified five with the WSRT. Flux densities at 1412 MHz ranged from 2.9 to 1 mJy. It is likely that fainter, as yet optically undetected, SNR may be found among the catalogued radio sources.

About 20 sources in the 37W catalogue seem to correspond to optical H II regions, but they are very weak sources compared, for example, to M33's H II regions. At 1412 MHz all but two have flux densities less than 5 mJy.

As described above, Braun's (1990) massive VLA survey of the NE half of M31 greatly increased the number of discrete sources that could be identified with optically-visible objects, especially with HII regions.

6.3 Spectra and Nature of the Sources

It was recognized early in the study of M31's radio spectrum that most of the radio radiation is non-thermal in origin. Figure 6.10 shows a spectrum plotted by Durdin and Terzian (1972), based on their Arecibo observations, as well as other global measurements.

Beck and Gräve (1982) analyzed the spectral index between 408 and 2695 MHz in an attempt to study the distribution of thermal and non-thermal radio emission in the disk of M31. They found the spectrum to be flattest in the 10 kpc ring. Subtracting out the presumed thermal component by using optical data on H II regions allowed them to establish that the spectral index of the non-thermal component from the ring is nearly constant at a value of 2.88. Beyond a distance from the center of about 12 kpc, the spectral index increases to about 3.0 (Figure 6.11).

Table 6.4
Radio Continuum Surveys of the Andromeda Nebula
(From Bystedt *et al.* 1984; see also van der Kruit and Allen, 1976 and Braun 1990)

Author	Year	Freq. (GHz)	Beam Size
Kraus	1955	0.250	1°.3 × 8°
Leslie	1960	0.178	13′.6 × 4°.6
Lynds	1961	1.400	35′.4
Kraus	1963	1.400	11′ × 40′
Kraus	1964	0.600	23′ × 85′
de Jong	1965	0.750	18′.3
		1.410	10′0
Pooley and Kenderdine	1967	0.408	80" × 120"
		1.407	23" × 35"
Dickel	1968	0.750	18′.3
		1.410	9′.8
van der Kruit	1972	1.415	23" × 35"
van der Kruit and Katgert	1972	1.415	23" × 35"
Spencer and Burke	1975	2.695	9"
		8.805	3"
Segalovitz, Shane, and de Bruyn	1976	0.610	3′.6 × 5′.4
Speed	1976	0.408	1".2
Wielebinski	1976	1.420	9′
Volodin and Dagkesamanskij	1977	0.1025	49′ × 24′
Beck, Berkhuijsen, and Wielebinski	1978, 1980	2.700	4′.4
Gillespie	1979	1.421	92" × 140"
Gräve, Emerson, and Wielebinski	1981	0.842	15′.7
Beck and Gräve	1982	0.408	5′
		2.700	5′
Berkhuijsen, Wielebinski, and Beck	1983	4.850	2′.6
Bystedt *et al.*	1984	0.610	54" × 82"
Braun	1990	1.465	5"

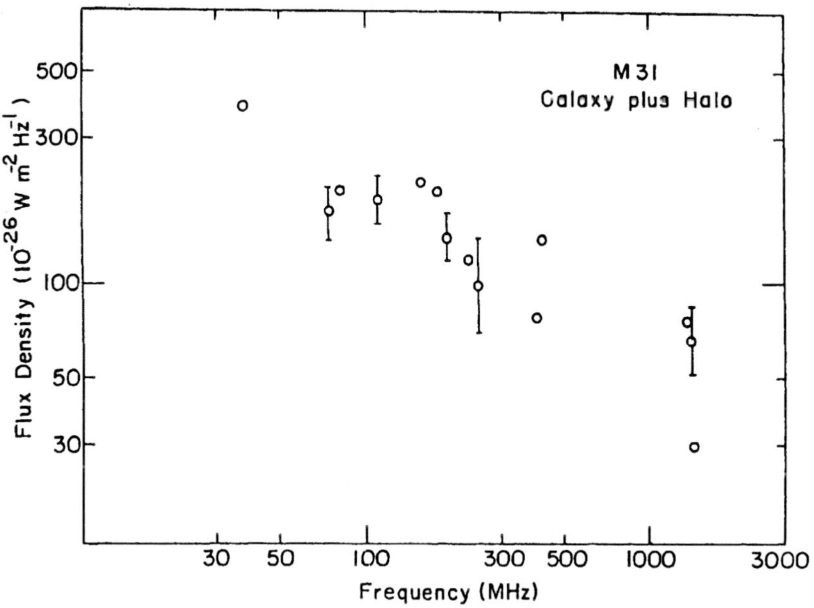

Figure 6.10 The radio spectrum of the M31 image (from Durdin and Terzian, 1972).

Beck and Gräve (1982) interpret these data in terms of models of cosmic rays constrained to the disk by the magnetic field embedded in the arms. The non-thermal radiation in the arms at 10 kpc comes presumably from supernovae and their remnants. The cosmic ray electrons are probably accelerated by the first order Fermi mechanism during the passage of a shockwave propagated by the supernova explosion. For the 10 kpc ring, where massive star formation is occurring, the rate of Type II supernova formation should be relatively high. The steepening of the spectral index outside of the ring can be understood as being the result of energy losses of the electrons as they diffuse out of the region of the ring in the disk.

The spectrum of the central continuum source is shown in Figure 6.12. This spectrum is not different at different locations, as can be inferred from Figure 6.4, where the three maps look quite similar. A mean spectral index of −0.75 is found. If all the emission is non-thermal, then Walterbos and Gräve (1985) find that it is quite reasonable for the expected supernova rate to explain the observed radio flux. Walterbos and Gräve (1985) have attempted to determine whether some of this radiation can be thermal, emitted by the optically identified H II filaments in the center. Although the intensities could be consistent with such a possibility, they conclude that the *optical* emission may not be thermal, but could be shock-induced, and thus the radio might be, too.

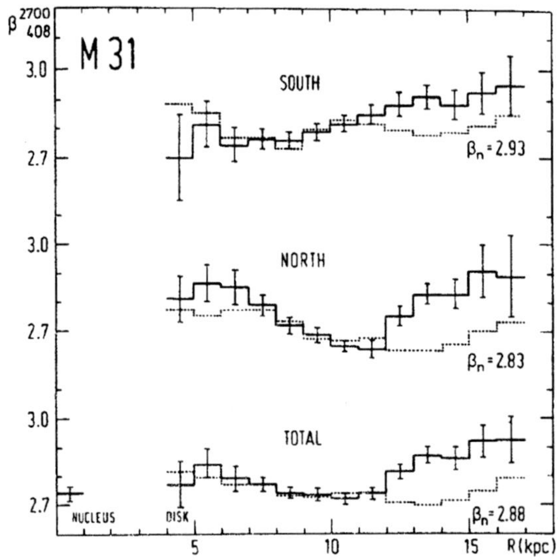

Figure 6.11 Variation of brightness temperature index as a function of distance from the center, showing the increase in the outer portions (from Beck and Gräve, 1982).

6.4 Radio Polarization and the Magnetic Field

The 100-meter radio telescope at Effelsberg has been used to map the radio polarization in M31 (Beck *et al.* 1980, Beck 1982, Beck *et al.* 1989, Beck 1990). The linearly polarized component of the 2700 MHz continuum shows a beautiful global pattern (Figure 6.13). The large-scale magnetic field is aligned along the spiral arms that form the 10 kpc ring. The direction of the magnetic field lines is the same as the direction of M31's rotation. There are no reversals seen with a size scale larger than ~ 3 kpc within the area where the signal was strong enough to look for them.

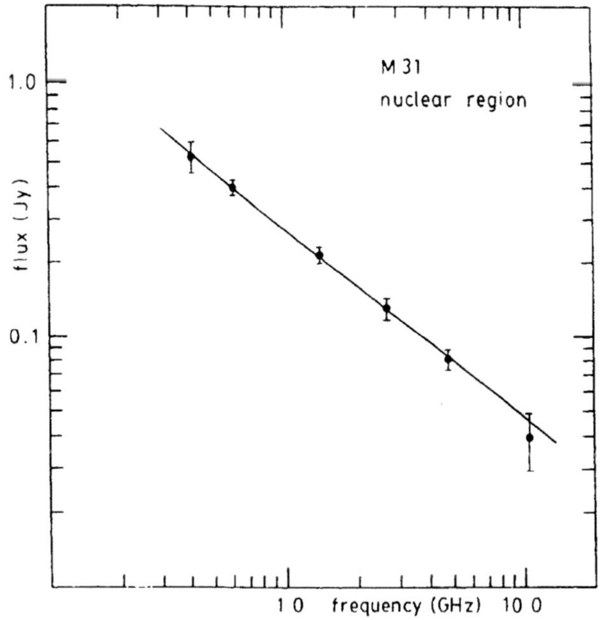

Figure 6.12 Radio spectrum of the nuclear complex within a 1 kpc radius (from Walterbos and Gräve, 1985).

Beck (1982) calculates that the maximum strength of the field is $5.2 \pm 1.4\mu G$. In the ring the mean value is $4.4 \pm 1.2\mu G$. The Faraday rotation in the ring can be accounted for by the following model: a $2.4\mu G$ uniform field in the plane of M31 and thermal electrons with a density of 0.03 cm^{-3} distributed in a disk 900 pc thick (full width to half power points). Beck explains the very striking uniformity of the field as being the result of compression of a pre-aligned field by the passage of density wave shock fronts.

6.5 Summary

Continuum radio radiation from M31 is concentrated in a nuclear source, the 10 kpc "ring", and a number of discrete sources in the disk. Some of the sources are H II regions and supernova remnants. The diffuse radiation of the disk is polarized, indicating that it is caused by relativistic electrons moving in a 4μG field aligned along the spiral arms.

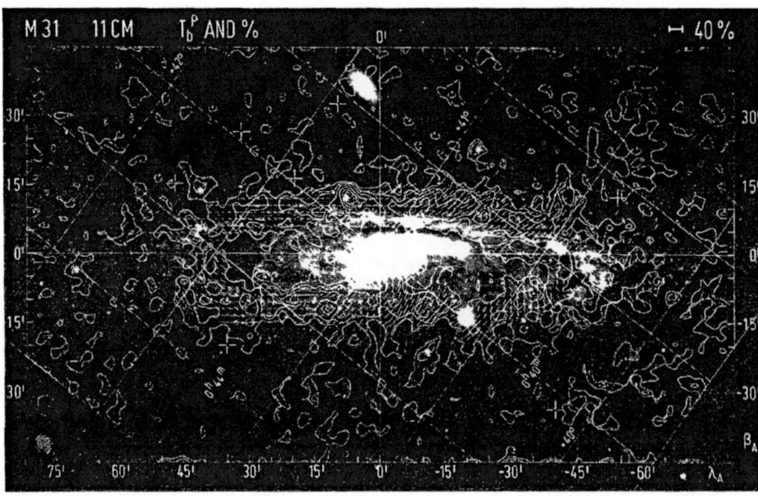

Figure 6.13 Contour map of main beam polarization temperature at 11.1 cm. The first contour is 4 mK in T_b^p *and intervals are 4 mK. Vector lengths for the brighter, more polarized areas indicate the percentage polarization. Asterisks are polarized point sources (from Beck et al. 1980).*

CHAPTER 7

ROTATION

Rotation curves for M31 have been obtained from both optical spectra and 21-cm velocity measurements. Even the earliest optical spectra were useful for understanding the gross rotational parameters of the galaxy, and subsequent optical data have greatly refined the details of what turns out to be a rather complicated rotation curve. The hydrogen gas motions agree, in general, with the optical data, but extend out to greater distances, giving important knowledge about the outer mass distribution (unfortunately, not enough knowledge, however, to answer the questions that are raised there). Very recently, the rotation of the nucleus has been scrutinized with modern techniques, to reveal a rapidly rotating object that may be driven by a massive central black hole.

Normally, three assumptions are made about a galaxy's rotation curve when it is obtained from observations of radial velocities of its components:

1. All motions are circular;
2. All motions are confined to a single plane;
3. The galaxy possesses circular symmetry.

For M31, there is clear evidence that the first two assumptions are not correct, and, at least locally, the third is also not perfectly true. Much of the recent progress in understanding M31's rotation has involved attempts to disentangle these imperfections and to separate the global from the local anomalies.

Although there has been considerable progress in obtaining stellar absorption line radial velocities, especially in the inner parts of M31 (Section 7.6), most rotation curve data come from emission line measurements, from both optical and radio lines. Table 7.1 lists many of the sources of emission line radial velocity data obtained for M31.

7.1 Early Optical Results

The rotation of M31 was one of the first features to be detected early in the 20th century, when spectra of objects could be obtained to sufficiently faint levels. Even as early as 1914, Slipher (1914) noticed the slanted nature of the object's spectral lines. Pease (1918) showed a dedicated interest in the subject when he obtained a spectrum of M31 in 1917, using the Mt. Wilson 60-inch telescope on several nights during the months of August, September and October. The single spectrum had a total exposure time of 79 hours! He was able to measure the rotation curve out to distances of 150″ from the center.

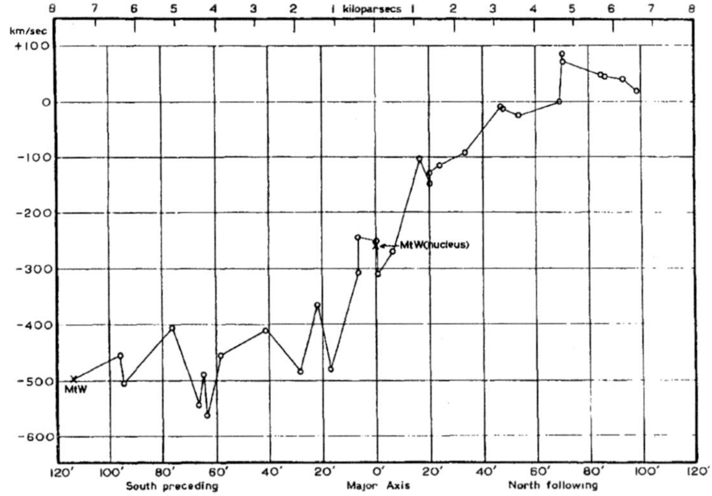

Figure 7.1 The rotation curve of M31, obtained by Mayall from spectra of emission regions in the disk.

A landmark study of the rotation curve was that of Babcock (1939), who used the Lick Crossley telescope to measure a more complete curve, including data for both the central bulge, where it was based on the absorption lines of the stellar component, and the disk, based on measurements of velocities of individual H II regions. Babcock's velocity curve led to the first attempts to derive the mass of M31 from a detailed matching of model and observed data, as explained in Chapter 8.

An even more complete study of the rotation resulted from another Crossley telescope project, when Mayall (1951) extended Babcock's survey to include many more H II regions in the disk, providing a more complete and accurate rotation curve (Figure 7.1).

Table 7.1
Emission Line Radial Velocity Observations in M31

Source	Wavelengths
Babcock (1939)	[O II], [O III], Hα
Mayall (1951)	[O II]
van de Hulst *et al.* (1957)	21 cm
Münch (1962)	[O II]
Argyle (1965)	21 cm
Gottesmann *et al.* (1966)	21 cm
Roberts (1966)	21 cm
Deharveng and Pellet (1969)	Hα
Gottesmann (1970)	21 cm
Rubin and Ford (1970, 1971)	Hα, [N II], [S II], [O III]
Deharveng and Pellet (1975*a*, *b*)	Hα, [N II], [S II]
Roberts and Whitehurst (1975)	21 cm
Emerson (1976)	21 cm
Pellet (1976)	21 cm
Combes *et al.* (1977)	2.6 m
Unwin (1980)	21 cm
Cram *et al.* (1980)	21 cm
Bajaja and Shane (1982)	21 cm
Unwin (1983)	21 cm
Brinks (1984)	21 cm
Kent (1989)	Hα, [N II]
Braun (1990)	21 cm

7.2 Modern Optical Rotation Curves

Mayall's data were sufficiently good that it was twenty years before an improved rotation curve was obtained from optical data. With the introduction of electronic amplification of photographic detectors, it became possible, by the 1970s, to measure velocities of much fainter H II regions and to register spectra of regions of the bulge too faint for earlier techniques. While Mayall had to expose for as much as two nights for a single velocity, Rubin and Ford (1970, 1971) using an image tube could obtain good spectra of the same faint objects in less than an hour. The results of their work were published in two papers; the first (Rubin and

Ford 1970) deals with measurements of the outer parts (the disk), based mostly on emission line spectra of H II regions, and the second gives a rotation curve for the bulge that is based on emission lines detected in centrally-placed long slit spectra.

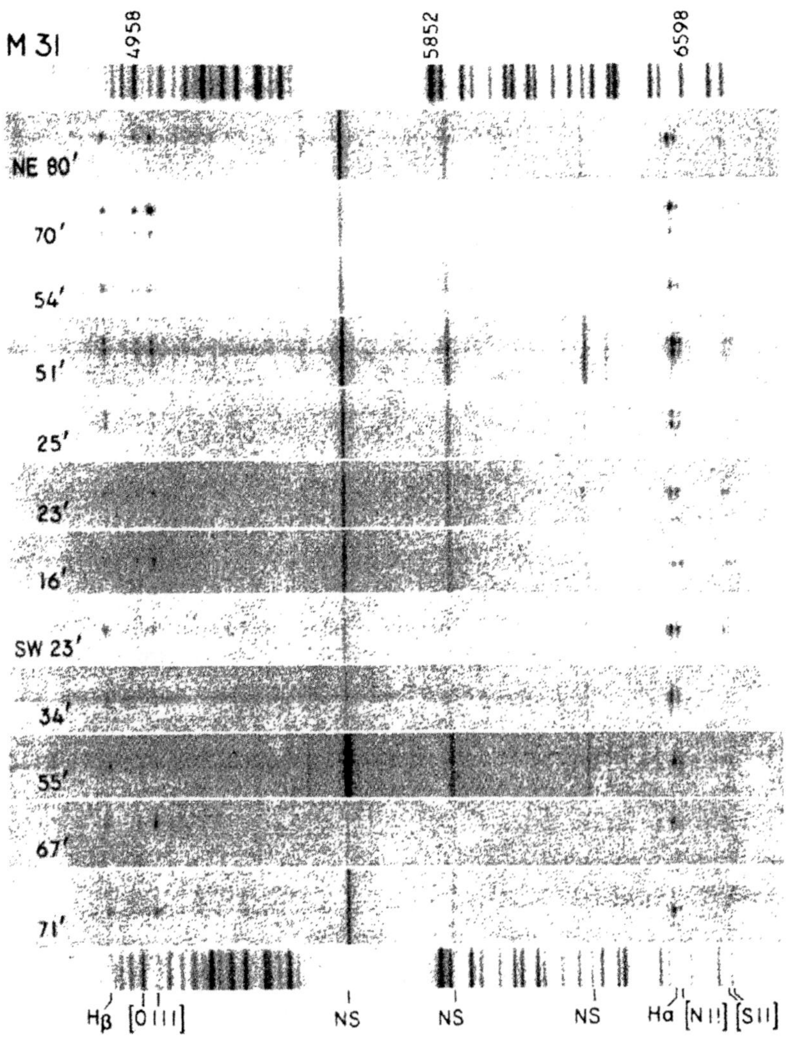

Figure 7.2 Spectra of some of the emission regions observed by Rubin and Ford (1970). Distances from the center are indicated at the left. Lines that extend entirely across the spectra are night sky lines. The main emission lines are labeled at the bottom.

Figure 7.2 shows several examples of the spectra Rubin and Ford (1970) obtained of H II regions in M31. They were able to take advantage of the extensive catalog of H II regions published by Baade and Arp (1964) to select 67 emission regions, spread out over the disk at distances from the center ranging from 16 to 120 arcmin. Accuracy of the velocities was ≈ 10 km/sec. On the basis of an assumed circular rotation in a plane, the rotation curve was derived from the relation

$$V_{obs} - V_c = V(R)\sin\psi\cos\theta, \qquad (eq.\ 7.1)$$

where V_c is the systemic velocity, ψ is the inclination angle, and R and θ are the coordinates of the emission object in the galaxy's plane. Circular velocities so calculated ranged from 66 km/sec in the inner portion to 433 km/sec at a point 4 kpc to the northeast.

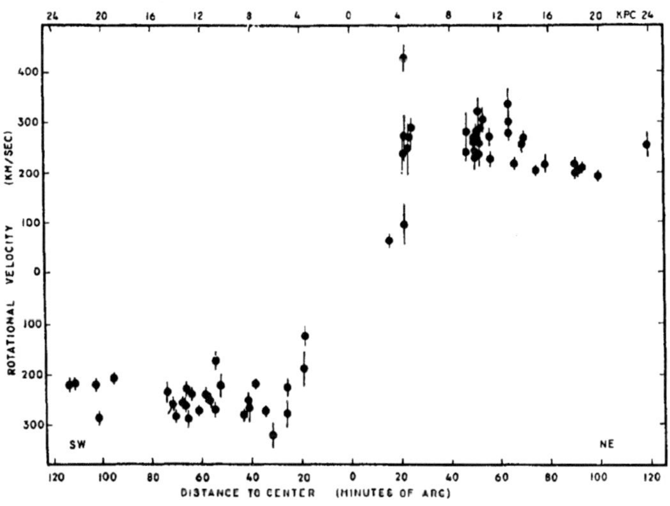

Figure 7.3 Velocities for 67 emission-line objects in M31 observed by Rubin and Ford (1970).

In their 1970 paper, Rubin and Ford also looked at the velocities in the bulge by using spectra that showed a narrow emission line of [N II]. These measures showed that the nuclear area has a rapid rotation rate and that there is a deep minimum in the rate at a distance of about 2 kpc from the nucleus (Figure 7.4). The maximum is at a distance of 400 pc and reaches velocities of V = 225 km/sec. It should be noted that this inner maximum in the rotation curve is not the one observed for the nucleus proper, which has a maximum value at a distance of only 7 pc (Section 7.6).

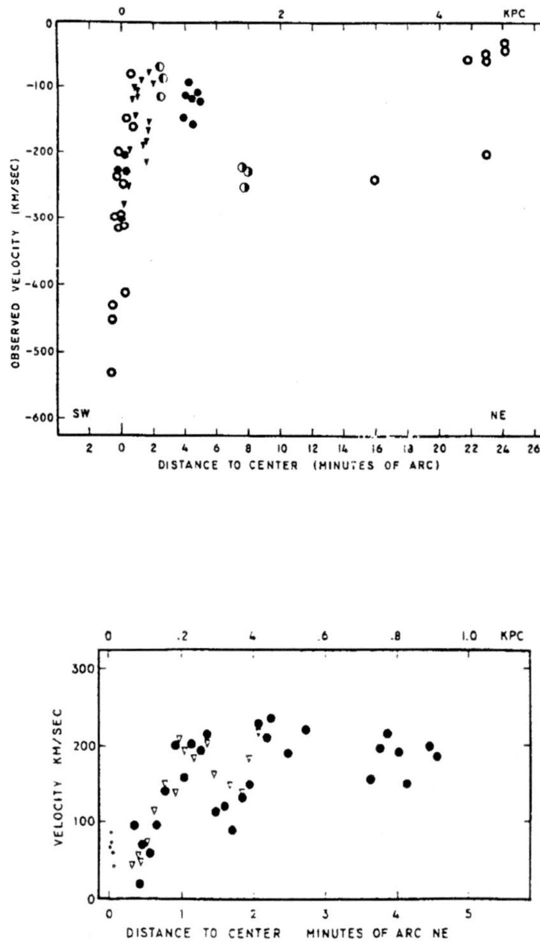

Figure 7.4 Velocities along the NE major axis of M31, based on the [NII] λ6583 line (from Rubin and Ford, 1970). Different symbols indicate different plates. The lower panel shows velocities near the nucleus.

The combination of the data for various realms of the galaxy leads to a fairly complicated rotation curve, even when non-circular motions are ignored. Figure 7.5 shows a curve based on the optical velocities (outside of the nearly unresolved nucleus) from Rubin and Ford's measurements.

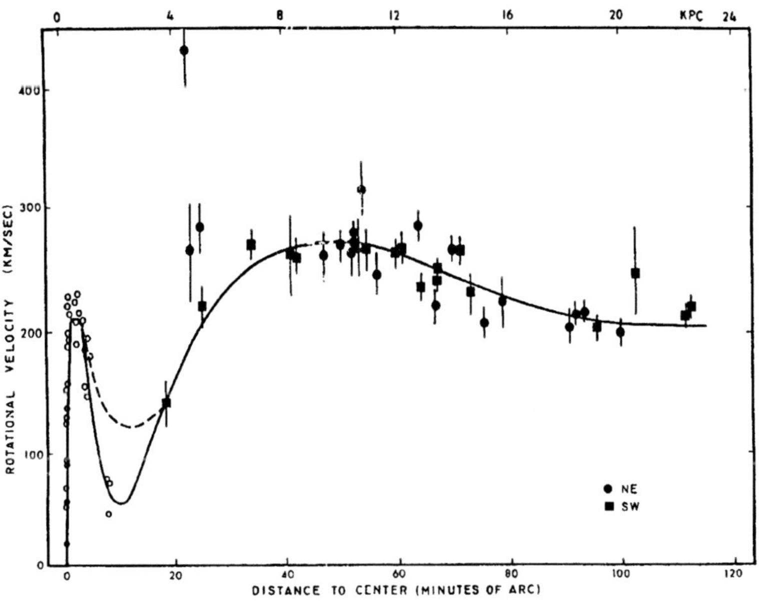

Figure 7.5 A rotation curve fit to the optical emission-line velocities measured by Rubin and Ford (1970). The dashed curve is an alternative fit for the inner disk, with higher minimum velocities.

In a more detailed study of the inner 400 pc, Rubin and Ford (1971) obtained spectra for 16 different position angles across the nucleus. Using image tubes allowed them to detect and measure several emission lines, including Hα, [N II] 6548, [N II] 6583, [S II] 6717, [S II] 6731, He I 5876 and [O III] 5007. Figure 7.6 shows examples of some of their results. The most interesting outcome of the coverage of so much of the bulge was their conclusion that much more than simple symmetric circular rotation must be taking place. There is a very thin rotating disk, with the rotation not being axially symmetric. There are expansion velocities superimposed on the rotation, with velocities in certain position angles reaching values as high as 100 km/sec. A correlation was found between positions of areas of excess positive velocities and positions of dust clouds, leading them to conclude that these may indicate infalling motions of low-lying clouds of gas and dust.

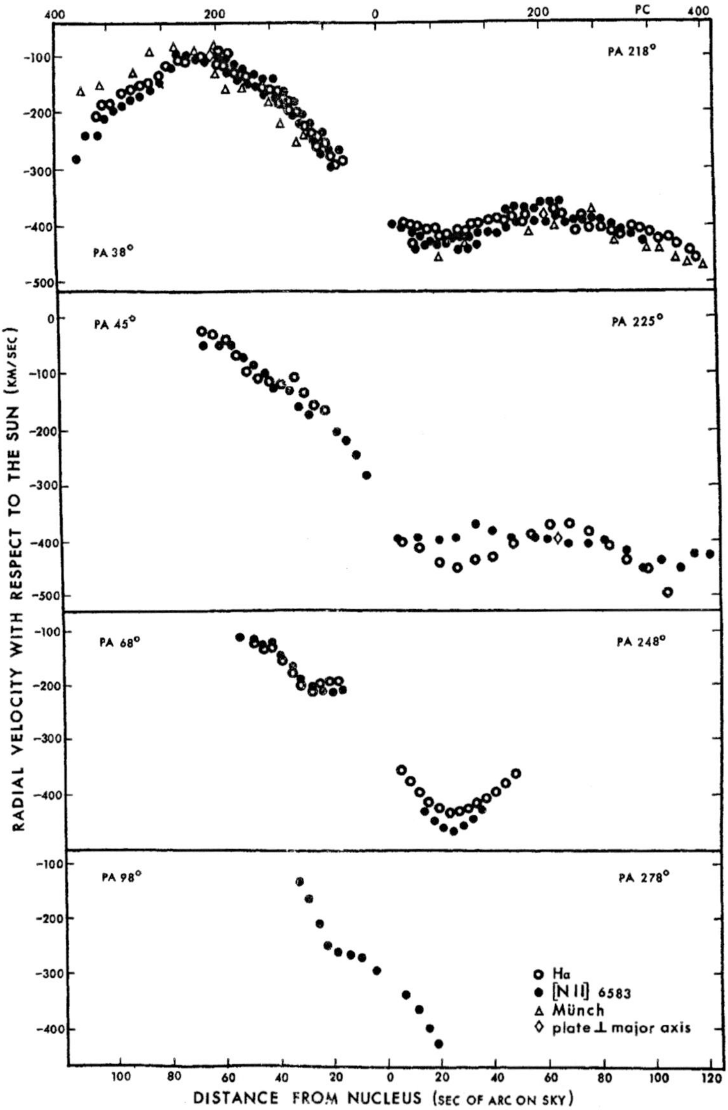

Figure 7.6 Observed radial velocities in the inner bulge of M31, as measured at various position angles. The scale refers to the major axis data only. From Rubin and Ford (1971).

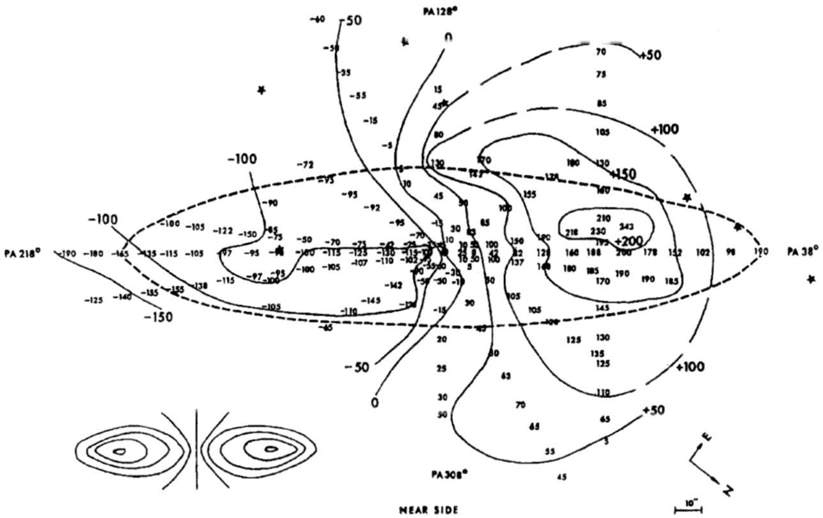

Figure 7.7 Radial velocities in the inner parts of M31, based on emission lines and with the central velocity subtracted. Velocities are given in km s^{-1}. The dashed line is a 400 pc disk in M31's plane and the diagram at the left is a predicted isovelocity diagram for a simple rotation law. (From Rubin and Ford, 1971).

More recent studies of the bulge region (see especially the extensive absorption line velocity study of McElroy, 1983) essentially confirm the Rubin and Ford results in general. McElroy's results suggest a rather less smooth velocity field than expected, which may be the result of the optical effects of dust clouds or, possibly, the residual effects of small-scale dynamical events, such as cluster tidal breakup, though the analysis carried out by Teuben *et al.* (1984) suggests that neither of these effects is the likely cause.

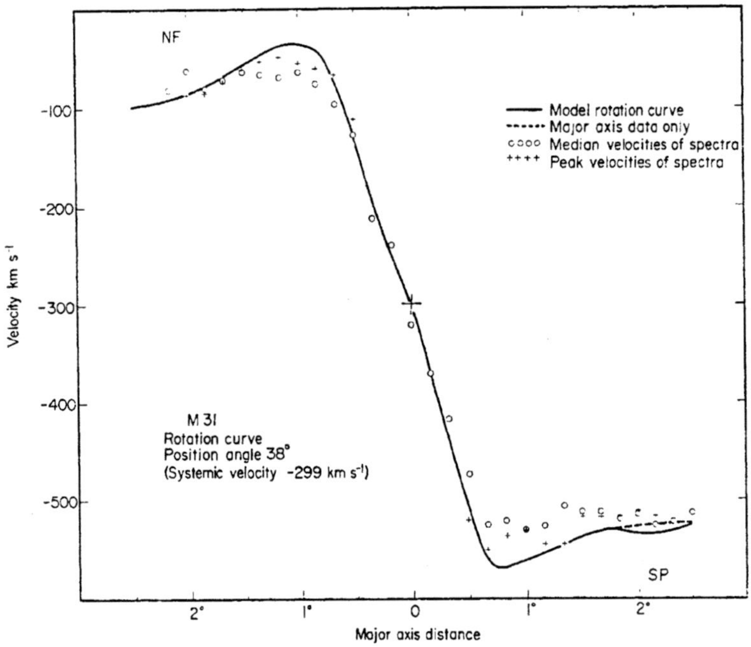

Figure 7.8 An H I rotation curve for M31 based on single-dish observations made with the Jodrell Bank 150-ft telescope (from Gottesman and Davies, 1970).

7.3 H I Rotation Curves

A complete, filled-aperture survey of the H I in M31 was carried out by Cram *et al.* (1980) using the Bonn 100-meter radio telescope, giving a beamwidth of 8′.8. The major axis velocity contours obtained are shown in Figure 7.11, which has the characteristic shape of data from a highly-inclined, rotating disk. Note, however, that there are multiple values for the velocities at many major axis positions, a manifestation of the inappropriateness, in detail, of a simple, flat, purely-rotating disk model. Cram *et al.* (1980) pointed out that the departures from the expected curve can be attributed to the warp of the disk and to areas of infall motion towards the center, as previously concluded from Westerbork data by Shane (1975) and by others (Section 7.5).

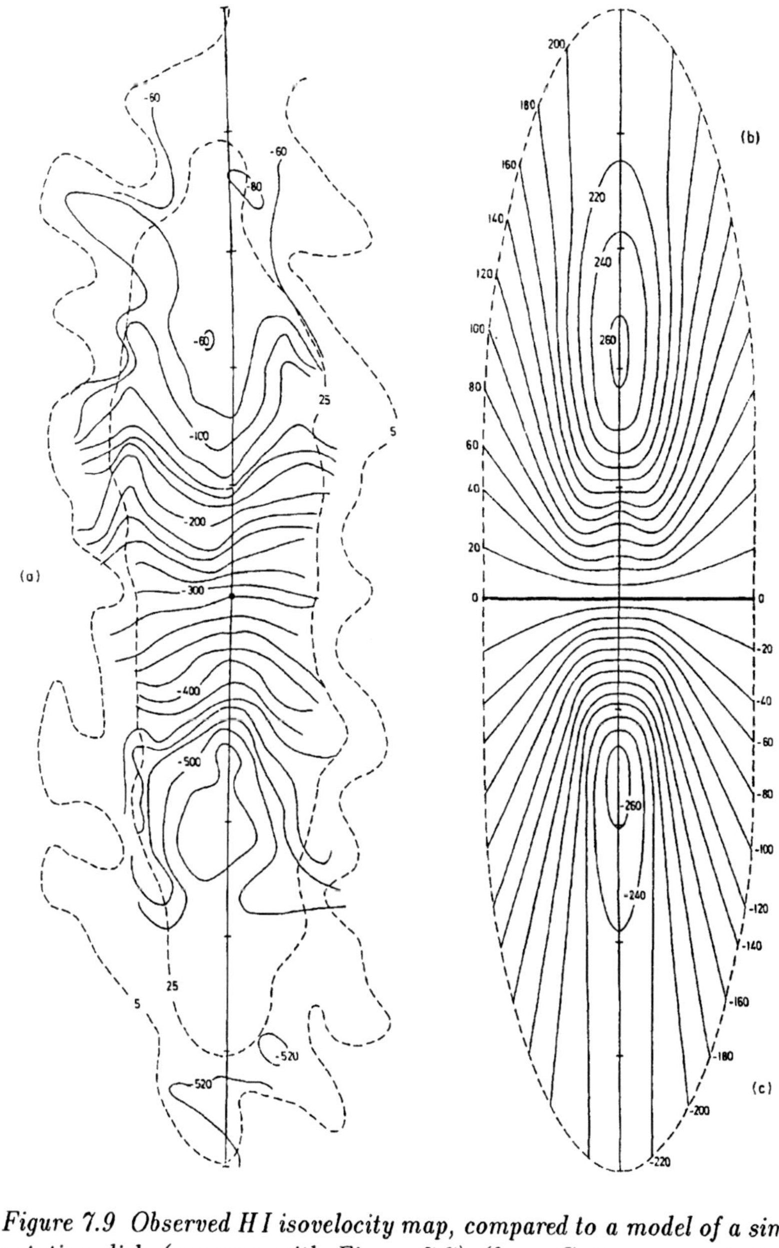

Figure 7.9 Observed H I isovelocity map, compared to a model of a simply-rotating disk (compare with Figure 7.7) (from Gottesman and Davies, 1970).

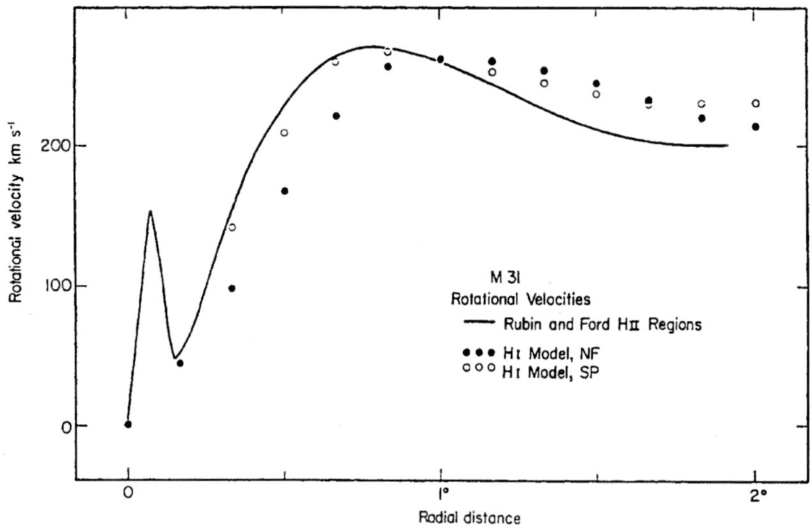

Figure 7.10 HI rotation curves for the north-following (NF) and south-preceding (SP) major axes of M31, compared to the optical curves of Rubin and Ford (1970) (from Gottesman and Davies, 1970).

Further progress in disentangling the rotation curve from the various interfering complications was made by Haud (1981), who primarily used HI data, but who also included a comprehensive collection of optical and mm-wavelength (Combes *et al.* 1977) data. Haud's results were based on an analysis of 4077 individual radial velocities, which were processed simultaneously to provide values for rotational velocities and radial velocities, adopting a modified version of Henderson's (1979) bending model. Figure 7.12 reproduces his curves of the rotation and radial motions. The rotation curve is similar to those produced by others shown in this chapter. Haud (1981) notes that the rotation curve shows evidence for fine structure, which he calls "waves" and which correspond in location to the crossings of the spiral arms (marked by arrows in Figure 7.12), indicating a distortion of motions somewhat like that predicted by the density wave model of spiral structure, an effect confirmed in part by more recent, high resolution studies (see below).

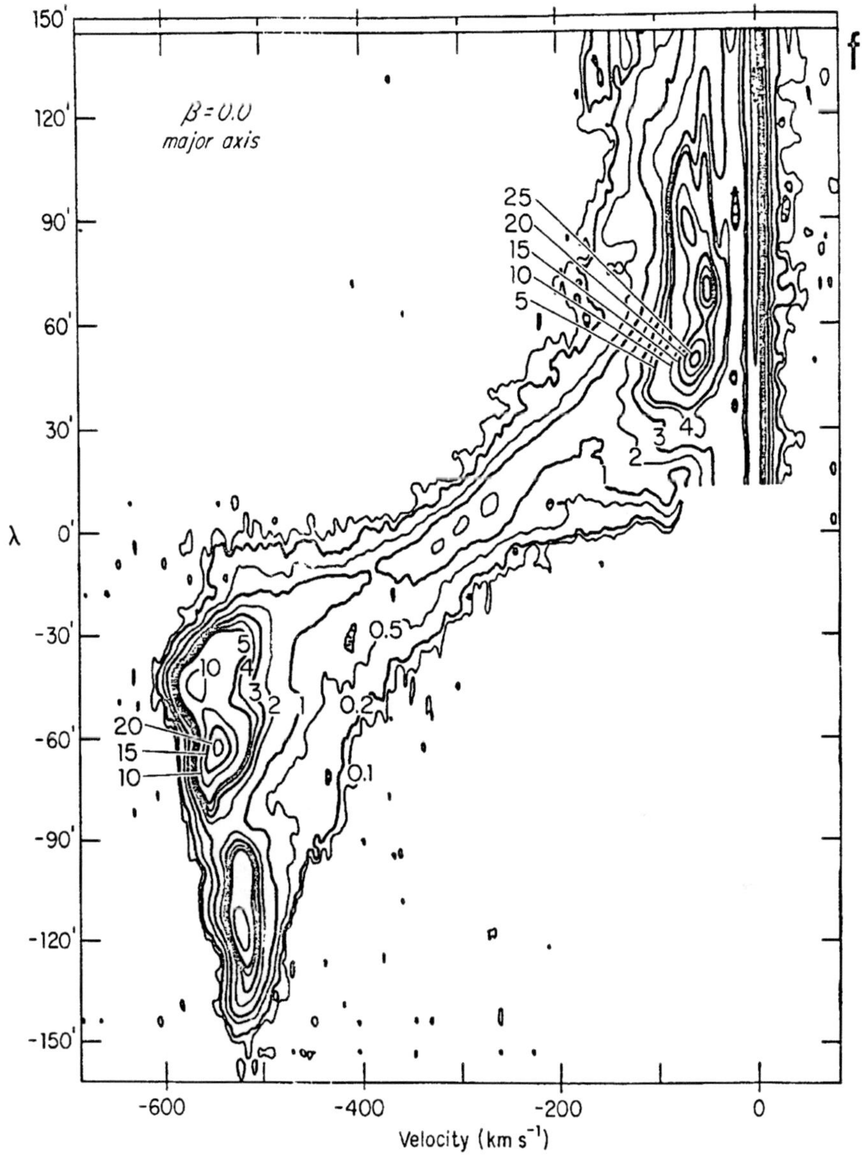

Figure 7.11 Isovelocity contours for the major axis of M31, based on filled-aperture measurements obtained with the 100-m radio telescope at Effelsberg. Contour units are K (antenna temperature) (from Cram et al. 1980).

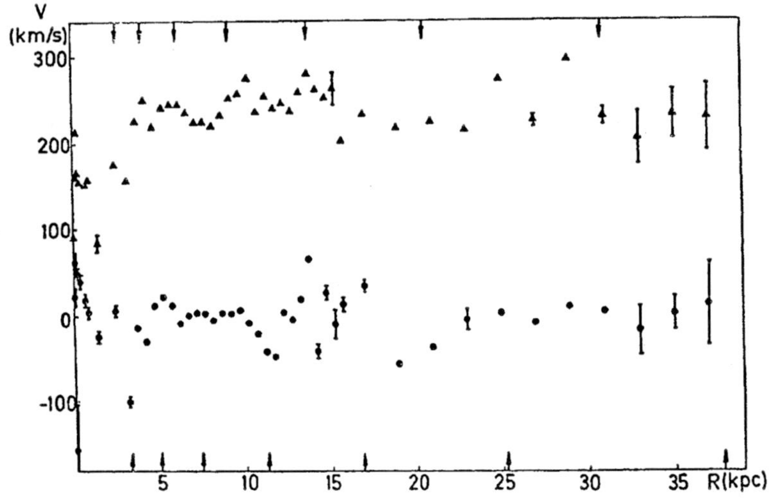

Figure 7.12 The rotation curve (upper) and expansion curve (lower) based on the bending model developed by Haud (1981). Arrows mark the positions of spiral arm crossings.

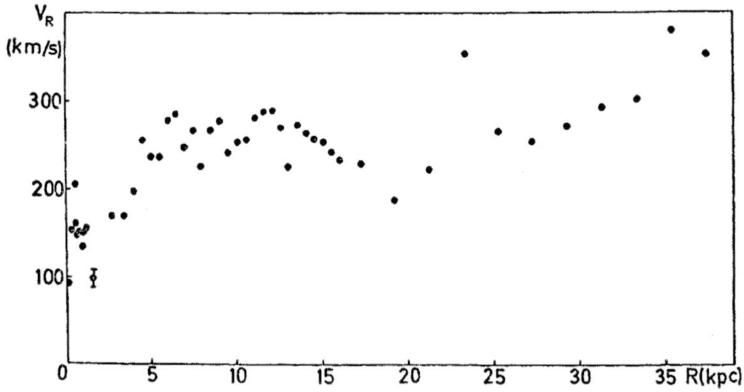

Figure 7.13 Haud's (1981) rotation curve plotted without taking into account the expansion and warp of the plane of the disk.

The main difference between solutions that include the effects of radial motions and warping and solutions that do not can be seen by comparing Figure 7.13, which is Haud's (1981) solution for a plane, purely-rotating disk, with Figure 7.12. Principally, the warping is reflected in the outer parts (Section 7.4), where a flat-disk model gives a rather steeply-rising velocity curve exterior to R = 20 kpc. This behavior is not shown in major-axis-only plots (*e.g.*, Roberts and Whitehurst, 1975), which are not affected by the bending of the disk.

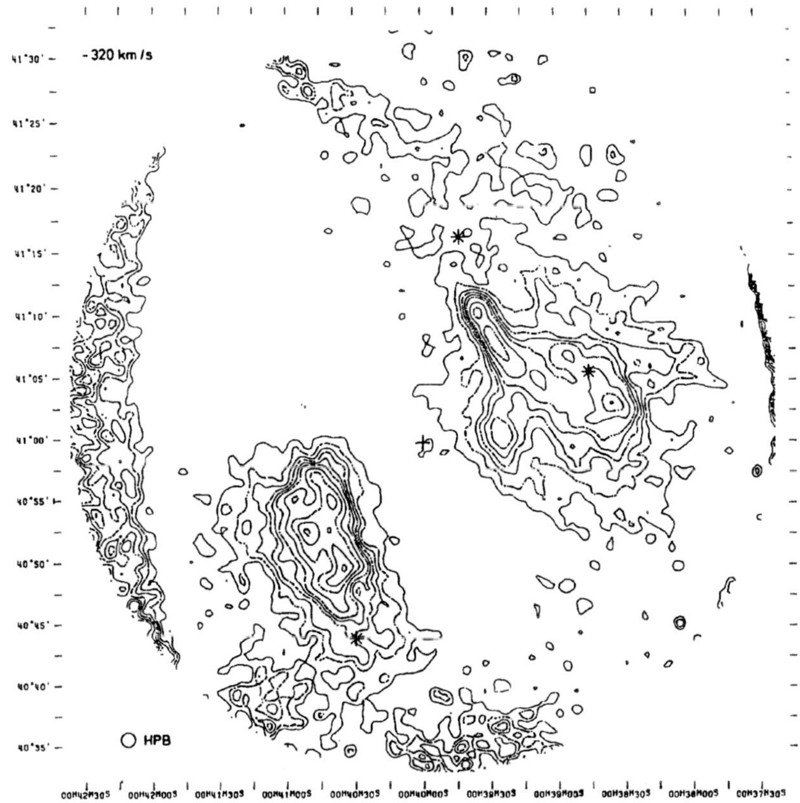

Figure 7.14 An example of a high-resolution Westerbork map of the H I in M31, in this case for gas with a velocity of -320 km s^{-1} (from Bajaja and Shane, 1982).

A pilot study to map the H I velocities at high resolution was undertaken in the 1970s at Westerbork (Bajaja and Shane, 1982, and previous references cited therein). These observations of the central part of M31 had a resolution of 24 × 37 arcsec^2, and, although limited in areal extent and undersampled in velocity, showed the great promise of such a survey of M31 with the WSRT. Figure 7.14 gives a sample of one of their velocity maps, for $v = -320$ km/sec (near the systemic velocity), showing the great detail in the H I distribution attainable with this instrument.

Unwin (1983) used the Cambridge Half-Mile Telescope to obtain a survey of the H I in M31 with a spatial resolution of 0.8 × 1.2 arcmin^2 and a velocity resolution of 16 km sec^{-1}. One of the more significant findings of this study was the fact that double-valued velocities were found inside R = 10 kpc; previous models of the warp had been restricted to larger distances from the nucleus. Also, Unwin found evidence for non-circular motions associated with the spiral arms (Section 7.5).

A higher resolution, complete survey of H I with the WSRT was subsequently published by Brinks and Burton (1984), whose higher resolution data and greater spatial coverage has produced the best information so far on M31's rotation. With a spatial resolution of 0.5 arcmin Brinks and Burton's (1984) survey resulted in an immense amount of data, which could only be displayed in its completeness in the form of a rather spectacular movie. The fine detail made available did not greatly alter our knowledge of the rotation curve of M31, except by showing more clearly and accurately the different departures from a simple rotational model fit. It did demonstrate the need for a detailed and comprehensive kinematical model of the gas, it allowed an improvement of the geometrical description of the warp (Chapter 5), and it provided velocity data that is still profitably being examined.

Braun (1991) published an exhaustive analysis of the rotation curve, largely based on recent VLA H I velocity measurements. His derived rotation curve agrees well with earlier curves and with the best optical data (Kent, 1989).

7.4 Rotation in the Outer Disk

A major development in our understanding of galaxy rotation (and the mass distributions determining it) occurred when Roberts and Whitehurst (1975) used the NRAO 91m radio telescope to probe the H I motions in the outer parts of M31, beyond all optical studies and beyond most previous radio surveys. They studied the southwest portion of the major axis, extending data out as far as 30 kpc. Previous optical studies (Section 7.2) had reached the outermost known H II region at a distance of 24 kpc

(Rubin and Ford 1970). Roberts and Whitehurst (1975) found that velocities in the previously unexplored region were greater, with respect to the galaxy's systemic velocity, than was expected on the basis of the usual assumptions, which predicted that there should be a decrease following a Keplerian relation. Because there is little optical radiation in the area beyond 20 kpc, it had been assumed that little material existed here; thus, the H I at these distances should be responding gravitationally to mass that is virtually all inside its position.

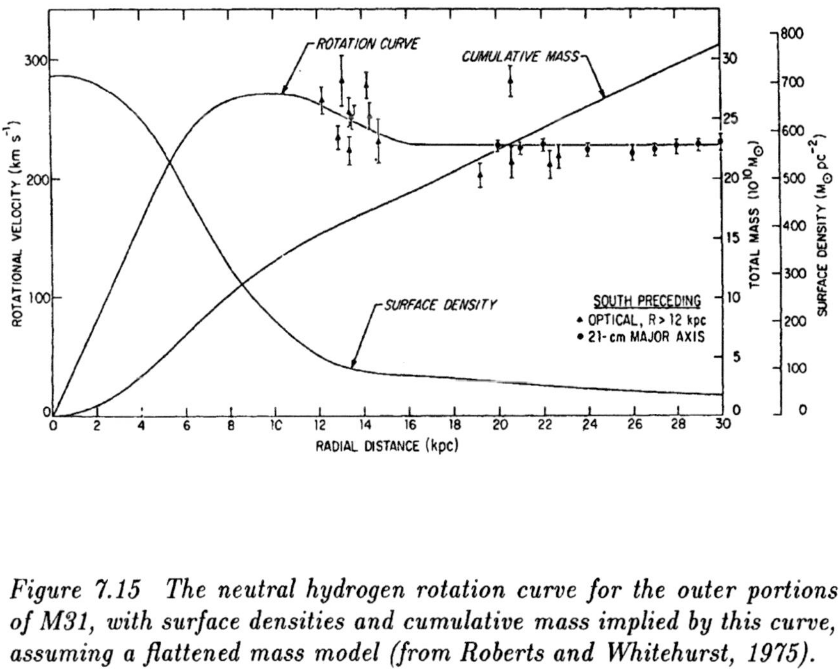

Figure 7.15 The neutral hydrogen rotation curve for the outer portions of M31, with surface densities and cumulative mass implied by this curve, assuming a flattened mass model (from Roberts and Whitehurst, 1975).

Instead, they found (after correcting for the effects of the warp) that the outer rotation curve is flat (Figure 7.15). This implies a mass that increases linearly with R and a mass-to-light ratio that becomes enormous in these outer parts. In their rather simple model, a surface density of 50 solar masses pc^{-2} is derived for R = 30 kpc, leading to a value of $M/L \approx 200$. The cumulative mass within this radius is 10^{11} solar masses, but, of course, a much larger total mass is implied by the fact that the rotation curve is still flat at this distance (see Chapter 8).

7.5 Radial Motions

After (and if) the effects of the warp of the disk have been taken into account adequately, any remaining anomalous velocities, consisting of departures from a smooth rotation curve or of deviations from simple isovelocity curves, can be isolated and studied. Unwin (1983), for instance, noted that the most common anomaly in his isovelocity contours is a turning in towards the minor axis. Also, many arms showed a nearly pure rigid body rotation and others showed, instead, a turning toward the major axis. He pointed out that the former effect can be caused by one of three possible situations: (1) a rotation curve with a local steep velocity increase, (2) an inward radial motion on the far side of the disk, or (3) an outward radial motion on the near side.

A demonstration of the clear presence of radial motions is given in Figure 7.16, which is a WSRT velocity curve for the minor axis, showing significant departures from circular motion that correspond to Baade's spiral arms 3 and 4; the bump near the nucleus corresponds to an "optically indistinct" inner feature (Bajaja and Shane 1982). That these are connected to the arms, rather than being geometrically superimposed on the minor axis from some more distant, distorted segment of the disk, is indicated by the fact that CO emission shows the same velocity deviations (Stark 1979).

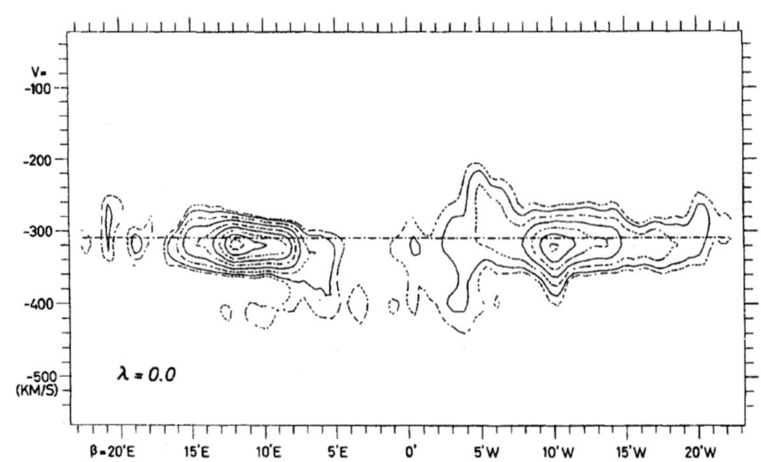

Figure 7.16 H I isodensity contours for the minor axis of M31. The dashed line shows what the velocity would be for perfect, simple circular rotation (from Bajaja and Shane, 1982).

There have been several attempts to disentangle the velocity components and geometry of these radial motions. For example, for the WSRT observations of Baade's arm 3, the velocity departs as much as −70 km/sec from model predictions at its extreme value about 8 arcmin north of the minor axis. Although Bajaja and Shane (1982) admitted that a unique model is not possible from the data, a kinematical picture that does fit includes an inward component of 30 km/sec (the value observed on the minor axis) and an excess rotational velocity of 40 km/sec. For arm 4, on the other hand, a simple fit was attained for an *outward* velocity of 20 km/sec, with no departure from the model rotational velocity. The third component shown in Figure 7.16 (at $\sim$ 3.5 arcmin west of the nucleus) shows a 70 km/sec outward velocity.

Several attempts have been made to explain individual examples of these anomalous velocities. The possible causes of such motions are varied: streaming to or from the nucleus, density wave effects, the dynamical effects of a bar, supernovae and stellar wind gas pressures, extragalactic infall, tidal effects by companions, and the dynamical breakup of star clusters and associations. It has not yet been possible to disentangle all of the complications, but possible causes of some local examples have been put forward. For example, for arm 3, Shane (1978) found that the velocities could, at least formally, be explained as being the result of an explosive event in the galaxy's nucleus 30 million years ago. Its similarity to our Galaxy's 3 kpc arm has been noted; they are at about the same distance from their galaxy's nucleus and the morphology is similar. Their radial motions are about equal, but the similarity ends when it is noted that one is expanding and the other falling in. With respect to density wave theory effects (Lin *et al.* 1969), Unwin attempted to compare his data with predictions, and, although some of the observed departures are consistent with the presence of density waves, which alter the values of the velocities in the plane, most are not clearly identifiable as such. We must conclude that the causes of the anomalous velocities remain uncertain. Further discussion of this important topic is given in Chapter 8.

7.6 Rotation of the Nucleus

The nucleus of M31 has been recognized as an optically discrete component for over 100 years (Chapter 1). However, it was something of a surprise when the first high resolution spectra were obtained to discover that it is also a discrete dynamical component (Lallemand *et al.* 1960). It rotates essentially independently of both the bulge and the disk. Lallemand *et al.* (1960) found a rotation curve with a maximum velocity of $\sim$ 80 km/sec at a distance of only 2″ (7 pc). It is important to clearly distinguish this maximum from that found in the bulge at 400 pc or the one for the disk, which has a maximum at $\sim$ 10 kpc. All represent rotations of distinct subcomponents on very different scales.

Further work on the rotation of the nucleus was carried out by Walker (1974, 1975) and Peterson (1978). More recently, a detailed study was published by Tonry *et al.* (1987) and others (see Chapter 8), who used high enough resolution and sensitivity to be able to establish the shape of the rotation curve and then determine the most probable mass distribution in the nucleus. As explained in Chapter 8, the best mass distribution fit implies the presence at the center of a massive, dark object, the most likely candidate for which is a black hole.

7.7 Summary

The rotation curve for M31 is complex. There are three principle components: a highly-rotating nucleus with a maximum velocity of $\sim$ 150 km/sec at a distance from the center of 3.5 pc, a central disk with a maximum rotational velocity of 225 km/sec at a distance of 400 pc, and an extended, warped, and flared disk with a maximum rotational velocity of about 280 km/sec at a distance of about 10 kpc. Superimposed on these rotational velocities are complex large- and small-scale anomalies, which may be related to gas-streaming motions, density-wave effects, or other causes. These velocities are still poorly understood and much further research is needed to disentangle their properties and to ascertain their causes.

CHAPTER 8

DYNAMICS AND MASS

From the measurements of the velocities of the various components of M31, it is possible to infer its dynamics and to calculate the mass of material that can explain its kinematics. This field is a complicated one; although in principle there should be no difficulty in deriving the mass distribution and total mass from the rotation curve, in practice there are many details that make a simple solution impossible. For example, there is the composite rotation curve in the inner parts, there is the asymmetry of the motions found at various radial regimes, and there is the problem of the apparent lack of any Keplerian decrease in the velocities in the outermost parts of the galaxy. All of these mean that we do not yet have a complete, detailed model for the dynamics or the mass distribution of M31. We do have an increasingly realistic picture of the complex kinematics and of their implications for our understanding of a galaxy's structure and motions.

8.1 Methods

For M31 there are two basic methods for using observed velocities to determine the mass distribution and to construct dynamical models. For the relatively flat disk, most of the motion is in the form of circular rotation, so we can use rotation curves, fitting them to various mass-distribution models to find a best fit. For the central bulge and the halo, where orbits are more nearly radial, we can use the dispersion in the velocities, together with an appropriate dynamical model, to determine the mass within a certain radius. Velocity dispersion measures can come from measurements of a number of types of sources, including bulge stars measured by means of their absorption lines in long-slit spectra, planetary nebulae, and globular clusters.

8.2 Dynamics of the Disk

The first attempt to interpret the rotation curve of M31 in terms of a dynamical model was made by Babcock (1939), who based his analysis on the rotation curve that he obtained from H II region radial velocities. Babcock made the assumptions, since used also by most other studies, that all of the motions that show up in the rotation curve are circular motions in a plane, that the stars and gas are moving only because of the balance between the galaxy's gravitational field at each point and the centrifugal force, and that pressure is not important. He constructed a model that consisted of four parts: a small spherical nucleus plus three concentric, coaxial, highly-flattened spheroids. Each different component was assumed to have a uniform density, the run of which decreased outward. The result was a reasonably good fit to the rotation curve. He derived mass-to-light ratios and a total mass for the system.

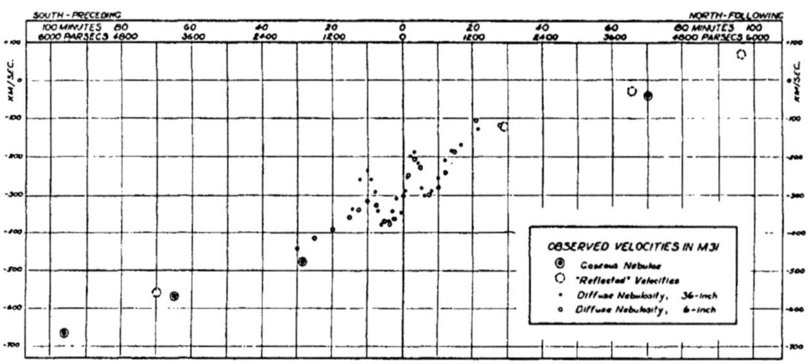

Figure 8.1 The observed rotational velocities of M31 used by Babcock (1939) for his analysis of its rotation.

Wyse and Mayall (1942) used a somewhat more analytical procedure to fit Babcock's observations. They employed a series of non-homogenous circular disks and solved for the properties by finding a best fit. A satisfactory fit was found for two such disks, and they derived a total mass of 2.7×10^{11} solar masses (using the 1942 distance, of course).

As more data became available and more understanding of galaxy dynamics developed, additional attempts to determine the mass and mass distribution in M31 were published (Table 8.1). Notable among these are the papers of Schwarzschild (1954) and Lohman (1954). With the advent of H I measurements, the rotation curve could be extended and further improved, leading to Schmidt's (1957) solution, which was similar to his multiple-spheroid model for our Galaxy.

Table 8.1
Determinations of the Mass of M31*

Source	Method	Mass (in units of 10^{11} solar masses)
Babcock (1930)	Optical rotation curve	3.3
Wyse and Mayall (1942)	Optical rotation curve	3.1
Schwarzschild (1954)	Optical rotation curve	2.1
Lohman (1954)	Optical rotation curve	3.7
Schmidt (1957)	H I rotation curve	2.8
Brandt (1960)	Optical and radio rotation curve	3.2
Rubin and Ford (1970)	Optical rotation curve	1.8
Einasto and Rummel (1970)	Optical and radio data	2
Gottesman and Davies (1970)	Radio rotation curve	2.2
Hartwick and Sargent (1974)	Globular clusters	3
Deharveng and Pellet (1975)	Optical rotation curve	1.6
Roberts and Whitehurst (1975)	H I outer rotation curve	3.7
Gunn (1975)	Timing of orbits	40
Hodge (1975)	Tidal cutoff of companions	60
Bosma (1978)	H I rotation curve	2.9
Rood (1979)	Virial mass of M31 group	2.5
van den Bergh (1981c)	Viral mass from globulars	2.4
Braun (1991)	H I rotation curve	2.0

*All values are reduced to a standard distance of 725 kpc, and generally refer to the mass within the observational limit of the data used.

An even more elaborate treatment of the problem was published by Brandt (1960), who showed that it was possible to represent the rotation curve in terms of a series of spheroidal homoeoids, which lend themselves to a tractable analytical solution. Because the treatment uses analytical expressions rather than ad hoc combinations of empirical structures, several different interesting quantities can be calculated directly, such as the surface density and the mass interior to any point. Figure 8.2 shows Brandt's fit to the rotation curve for various observational data and Figure 8.3 shows the corresponding mass fraction function. In a subsequent paper, Brandt and Scheer (1965) developed a more elaborate fit to take into account the double-peaked rotation curve that had then become available. Shortly thereafter, Takase and Kinoshita (1967) and Takase (1967) extended Brandt's method and applied their version to both M31 and the Galaxy.

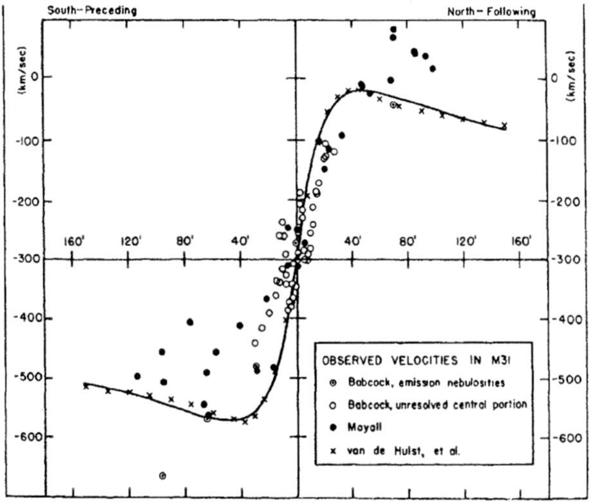

Figure 8.2 The rotation curve, including both HI velocities and optical values, used by Brandt (1960) for his determination of the mass distribution.

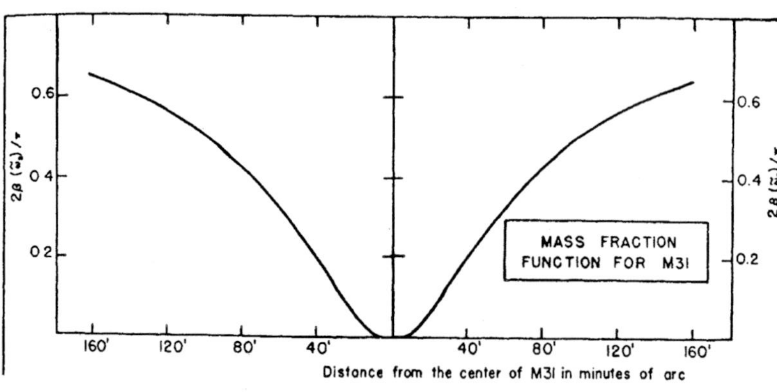

Figure 8.3 The mass fraction distribution in M31 determined by Brandt (1960).

A large number of papers on the subject of galaxy dynamics in general and of M31 in particular have been published by the Tartu group (*e.g.*, Einasto 1969, Einasto and Rummel 1969, Einasto 1970). For M31, their approach has been to divide the galaxy into four components: the nucleus, the bulge, the disk, and a flat component. Their results give somewhat different values for the parameters that describe the best fit models. In particular, they find a smaller total mass (2×10^{11} solar masses) and a larger central concentration than found previously. The mass-to-light ratio that they calculate for the nucleus is about 170 (see Section 8.4 for further discussion of this interesting point).

In 1970, when Rubin and Ford (1970) published their new velocities based on a large number of H II regions (see Chapter 7), new solutions to the mass and mass density could be found. Their solution, which includes consideration of all velocity data available at that time, consists of a dense, rapidly rotating central region with a mass of 6×10^{9} solar masses and a radius of about 2 kpc. (This is not the nucleus, which is much smaller. See Section 8.4). The density is low at a distance of 2 kpc, reaching values that become negative (of course, unrealistically), according to their solution (Figure 8.4). Beyond distances of 2 kpc, the density increases sharply to a maximum value of about 500 solar masses per projected square parsec at a distance of 5 kpc. Then there is a gradual decline, reaching a value of "a few" solar masses at distances of about 15 kpc, beyond which their model remains approximately uniform in density. Their mass distribution curve (Figure 8.5) shows a gradual increase outwards from the center to about 4 kpc, after which it rises more steeply, approximately linearly to distances of ~14 kpc, and more slowly thereafter. The total mass out to their most remote H II region is 1.8×10^{11} solar masses.

Deharveng and Pellet (1975) fit several models to their very extensive velocities, most obtained with a Fabry-Perot interometer, and concluded, among other things, that the mass-to-light ratio varies from a low value (~10) at a radius of ~2 arcmin to a high value (~30) at a radius of 25 arcmin. The mean mass-to-light ratio is about 12, in agreement with the value found by many other studies.

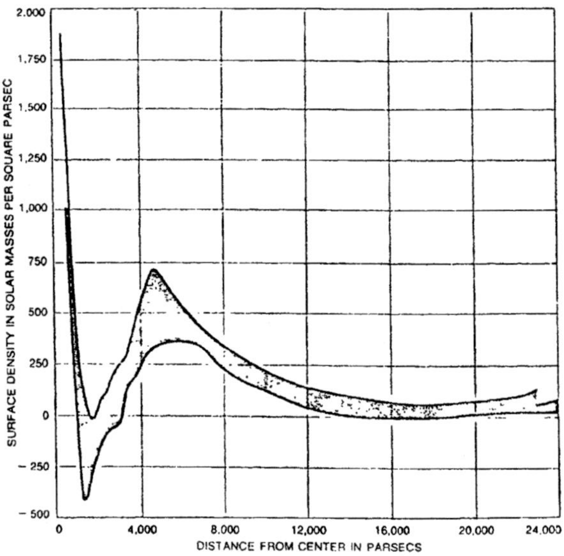

Figure 8.4 The distribution of the surface density in the plane of M31 as determined by Rubin (1973), based on a thin disk model. The thickness of the line represents the uncertainty that results from observational errors in the velocities.

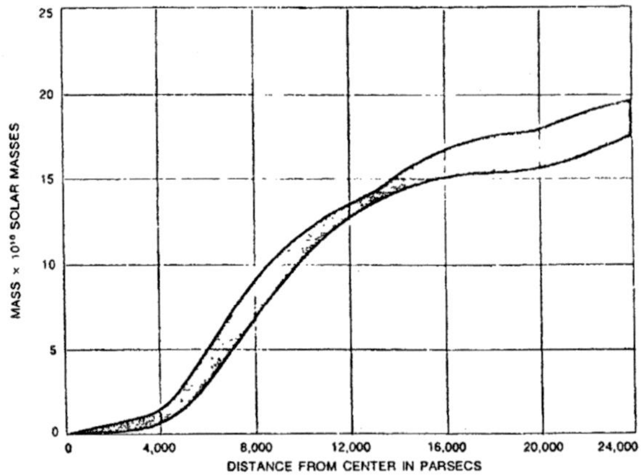

Figure 8.5 The mass distribution from Rubin (1973), determined as for Figure 8.4.

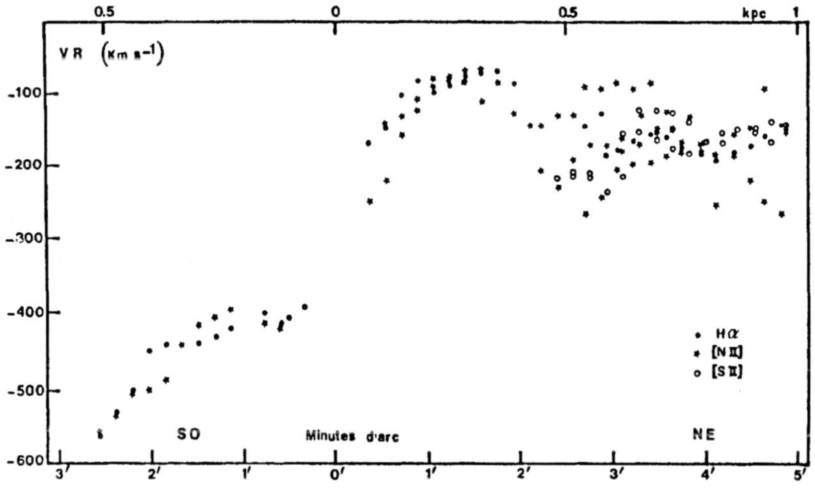

Figure 8.6 The rotation curve measured from emission line regions in M31 by Deharveng and Pellet (1975), using a Fabry-Perot interferometer.

8.3 Dynamics of the Bulge and Halo

The motions of the stars, clusters and gas in the bulge of M31 have been used to deduce the mass of material in that component. Most of our information about the mass of the inner bulge comes from the stellar velocity dispersion, while our information about the mass of the halo, which is usually interpreted as an extension of the bulge to large radii, comes from radial velocities of the outer globular clusters and companion galaxies. The halo is usually modelled so that it includes the dark matter of the galaxy, although the exact distribution of dark matter in M31 is largely conjecture.

Early attempts to model the distribution of mass in the bulge were made by Rubin and Ford (1970, 1971), who based their work on emission-line spectra of the excited gas in the bulge. Later (Rubin, Ford and Kumar 1973), they used the same spectra to measure absorption-line velocities of the brighter parts of the bulge. They found good agreement between the gas and stellar motions. Along the major axis they found that the velocity has a minimum value at a distance of 1.5 kpc, and a sharp rise from the nucleus to around 500 pc, where the rotational velocity is on the order of 130 km/sec. Along the more distant minor axis there were found anomalous positive velocities, suggesting streaming motions out from the nucleus area. Model-fitting indicated that the probable total mass of material inside a radius of 400 pc must be $\sim 3 \times 10^9$ solar masses.

Ruiz (1976) constructed dynamical models of the central region of M31 and compared them with the observations then available (Figure 8.7). She used two nonhomogeneous axisymmetric spheroids to represent the nucleus and the bulge separately, and obtained reasonable fits, though claimed that they were not unique. Her models predict that the ratio of the velocity dispersion in the bulge to that in the nucleus should be ~1.7.

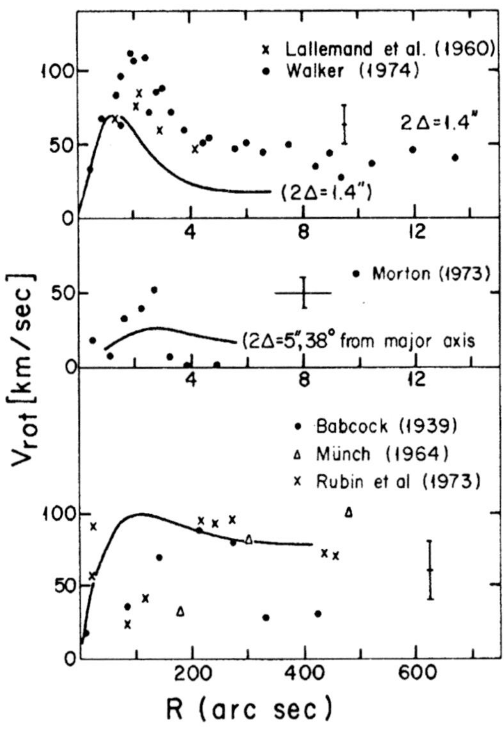

Figure 8.7 A comparison between observed and model rotation curves for different regimes in M31, published by Ruiz (1976).

Morton, Andereck and Bernard (1977) carried out observations to check this prediction, using artificially broadened spectra to compare with their spectra of the bulge and nucleus (Figure 8.8). They found that the ratio must be smaller than Ruiz (1976) predicted, with a most probable value near 1.0.

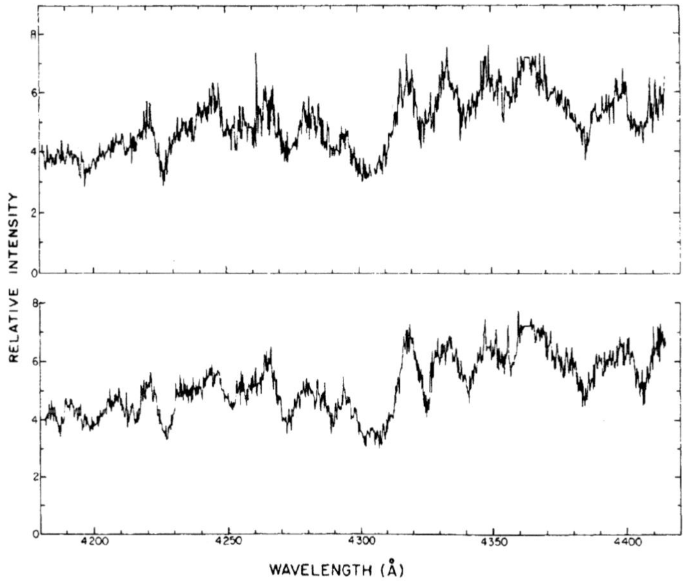

Figure 8.8 Spectra of the nucleus (top) and the bulge (bottom) of M31, used to measure the velocity dispersions (from Morton et al. 1977).

A second modern study was made by Whitmore (1980), who obtained new, high-resolution spectra of the bulge and the nucleus and who used a Fourier quotient method to determine the velocity dispersions. He attempted to make separations of the effects on the result of mutual contamination and rotational mixing. He found a velocity dispersion in the bulge of 151 ± 16 km/sec, significantly less than that found for the nucleus (181 ± 12 km/sec), and concluded that the mass-to-light ratio might be approximately four times larger in the nucleus than in the bulge to explain this difference.

A more complete mapping of the velocities in the bulge was carried out by McElroy (1983), who obtained fairly complete coverage of the entire bulge, using spectra at many position angles. He also examined the possible effects of dust on the results and found that the effect of dust can mimic asymmetries in the velocity field, probably explaining the asymmetries that he found to be especially noticeable near the major axis. The velocity dispersion in the bulge was found to reach a value of 150 km/sec at r = 30 arcsec to 60 arcsec, decreasing outwards from there to values of about 100 km/sec at r = 3 arcmin. The fact that he could discern no difference between the major and the minor axis dispersions suggests that the velocities are isotropic.

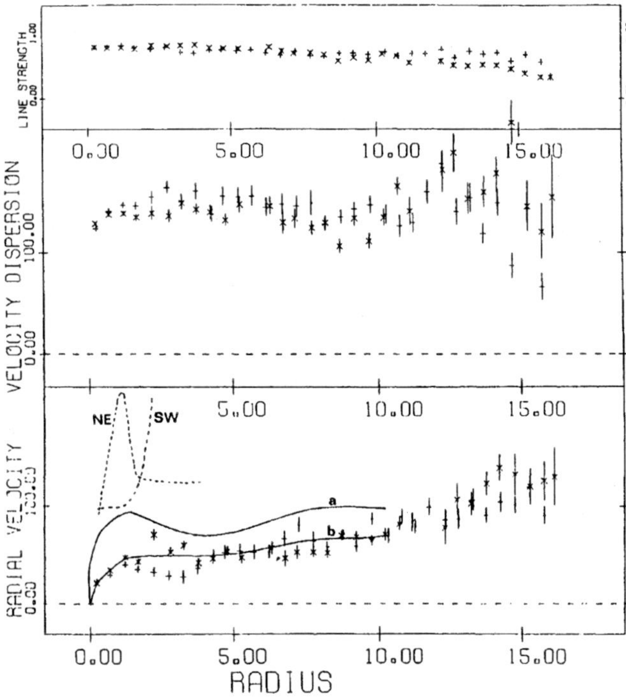

Figure 8.9 McElroy's (1983) results for the bulge of M31, showing the radial variation of the line strengths (top), the velocity dispersion (middle) and the radial velocity (bottom).

McElroy's dynamical models for the bulge lead to values for the mass within 1 kpc of $\sim 7 \times 10^9$ solar masses. He found a global mass-to-light ratio of 3.5, which appears to be nearly constant within this radial limit. The velocities of the gas and the stars, while both indicate rotation, seem to show little detailed similarity in their rotation curves; thus there is little dynamical coupling between these two components. Table 8.2 lists these and other measurements of the velocity dispersions of the bulge and nucleus of M31.

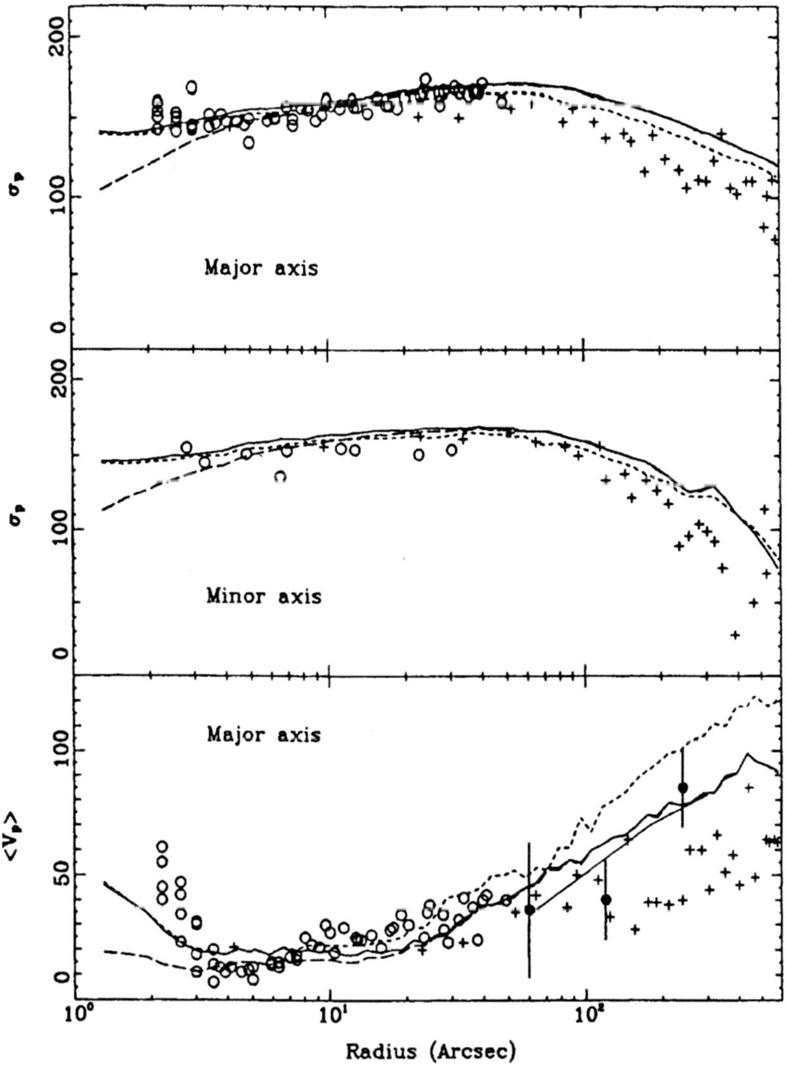

Figure 3.10 Kent's (1989) analysis of the M31 bulge. The projected velocity dispersions along the major (top) and minor (middle) axes are based on an assumed isotropic velocity distribution. The corresponding mean projected major axis rotational velocity is also shown (bottom). The solid line is for a mass model that includes a central point mass of 5 million solar masses; the long-dashed line is for a similar model, but without a point mass, and the short-dashed line is for a large-bulge model.

Table 8.2
Velocity Dispersion Measurements for M31 (after Whitmore 1980)

Author	Year	M31 ($km\ s^{-1}$	Method
Minkowski	1962	225	visual
Burbidge, Burbidge, and Fish	1961		
Richstone and Sargent	1972		least squares
Morton and Thuan	1973	120 ± 30	visual
deVaucouleurs	1974	150	least squares, composite
Faber and Jackson	1976	190	power spectrum
		180	visual
Morton and Elmergreen	1976	200	Fourier quotient
		180	least squares
		130 ± 20	visual
Morton, Andereck, and	1977	120 ± 20 (nucleus	visual
Bernard	1977	110 ± 20 (bulge)	visual
Williams	1977	125 ± 20	visual, composite
Sargent *et al.*	1977	165 ± 8	Fourier quotient
Monnet, Pellet, and Simien	1977	140 ± 20	visual
Pritchet	1978	145 ± 10	power spectrum
		150 ± 20	visual
Schechter and Gunn	1979	190 ± 5 (nucleus)	Fourier quotient
Whitmore	1980	181 ± 12 (nucleus)	Fourier quotient
		151 ± 16 (nucleus and bulge)	Fourier quotient
McElroy	1983	150 30" <R< 1′)	
		100 (R = 3′)	Fourier quotient

Kent (1989) constructed a new set of mass models for the inner parts of M31, taking advantage of his new photometry (Kent 1987) and the several new measurements of velocities cited above and in Section 8.4 below. He found that the bulge behaves like a rotating, oblate, isotropic spheroid with a mass-to-light ratio of 5, compared to a disk ratio of 10. Figure 8.10 shows his fits to the M31 kinematics and Table 8.3 lists the characteristics of the best fit models.

Table 8.3
Parameters of Kent's (1989)
Model for M31

Parameter	Value	Units
$(M/L)_B$	5.00 $\pm$0.5	
$(M/L)_D$	10.4 $\pm$1.0	
ρ_H	9.5 $\pm$1.9	10^{-26} g cm^{-3}
Bulge Mass	4.0	10^{10} solar masses
Disk Mass	1.6	10^{11} solar masses

Globular clusters' velocities can be used to further explore the mass distribution in the bulge and halo. Using data for a small number of globulars found at moderately large distances from the main body of M31, Hartwick and Sargent (1974) concluded that the total mass of M31 for a mean distance of 17 kpc is on the order of 3×10^{11} solar masses, in good agreement with the total determined for this portion of the galaxy from its rotation curve.

Using similar data but a different analysis technique, van den Bergh (1981) calculated that the mass of M31 inside 6 kpc is $\sim 9 \times 10^{10}$ solar masses, while six outer globulars, with distances on the order of 20 kpc, gave a mass of $\sim 2 \times 10^{11}$ solar masses.

Kent, Huchra and Stauffer (1989) analyzed the radial velocities of 149 globular clusters and concluded that little could be said with confidence without knowledge about the isotropy of the cluster velocity distribution. If it is assumed to be globally isotropic, then the cluster velocities, which indicate a low rotation velocity, present evidence against a model that has all of the mass in a spherically symmetric halo. If the velocity distribution is also locally isotropic, then the data argue for the presence of a massive halo.

8.4 Dynamics and Mass of the Nucleus

As described in Chapters 4 and 7, the rotation and structure of the nucleus of M31 indicates that it is a separate unit, not necessarily coupled at all with the rest of the galaxy. In the 1980s, as better and better resolution of both photometry and velocities were obtained, the astronomical community became more and more convinced that the nucleus must contain an unusually massive, small object. The most frequently suggested kind of object is a black hole, though other possibilities, such as a massive cluster of low-mass stars or stellar remnants may be possible (Goodman and Lee 1989).

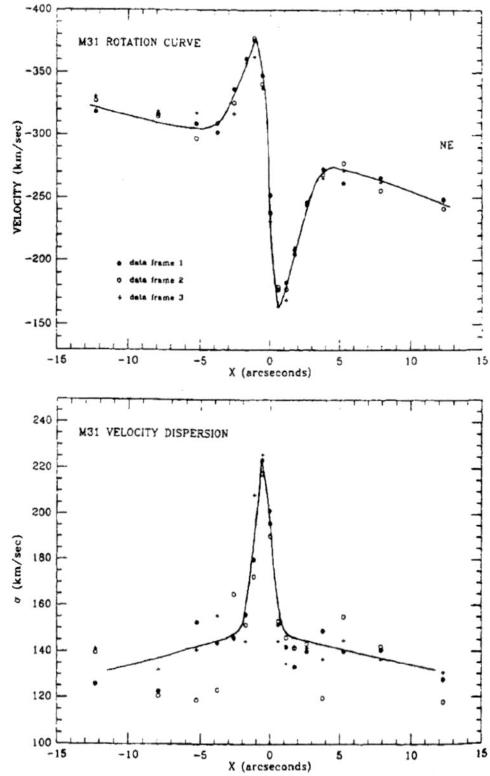

Figure 8.11 The rotation curve (top) and velocity dispersion (bottom) for the nuclear region of M31 as determined by Dressler and Richstone (1988) and used to demonstrate the presence of a mass concentration in the nucleus.

The suggestion that the nucleus of M31 has an exotic massive object in it was put forward by Dressler (1984) on the basis of the sharp rise in the rotation rate near the nucleus and the strong increase in the velocity dispersion at the center. Subsequently, two more detailed examinations of the problem, including better-resolution data, have been published (Dressler and Richstone 1988, Kormendy 1988). Both conclude that the new data strongly indicate the presence of a black hole. Dressler and Richstone's data, taken with the Palomar 5-m telescope, are shown in Figure 8.11. Their conclusion is that the black hole in the nucleus has a mass of $3 - 7 \times 10^7$ solar masses. (In a reanalysis, Richstone, Bower, and Dressler (1990) later revised this value upward to 10^8 solar masses for a mass that lies within 3.4 pc of the center). They note a peculiar phenomenon when the positions of the peak of light brightness is compared with the kinematic center: they are not in the same place, but are separated by approximately 0.6 arcsec (projected). The reason for this is not known, but it may be the result of some non-symmetric distribution of material around the nucleus.

Kormendy (1988) took advantage of the excellent seeing at Mauna Kea Observatory and obtained data with spatial resolution of ~0.5 arcsec. At 1 arcsec from the center, he found a rotational velocity maximum of 149 km/sec; the velocity dispersion at the center was 245 km/sec. The mass-to-light ratio is found to rise sharply to values of order 100 within 0.5 arcsec from the center. This strongly suggests a non-stellar, massive dark object, a black hole being the most likely. Kormendy's models suggest that the mass of the supposed black hole is 10^7–10^8 solar masses, in excellent agreement with the Dressler and Richstone (1988) conclusion.

8.5 Dynamical Disturbances

Because M31 is so close and well-resolved, we see many details of the motions in the disk that suggest that a simple, symmetric dynamical model does not tell the whole story. Small-scale disturbances in the rotation curve show up in both optical and H I high-resolution data (Chapter 7).

One of best-studied disturbances of M31 dynamics is that ascribed to the presence of M32 (Schwarzschild, 1954; Byrd 1976, 1978). Byrd was able to reproduce many of the asymmetries seen in the velocities and positions of H I in the M31 disk. Contours of H I density along the northeast semi-major axis are displaced away from the major axis, while the contours along the southwest semi-major axis are displaced towards the major axis. Byrd's models of M32's encounters with M31 suggest that the smaller galaxy may well have passed through the larger galaxy, causing very complicated disturbances to the disk.

If M32 is truly interacting strongly with M31, then dynamical friction should cause its orbit to decay and it should eventually be accreted. Tremaine (1976) examined this problem and estimated that the time for M32 to be accreted by M31 is on the order of 3×10^8 years, a very short time compared to its lifetime. However, Byrd (1979) re-examined the problem, using better defined orbital characteristics and masses, and obtained a much longer decay time, on the order of 2.6×10^9 years.

Smaller-scale disturbances have been examined from the high-resolution density and velocity H I maps obtained, for example, by Unwin (1980) and Brinks (1984). One example is the spur found near the giant OB association NGC 206, which Byrd (1983) explained by models of the gravitational effects of the mass of NGC 206 on the neighboring disk. Further modeling of the situation was carried out by Byrd, Smith, and Miller (1984), who generalized the conclusions that objects as massive as NGC 206 can show conspicuous disturbances, both near the disturbing mass and at distances as large as kiloparsecs, both inside and outside the orbit of the disturbing mass.

8.6 The Total Mass of M31

The total mass of M31 has steadily grown since it was first measured. Much of this growth has resulted from refinement of its distance, some from application of increasingly more realistic mass models, and some from the detection of mass at greater and greater distances (Chapter 7). It is clear from an examination of Table 8.1 that the mass of the optically visible portion of M31 must be on the order of $2 - 3 \times 10^{11}$ solar masses. However, it is clear from the H I rotation curve that much mass exists far beyond the optical image. As one looks at more and more distant regions, the density of radiating H I drops to an undetectable level before the mass density clearly begins to decrease.

The best way to measure the total mass of M31 should be by means of a method that senses its gravitational field at very large distances, external to M31 itself. Gunn (1975), for instance, used orbit timing arguments and derived the very large value of 4×10^{12} solar masses for M31. Hodge (1975), on the other hand, used the measured tidal radii of the distant elliptical companions of M31, NGC 147 and NGC 185, deriving a total mass of 6×10^{12} solar masses. A much smaller value was obtained by Rood (1979), who examined the viral mass for the M31 group and calculated a total mass for M31 of only 2.5×10^{11} solar masses. A somewhat different approach was used by van den Bergh (1981), who applied an alternative dynamical mass estimator to the M31 *group* and found a mass of 7.5×10^{11} solar masses. The wide dispersion of these values illustrates the difficulty of the problem; the virial theorem and its equivalent suffer from the small numbers of objects and the other methods tend to give limits that are dependent on assumptions about several unknowns, especially about the orbits of the companion galaxies.

8.7 Summary

From the inside outwards, M31 has:

a. a very small, massive nucleus with a mass of $10^7 - 10^8$ solar masses, possibly a black hole;

b. a central bulge with a mass within 1 kpc of the center that totals approximately 10^{10} solar masses.

c. a flat disk with a mass out to the edge of the visible image that totals, including the mass of the inner dark halo, approximately 2×10^{11} solar masses; and

d. a halo, which is probably made up mostly of dark matter, giving a total M31 mass of, perhaps, as much as 5×10^{12} solar masses.

CHAPTER 9

GLOBULAR CLUSTERS

The globular clusters of M31 have provided a great deal of useful information, both directly about the nature of the galaxy's Population II and halo, but also indirectly about such questions as its history, dynamics and mass.

9.1 Identifications and Catalogues

Hubble's Pioneering Study

The first extensive list of globular clusters in M31 was published by Hubble (1932). He had not completed this work in time to include it in his massive 1929 paper on M31 (Chapter 2), so it is discussed here; it is a sufficiently important milestone in the development of the field that its results are presented in some detail here, rather than being dispersed at various places throughout this chapter.

Using Mt. Wilson 100-inch telescope plates, Hubble identified 140 objects in M31 that appeared to be non-stellar and unresolved, which he tentatively suggested to be globular clusters. This conjecture, which has since been fully confirmed for most of them, was supported at the time by Humason's spectrum of candidate number 62, which had a spectral type of F8 and a radial velocity of −210 km/sec, consistent with velocities in M31. The biggest problem with identifying globular clusters in a galaxy with such a large angular extent has been the possible confusion with distant, compact elliptical galaxies, and Humason's measures provided the needed reassurance on this issue.

Hubble measured the positions, the distances from the center of M31, the apparent integrated magnitudes, the apparent diameters, and the concentration classes for his candidates. He found that they ranged in position from near the center of the bulge to as far away as 204 arcmin. Their magnitudes ranged from 15 to 18 in the blue (pg), and their sizes ranged from 4 to 10 arcsec. The radial projected spatial distribution was found to resemble the distribution of light in elliptical galaxies (Figure 9.1).

The luminosity function of the globulars was found to show a peak value at $m_{pg} = 17$, with about a 1.5 mag standard deviation (Figure 9.2). As a historical curiosity, it is interesting to note that his conclusion that these objects might be globular clusters was partly based on the similarity of their absolute magnitudes with those of the Magellanic Clouds' globulars, but only one of the Magellanic Cloud clusters that he used for comparison are true globulars; the rest are now known to be massive young clusters.

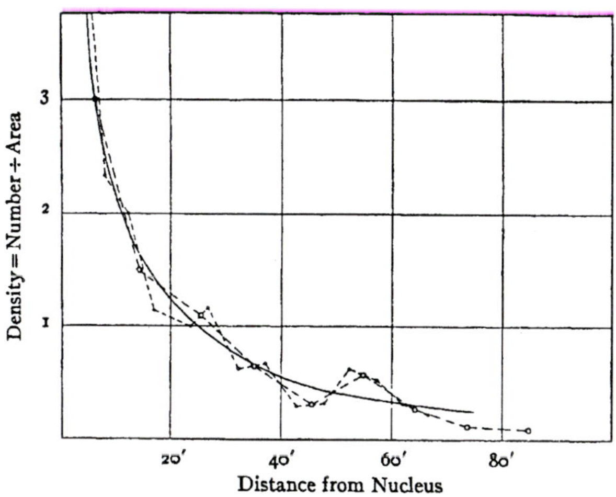

Figure 9.1 The radial distribution of Hubble's "nebulous objects", tentatively identified as globular clusters (from Hubble, 1932).

More Recent Surveys

In the process of his extensive survey of M31's content, Baade added 102 additional objects that he considered to be like globular clusters. A result of his reluctance to publish is the fact that, to find any record of these objects, one must look in a paper published by others (Seyfert and Nassau 1945). A few additional candidates were reported by Hiltner, Mayall and others. By 1962, the total number of catalogued clusters was 266 (Vetesńik 1962). An additional 17 were added shortly thereafter from Baade and Arp's catalogue of H II regions (without narrowband images or spectra, globular clusters can, from their shape and red color, be mistaken for compact H II regions).

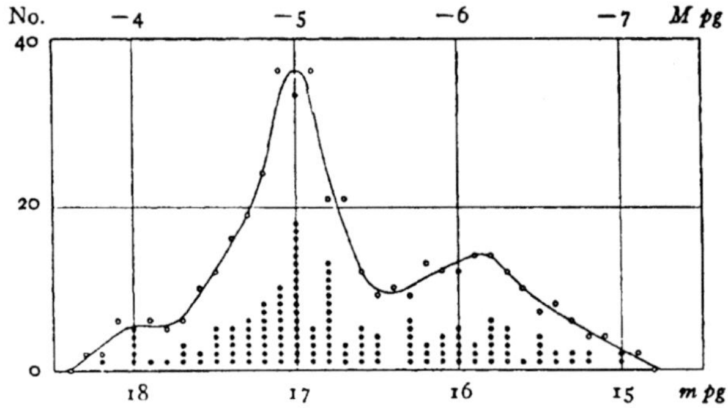

Figure 9.2 The luminosity function of Hubble's globular cluster candidates (from Hubble, 1932).

In the 1970s, several ambitious programs were begun at different observatories to discover and study more globular clusters, with the emphasis on finding fainter, more compact, and more distant objects than had been catalogued previously. Sharov began his extensive work on M31 globulars with a catalogue of 25 new candidates discovered with the Latvian Radio Observatory's Schmidt telescope in 1972 (Sharov 1973). In several papers, discussed below, he and his collaborators used a variety of large Soviet telescope to study the M31 globular objects further. Additional lists were published by Alloin *et al.* (1976, 1977).

The comprehensive new search was carried out by a group of astronomers who used the uniquely suited large field of the then new Kitt Peak 4-meter telescope (Sargent *et al.* 1977). A total of 29 different 50 arcmin-diameter fields were exposed, covering the main body of M31 and an extended area along the minor axis. All plates were in one color only (B), so identification was based primarily on the resolvability of the images. The total number of candidate clusters was 307, of which 183 were subsequently identified as previously catalogued clusters. Thus, the program provided 124 new objects. Later inspection showed that 48 previously known clusters had been missed, but could be confirmed, and 52 previous clusters were rejected. Thus, this paper was able to conclude that there were 355 confirmed globular clusters in M31.

In Italy, a group of astronomers were in the meantime embarking on an exhaustive search using more quantitative techniques (Battistini *et al.* 1980, 1982, 1987, Buonanno *et al.* 1982). Using PDS scans of the photographic images, they were able to distinguish compact clusters from stellar images in a quantitative way. Their master list, assembled in 1986 (Battistini *et al.* 1987), includes all objects identified to that date as globular clusters, within the central 3° × 3° field of M31. The objects are divided into three classes: A, which are objects very confidently identified as globulars; B, which are somewhat less certain; and C, which are doubtful. There are 254 A clusters, 99 B objects, and 152 C objects. Within the area covered, they find that even for 18th magnitude clusters, the sample is probably not complete. Their best estimate is that the completeness to that limit is about 85%, decreasing, of course, for fainter magnitudes.

The Battistini *et al.* (1987) catalogue includes most known data for each of the 353 class A and B clusters, including accurate positions, magnitudes and colors in the UBVR system, cross-listings to the Vetesńik (1962) and Sargent *et al.* (1977) catalogues, and remarks.

A different discovery technique was employed by Crampton *et al.* (1985), who used prime focus slitless spectra obtained with the CFHT 3.9-meter telescope on Mauna Kea. Their search covered about 4 square degrees and led to the recognition of 109 new candidates. Combining these with the previous lists, after rejecting several objects on the basis of the spectra, Crampton *et al.* compiled a listing of 509 probable globular clusters. This number is larger than that of Battistini *et al.* (1987), who rejected several of Crampton *et al.*'s (1985) new candidates, some because of faintness and some on the basis of evidence that they are distant galaxies.

Racine (1991) confirms that galaxies do contaminate the Crampton *et al.* catalog. Wirth *et al.* (1985) used a statistical argument to conclude that perhaps as many as half of the globulars in the bulge of M31 had been missed (they compared the number of stellar images there and compared it to the number expected on the basis of foreground star counts expected for M31's Galactic coordinates). The 20 or so missed clusters, if present, must be very compact in structure. Battistini *et al.* (1987) estimate that they would have to have core radii as small as 0.4 pc, according to their examination of the potential cluster images in the bulge. As there is a relation between size and radial distance for the globulars, with the size increasing outward, only small clusters are expected near the center.

9.2 Luminosities and the Luminosity Function

Following Hubble's (1932) initial look at cluster luminosities, the next major study was made by Vetesńik (1962), who compiled measurements, including many of his own, to assemble a listing of magnitudes and colors for 257 clusters. Other integrated photometry of the time included measures made by Hiltner, Eggen, Haro, Racine, and Kinman. Vetesńik (1962) published UBV data and derived a more complete luminosity function than Hubble's (Figure 9.3), finding a rather similar curve to that for the Galaxy, except for an extension for M31 to somewhat higher luminosities than found for our Galaxy's sample. There are five globulars in M31 with M_V brighter than -11, while our Galaxy only has one (ω Centauri). Both functions showed a maximum value of about $M_V = -8.5$. To do this work required, of course, that he derive reddening values and correct the observed magnitudes for extinction, which he was enabled to do from the three-color data (see below).

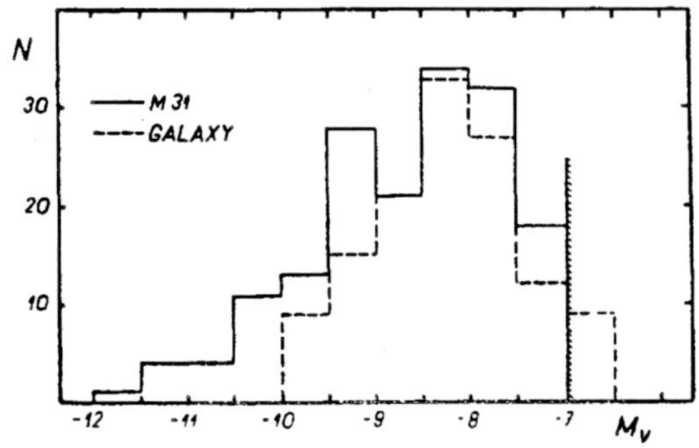

Figure 9.3 The luminosity function of the Vetesńik sample of M31 globulars, compared with that for the Galaxy's (from Vetesńik, 1962).

A few years later, van den Bergh (1969) used the Palomar 200-inch telescope to observe (photoelectrically) many of the clusters and to derive a new luminosity function (Figure 9.4). He concluded that the brightest globulars of M31 are brighter than those of our Galaxy, warning that this implies a danger in using them as distance indicators.

Sharov and his collaborators also used photoelectric methods to derive integrated magnitudes, as well as three-color data for many M31 globulars (Sharov 1975, 1983, 1984, Sharov and Lyutyi 1983, 1985, 1989). In a critical review, van den Bergh (1990) determined that no systematic difference exists between the voluminous photometries of Sharov and his collaborators and the photometric systems of earlier papers.

Turning their attention to only the outer clusters, where background effects are not important, Racine and Shara (1979) measured the magnitudes of 86 clusters in the halo and derived a luminosity function for them. They found a maximum at $V = 16.79 \pm 0.12$, corresponding to $M_V = -7.9$. Their main point was that, when only halo clusters are considered, there is no statistically significant difference in the luminosity functions of M31 and the Galaxy, a rather different result from those found earlier.

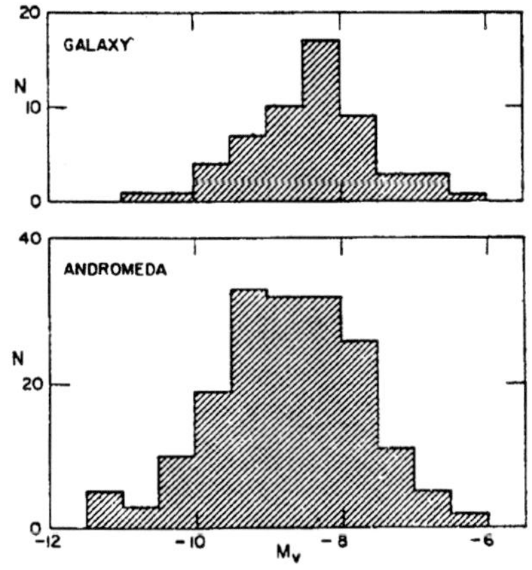

Figure 9.4 The luminosity function of the van den Bergh (1969) sample of globulars, compared with that for the Galaxy (from van den Bergh, 1969).

Crampton *et al.* (1985) found an apparent difference between the luminosity function for clusters inside a radius of 10 kpc and that for more distant clusters. The mean value of M_V for the inner clusters was -7.8 ± 0.1, while that for the more distant clusters was -7.0 ± 0.1. Although they were unsure to what extent the effect may be due to observational or selection effects, their conclusions do seem hard to explain away completely and the implications for the use of globular clusters as distance indicators might be serious. The differences found between M31's sample and the Galaxy's are shown in Figure 9.5. As theirs is the most complete list available, especially to faint magnitudes, Figure 9.5 is probably the best data set from which to generalize about any differences between luminosity functions for the two galaxies. The safest conclusion is that, although there is no significant difference in the luminosity of the maxima in the frequency curves (both have values of −7.0 for outer ($R > 10$ kpc) clusters, there does seem to be a wider dispersion in the M31 curves. The standard deviations for the outer clusters are 1.8 and 1.6 magnitudes for the Galaxy's and M31's, respectively, and corresponding values for the inner clusters are 1.0 and 1.5 mag.

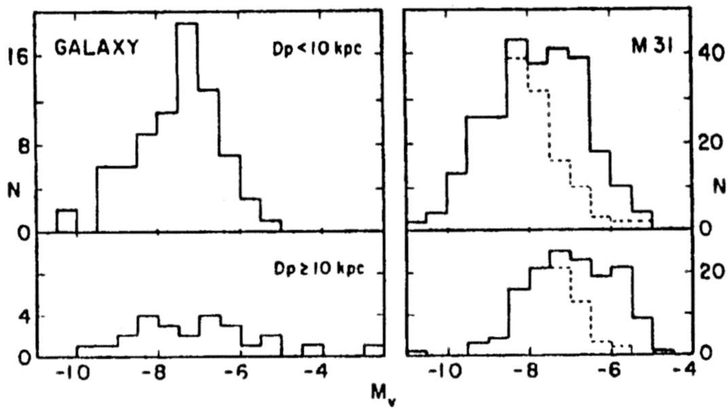

Figure 9.5 Another comparison of luminosity functions for M31 globulars, this one based on the Crampton et al. (1985) catalogue. the inner and outer globulars are compared (from Crampton et al. 1985).

We conclude that M31's globular clusters are more numerous than our Galaxy's (about 300 as compared to about 150), and they show more variety in their luminosities. However, their mean luminosities are almost identical.

9.3 Colors

Since Hubble's estimates, a great deal of effort has gone into measuring the colors of M31's globulars. The main reason for this is that colors can provide not only good corroboration of the nature of the objects, but also can lead to determinations of reddening and estimates of chemical abundances.

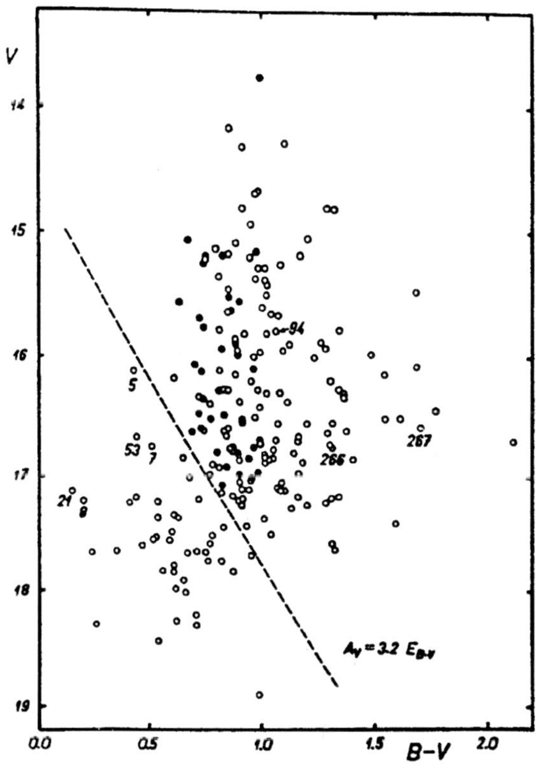

Figure 9.6 The color-magnitude diagram for M31's globular clusters, based on Vetesńik's data. The dashed line is for $A_V/E(B-V) = 3.2$ *(from Vetesńik, 1962).*

Vetesńik's (1962) study included four-color (UBVR) data for the clusters. From these he was able to plot a color-magnitude diagram (Figure 9.6), which showed a wide range of color, from values of $B-V = 0.2$ to 2.2. The former are clearly not normal globular clusters, but are probably fainter examples of the "blue globular clusters" found in the Magellanic Clouds (Hodge 1961), while the latter are probably highly reddened. Vetesńik determined the amount of reddening from his color-color diagram (Figure 9.7), which showed reddenings that varied up to values as large as 0.94 mag in $E_{(B-V)}$. The mean $B-V$, before correction, was calculated to be +0.83, about 0.2 mag redder than in our Galaxy, all of which was ascribed to reddening. The question of the amount of reddening, and particularly the question of whether the ratio of total to selective extinction is the same in M31 as in the solar neighborhood, will be taken up in Chapter 12.

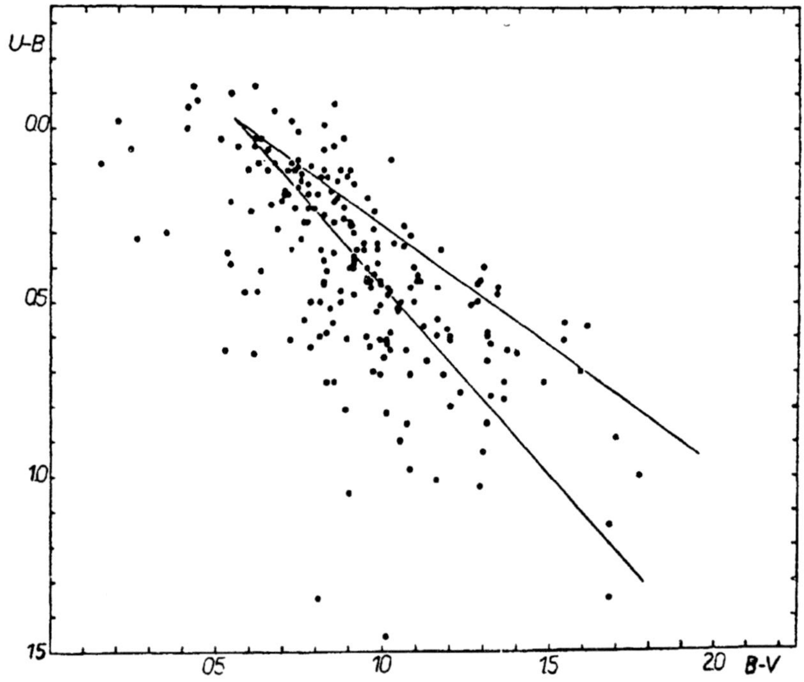

Figure 9.7 Two-color diagram from Vetesńik's photometry. The area occupied by Galactic globulars is indicated (from Vetesńik, 1962).

The color-magnitude diagram derived by van den Bergh (1969) looks very much like Vetesńik's, but with a smaller spread. Also, that of Buonanno *et al.* (1982) shows the same properties (Figure 9.8), with the majority of the globulars having measured colors of about B – V = +0.80.

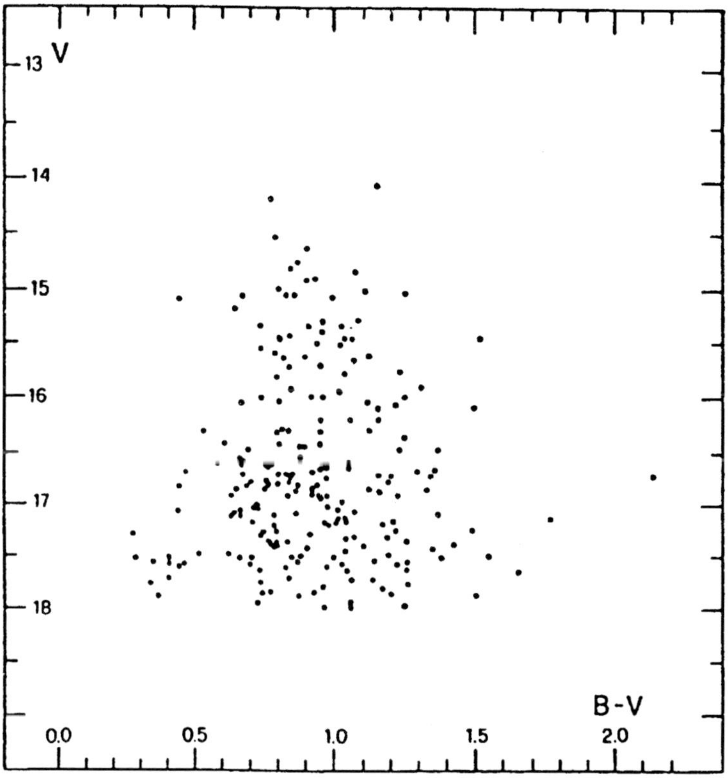

Figure 9.8 The color-magnitude diagram for the M31 globulars according to Buonanno et al. (1982).

Van den Bergh's (1967, 1969) studies included an analysis of the values of the reddening-free parameter Q (Johnson and Morgan 1953), which seemed to show a marked difference for our Galaxy and M31, with the latter being broader and less sharply peaked. Later, however, Hanes (1977) used the more extensive UBV color data given by Harris (1974) for 180 M31 clusters, concluding that the differences may instead be explained merely by observational errors. Figure 9.9 shows Hanes' plot of the distributions of Q values, broadened artificially by assumed errors. This issue is still very controversial.

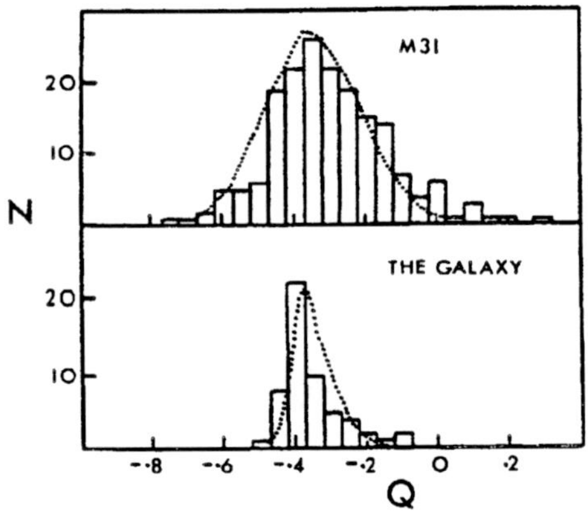

Figure 9.9 The reddening-free parameter Q for the M31 globulars shows more widely spread values than the local sample, probably because of larger observational errors (from Hanes, 1977).

The study of colors carried out by Elson and Walterbos (1988) includes a color-color diagram for a carefully selected sample of clusters, having what is considered the most reliable colors. The reddening-corrected version of this diagram is shown in Figure 9.10, in which the clusters are separated by age, the "true globulars" being those enclosed in the dotted box. Clusters bluer than $(B-V)_0 = 0.45$ are separated out as being too young to be classified as globulars.

Most of the above discussion has related to colors in the UBV system. Other color systems have also been used to study the M31 globulars, such as the Washington System (Harris and Canterna 1977), which was used to determine metal abundances (§ 9.4). Infrared photometry has been carried out by a number of astronomers. Kron and Guetter (1976) measured the near-infrared colors of seven M31 globulars, and Frogel *et al.* (1980*a*, *b*) determined $V-K$ colors for 40 globulars. In the latter study, it was determined that the $(V-K)_0$ colors are useful metallicity indicators (see § 9.4). The broadband infrared data were compared to predictions of spectral energy distributions based on stellar evolutionary models and it was found that the clusters' colors are consistent with a mass function of slope less than or equal to the Salpeter value.

Ultraviolet colors of 17 clusters, including 13 red globulars, were obtained with a rocket-borne telescope by Bohlin *et al.* (1988). They concluded that all but one of the globulars showed normal UV fluxes. The anomalous one, $B_o171 \equiv G222$, is a high-metallicity cluster with an ultraviolet excess.

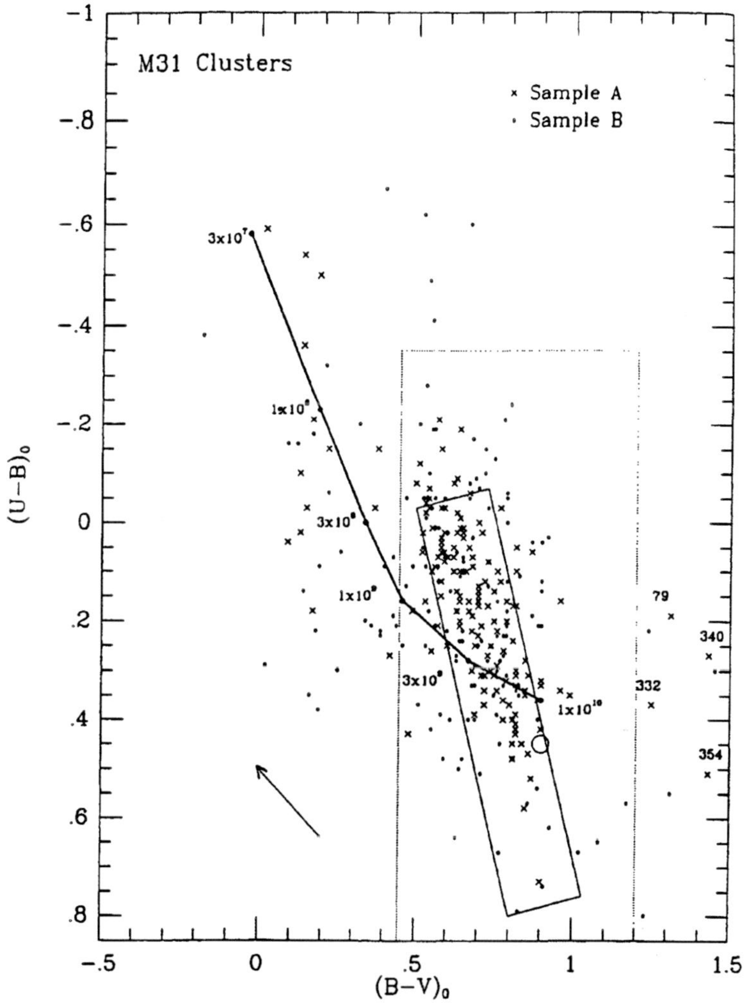

Figure 9.10 The reddening-corrected two-color diagram for two samples of M31 globulars, selected by Elson and Walterbos (1988). Only objects in the dotted box are considered to be true globulars. The solid box identifies the locus of Galactic globular clusters. The line represents a cluster evolutionary track.

Sitko (1984) extended the infrared coverage by measuring more M31 globulars, especially fainter and more distant ones. He corroborated many of Frogel *et al.*'s (1980*a*, *b*) findings, and found a strong relationship between the absolute magnitude in K and the infrared colors for the globulars in his sample (Figure 9.11). This relationship may be useful as a distance indicator, if it is found to be generally true for globulars.

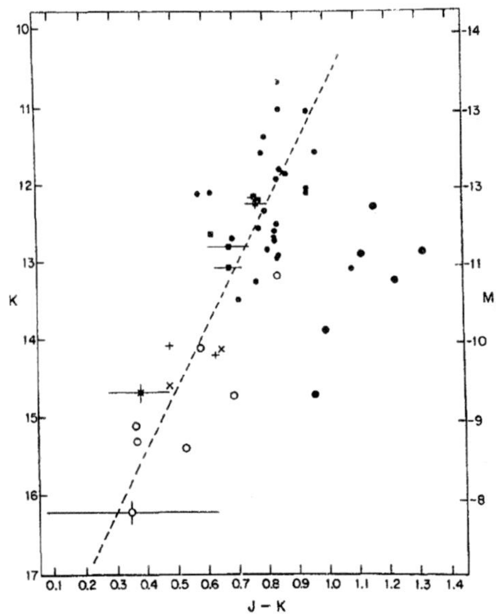

Figure 9.11 Infrared color-magnitude diagram for M31 globulars (from Sitko, 1984).

9.4 Spectroscopy and Abundances

The chemical abundances of the stars in M31's globular clusters are of interest, especially because of the way that they help us to learn about the early history of star formation at different places in the galaxy. Both color data and spectra have been used to determine the abundances, with various degrees of reliability. The first spectroscopic survey was that of van den Bergh (1969), who obtained image tube spectra of the brightest globulars with the Hale 200-inch telescope. Line strengths were estimated for the H and K lines, the G band, λ4226 of Ca II, and λ4325 of Fe I. The results were combined into a "line strength index", L, which was then used to study the abundances as a function of position and other parameters.

Figure 9.12 shows a comparison between the distribution of line strength indices in our Galaxy with that for M31. Values of L range from extremely weak-lined (L = 0) to strong-lined, roughly solar (L = 15). This diagram shows what is perhaps the most surprising result; the data indicate a very different distribution, with many more high metal content globulars being found in M31.

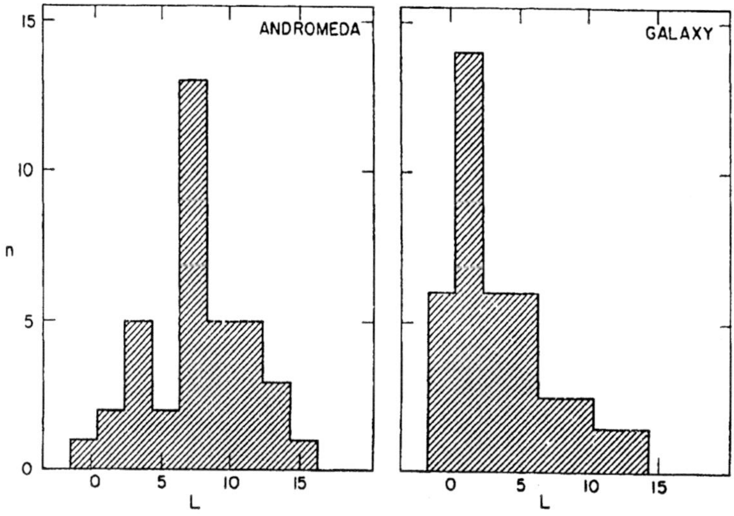

Figure 9.12 The differences in the distributions of line strength indices for M31 globulars and the Galaxy. Larger numbers represent stronger-lined spectra (from van den Bergh, 1969).

Another surprising result of van den Bergh's program was the apparent lack of any metallicity gradient with position. Instead of a trend of decreasing metallicity with radius, as found in our Galaxy, the M31 globulars seemed to show a similar and wide range of abundances for each radius. The reanalysis of these data by Hartwick and Sargent (1974), however, reopened the question, as they claimed to find a gradient very like that for our Galaxy.

Spinrad and Schweizer (1972) used scanning techniques to study the globulars' spectra and found results that confirmed van den Bergh's discovery. They extended the study by obtaining quantitative metal abundances from comparisons of the scans with synthetic spectra, concluding that a few of the very brightest clusters (specifically H 87 and B 282) are more metal-rich than the sun. They furthermore found that there is an ultraviolet excess in these clusters, indicating the presence of hot stars. Because the only significant numbers of hot stars in Galactic globulars are horizontal branch stars, which are blue only for the most metal-poor clusters, this result seemed to make very little sense. Considering the extremely luminous nature, and therefore the implied very large masses of these clusters, there may be a rather simple solution to this problem. We now know that many of the dwarf elliptical galaxies, some of which are less massive than these globular clusters, contain small amounts of young stars as recently formed contaminants. Perhaps M31's most massive globulars also have been able to retain enough gas to continue to form a small number of stars, so that a few recently formed stars exist to explain the blue and ultraviolet excess.

Harris and Canterna's (1977) conclusions, based on Washington photometry, were that two of the clusters that they observed, H 42 and B 282, are metal-rich, but possibly not more so than 47 Tuc (Figure 9.13). The other four are also metal-rich, relative to typical Galactic halo clusters, but not unlike many Galactic clusters (*e.g.*, NGC 6723 and ω Centauri).

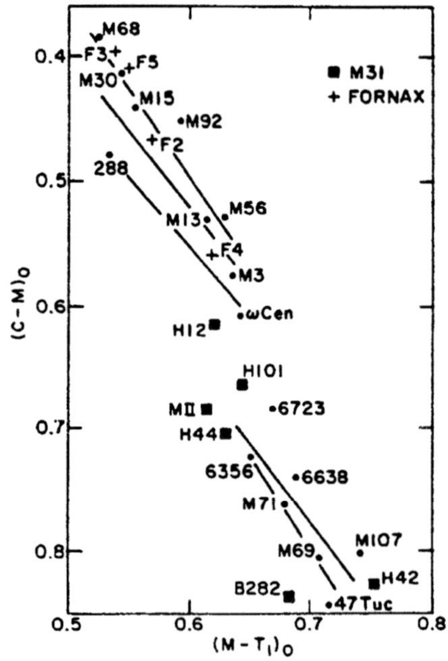

Figure 9.13 Washington system photometry of some M31 globulars, compared to Galactic and Fornax globulars (from Harris and Canterna, 1977).

Huchra *et al.* (1982) subsequently obtained digital spectra of 61 globulars in M31, using the MMT. They confirmed van den Bergh's conclusion that there is a wide range in metallicity for the globulars, and explored the kinematics (§ 9.5) and metallicity gradient. Rather than a correlation between distance and abundances, the data indicated that the range of metallicities (as well as the mean value) decreases outwards (Figure 9.14), with all very metal-rich clusters located at small radii.

More recently, Huchra *et al.* (1991) published metallicities for 150 globulars. From these more complete data, they found a considerable similarity between the M31 and the Milky Way globular cluster systems. A weak metallicity gradient as a function of radius was detected and a mean metallicity of [Fe/H] = −1.2 was deduced.

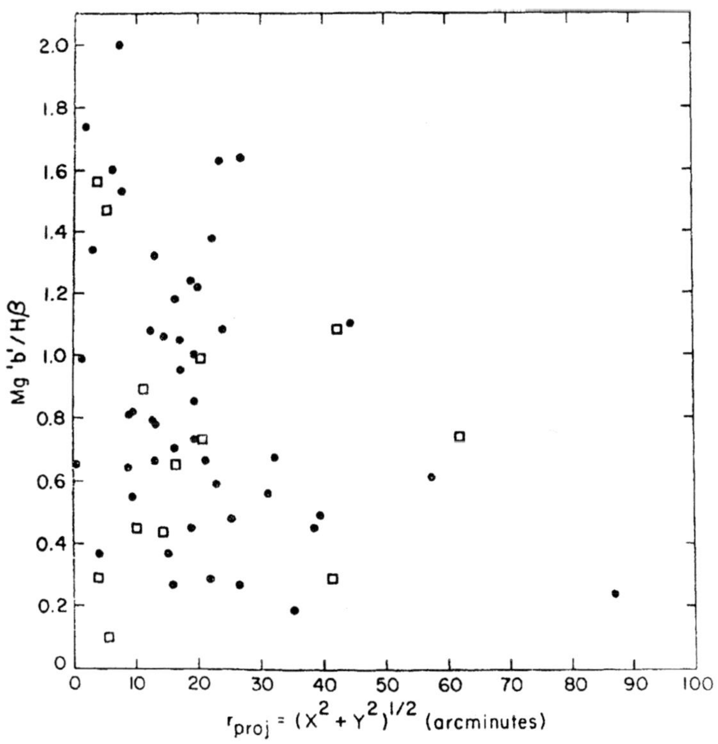

Figure 9.14 Line strengths as a function of distance from the center of M31, showing a trend for decreasing mean metallicity outwards (from Huchra et al. 1982).

Sharov and Lyutyi (1983) analyzed the photometric evidence and concluded, contrary to early work, that there is no significant difference between the distributions of reddening-free parameters (for example, Q) for the Galaxy and M31. They found evidence for a radial gradient in the metallicity and suggested that the gradient in [Fe/H] amounts to about -0.15 kpc^{-1} for distances from the center of less than 10 kpc, with a nearly uniform, low abundance exterior to that.

Tripicco (1989) examined high-resolution spectra of nine metal-rich globular clusters in M31. He found that certain differences could be discerned between his sample and similarly metal-rich globulars of the Galaxy. Specifically, the M31 clusters' light was the more dominated by dwarf stars (at λ4000), their CN band strengths were stronger, and any large contribution from blue horizontal branches was ruled out.

The strong CN band anomalies were confirmed in the infrared by spectroscopy in the $\lambda 1.5 - 1.8\mu$m and $\lambda 2.0 - 2.4\mu$m region by Davidge (1990). He found CN band strengths in four M31 clusters to be as strong as those in the anomalous Galactic globular cluster NGC 6304.

In a remarkable achievement for ground-based astronomy, Davidge *et al.* (1990) detected an absorption-line gradient in the optical spectrum of the highly-luminous globular cluster, G78 (=V42). Most metal lines in the spectrum weaken with distance from the cluster center. Possibly there is a systematic abundance gradient or possibly there is a population of younger stars near the center of the cluster.

An equally impressive accomplishment was the derivation of CMDs of globular clusters by Heasley *et al.* (1988) and by Christian and Heasley (1991). For G219 (=Mayall IV) and G1, they found that the giant branch morphology of the globulars could be ascertained, using V and I colors measured at Mauna Kea with the CFHT. A metallicity for G1 was derived, with [Fe/H] = -0.7. for G219, the determined that the giant branch indicates a metal-poor cluster, like M15.

9.5 Distribution and Kinematics

The spatial distribution of M31's globular clusters can be derived from the various compilations. For example, de Vaucouleurs and Buta (1978) used the Sargent *et al.* (1977) catalogue, correcting the positions to face-on (Figure 9.15). The result shows a strong central concentration, with a thinly populated outer halo, both nearly circularly symmetric. Some evidence of the major dust lanes shows up. They tested the fit of the radial distribution to a $r^{1/4}$ law, such as describes the galaxy's spheroidal component and concluded that the clusters are spatially distributed like the spheroidal component stars, with an identical effective radius. The inner clusters did not fit the $r^{1/4}$ law, however, a fact ascribed by the authors to interference by the presence of the bulge and disk.

Crampton *et al.* (1985) used their somewhat larger and somewhat different list of clusters to do a similar analysis. They concluded, in agreement with the previous study, that the distribution fits a $r^{1/4}$ law very well except in the innermost regions. They derive a fit to the data of the form

$$\log \sigma = 2.59 - 1.87\ D^{1/4}$$

where σ is the projected surface density in number of clusters per kpc^2 and D is the projected distance in kpc. Figure 9.16 compares the distribution with that for Galactic globulars; no statistically significant difference is evident.

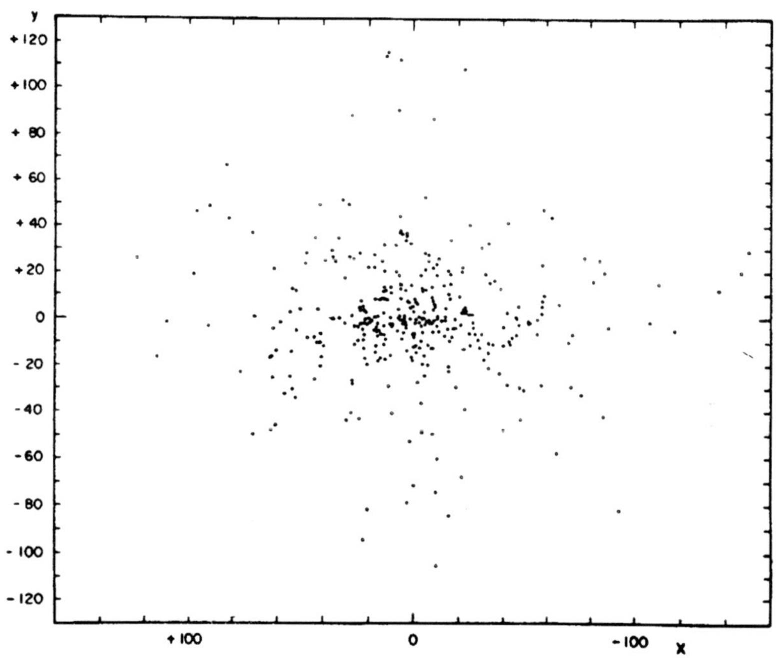

Figure 9.15 The spatial distribution of M31 globulars, plotted in the plane of the disk (from de Vaucouleurs and Buta, 1978).

The first look at the kinematics and dynamics of the M31 globular cluster system was carried out by van den Bergh (1969). He concluded that the velocity dispersion in the central regions is about 110 km/sec and that the clusters are primarily in a non-rotating halo. His sample of clusters did not strongly support the idea that the metal-rich clusters are dynamically distinct from the halo.

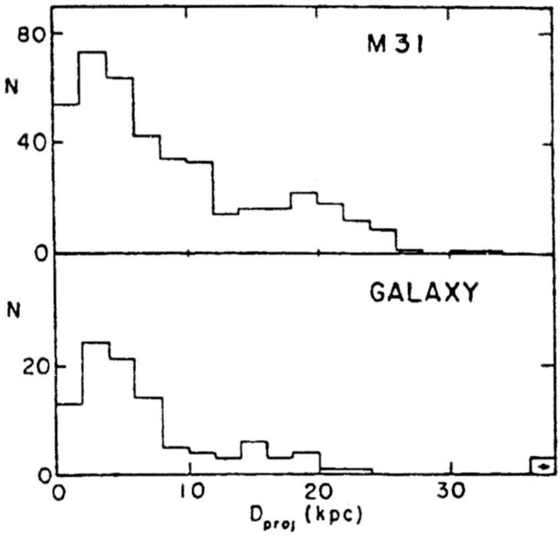

Figure 9.16 A comparison of the deprojected galactocentric distance distributions for M31 and Galactic globulars, showing a similarity (from Crampton et al. 1985).

Huchra *et al.* (1982), on the other hand, detected a significant rotational component, with a velocity amplitude of 160 ± 40 km/sec. When this is allowed for, a calculation of the central velocity dispersion gives 130 km/sec, which agrees with earlier data, including that used by van den Bergh to measure the mass of the core of M31. Elson and Walterbos (1988) characterize the situation by the following model: the most metal-rich clusters lie in a flattened, rotating disk, the intermediate metallicity clusters are more spherically arranged, while the metal-poor clusters are in a large, spherical halo with no (or slow) rotation. The flattening of the rotating component is similar to that of the stellar spheroid, as is the effective radius. The halo clusters, on the other hand, belong to a much more distended system.

9.6 Structure of the Clusters

Considering their large distance and small size, it might seem surprising that there has been a certain amount of information learned about the structure of individual globular clusters in M31. For example, a study of the structure of the cluster of M II, the brightest M31 globular, was published by Pritchet and van den Bergh (1984), who used a CCD on the CFHT. They found that the cluster is highly flattened for a globular cluster, with $\epsilon = 0.22$, even larger than for ω Centauri. The profile of the cluster (after correction for seeing and for interference by two adjacent bright stars) is well fit by King (1962) empirical models, with the best fit giving a core radius of approximately 0.5 pc.

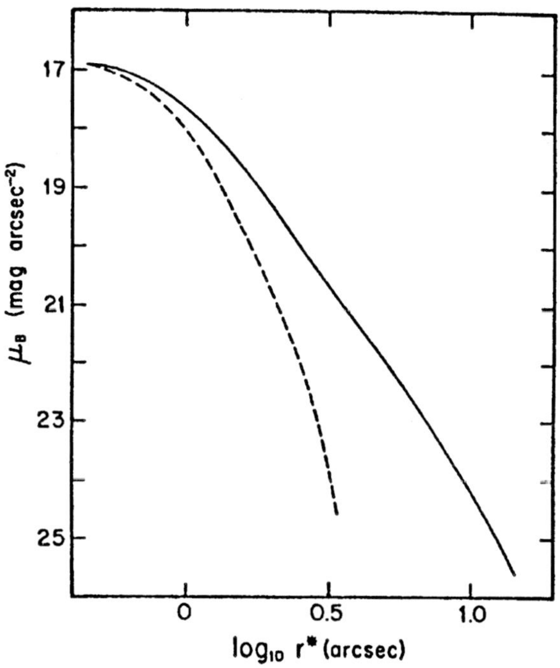

Figure 9.17 The structure of the M31 globular Mayall II, compared to a stellar profile from the same CCD image (from Pritchett and van den Bergh, 1984).

Spasova and Staneva (1984) measured the structure of five other globulars, measured on plates taken with the 2-meter telescope of the

Rozhen Observatory. They found ellipticities ranging from 0.05 to 0.16, spanning the range found for globulars in our Galaxy.

The automatic derivation of core radii that emerges from the method used by Battistini *et al.* (1982) to identify globular clusters is also a source of structural information. Crampton *et al.* (1985) also measured core radii for a large number of clusters, which agreed fairly well with the Battistini results, in spite of the obvious difficulties in measuring these quantities. They found core radii that ranged from near 0 to about 6 pc, a range very similar to that found for the globulars of our Galaxy (Peterson and King 1975).

9.7 Summary

M31 has approximately 300 globular clusters, similar in luminosity to our Galaxy's. They show a weak correlation in chemical abundance with position and exhibit a wider range in metallicity than our globulars. They form a rotating disk of metal-rich clusters, outside of which is an extensive, slowly-rotating spherical halo of globulars with generally more metal-poor abundances.

CHAPTER 10

OB ASSOCIATIONS

The study of M31's OB associations started with Hubble's (1929) work on NGC 206 (which he called a star cluster), as discussed in Chapter 2. More complete examination of this type of star group didn't occur until about 30 years later. Recently, with the advent of sensitive digital detectors, optical studies of the OB associations are rapidly progressing.

Before discussing them further, however, it is necessary to straighten out the relevant terminology. Similar objects in other galaxies, because of their various distances and scales, have been designated by different terms, without it always being clear what is meant. For example, NGC 206, which Hubble called a "star cluster", has also been called an "OB association", a "star complex" and a "star cloud". Efremov (1978, 1979) has suggested that these terms be carefully defined according to the linear dimensions and morphologies of the objects. In this chapter, I will use Efremov's terminology, as modified in Hodge and Lee (1988) and given in Table 10.1. A further discussion of the definition problem, especially as it relates to M31, is given in Hodge (1986).

10.1 Identifications and Catalogues

The first complete survey of OB associations in M31 was carried out by van den Bergh (1964a), who used plates taken in several colors with the Tautenburg Schmidt telescope. He blinked pairs of plates in different colors, ranging from ultraviolet to the near infrared and based his selection on the clumping of very blue stars. Using the red and infrared plates, he also searched for highly-reddened associations and associations of red supergiants, but found none.

Table 10.1
Terminology for Associations and Clusters

Term	Characteristics
Star clouds	Large ($\sim$ 1000 pc) assemblages of young stars
OB associations	Groups of young, massive stars resembling in size (100 pc) and stellar makeup the OB associations of our Galaxy
Open clusters	Low mass ($M < 10^4$ solar masses), compact groups of stars

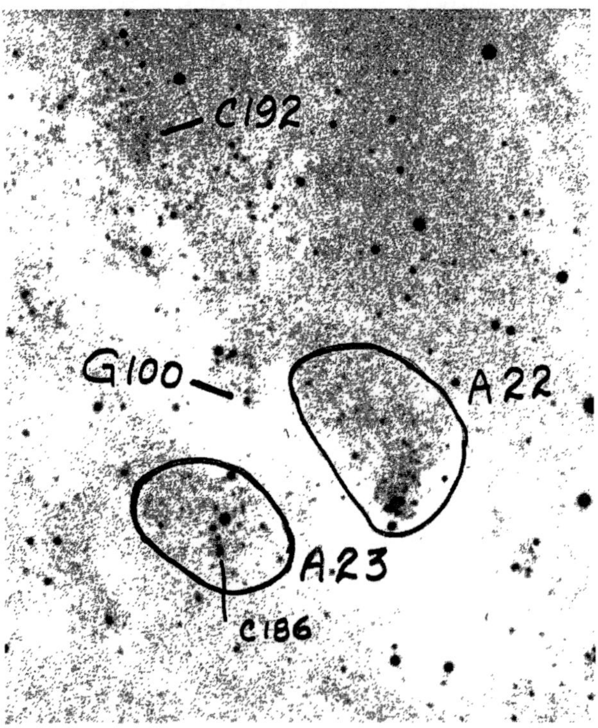

Figure 10.1 Examples of typical van den Bergh OB associations, A22 and A23. Also shown are a globular cluster (G100) and an open cluster (C192) (KPNO 4-meter photograph).

The result of van den Bergh's search was a catalog of 188 stellar associations, examples of which are shown in Figure 10.1. His list and its numbering system have been the basic reference used by most researchers in succeeding years. In this chapter, as adopted in the Atlas (Hodge 1981), I will use the prefix A to designate associations in van den Bergh's (1964a) numbering system. Some other authors (*e.g.*, van den Bergh himself and Efremov *et al.* 1987) have used the prefix "OB".

In his discussion of the identification criteria he used, van den Bergh pointed out that the boundaries drawn were somewhat arbitrary, and that others might delineate the structures differently. Such was the case when Efremov *et al.* (1987) re-examined the star complexes and associations, using plates taken with the Rozhen Bulgarian 2-meter telescope in U and B. They used the criterion that groupings with a density of more than 5 stars per square arcmin on the U plates (which reached to a magnitude limit of U = 20) are probably physically associated. The result was a list of 210 OB association candidates. Most of them are *inside* the van den Bergh associations, and Efremov *et al.* (1987) assert that their objects are the real counterpart of the traditional OB associations in our Galaxy. The van den Bergh objects, on the other hand, are considered to be too large to be so identified, and are designated as "star complexes". This question was addressed again by Massey *et al.* (1986), who looked at some van den Bergh associations in great detail, measuring color-magnitude diagrams and finding WR stars. They decided that some of the "associations" could be considered loose groupings of smaller ($\sim$ 100 pc) associations, based on measured ages. Hodge (1986) also re-examined the question and showed that identifications of associations in galaxies is very much a function of scale and magnitude limit; using plates of the LMC with the same linear scale and absolute magnitude limit as was used for M31, he found that large "star complexes" were chosen, rather than the smaller OB associations identified on larger-scale and deeper plates by Hodge and Lucke (1970).

An illustration of the relation between the Efremov *et al.* objects and the van den Bergh objects is shown in Figure 10.2, which shows the northeast region of the disk with the associations identified. It can be mentioned that Efremov *et al.* (1987) also identified several of the "open clusters" of the Hodge (1979) catalogue (Chapter 11) as being associations, as they are considered too large to qualify as normal star clusters.

Efremov *et al.* (1987) also searched their plates for what they call "star complexes" and found that most of van den Bergh's objects coincided with these larger objects. When both kinds of objects are counted up, the tally is as follows: there are 188 van den Bergh objects, plus 7 more found by Richter (1971), and 124 associations listed by Efremov *et al.*

(1987). These lists are probably relatively complete; there are probably only a few more objects in M31 that are OB associations *according to the criteria adopted by the authors.* Even in the areas along the minor axis that are marked by thick dust lanes, it is felt that a reasonably complete census was possible and that the existing catalogues include at least 90% of the OB associations that exist in the galaxy. The figure of 210 OB associations given by Efremov *et al.* (1987) is adopted here.

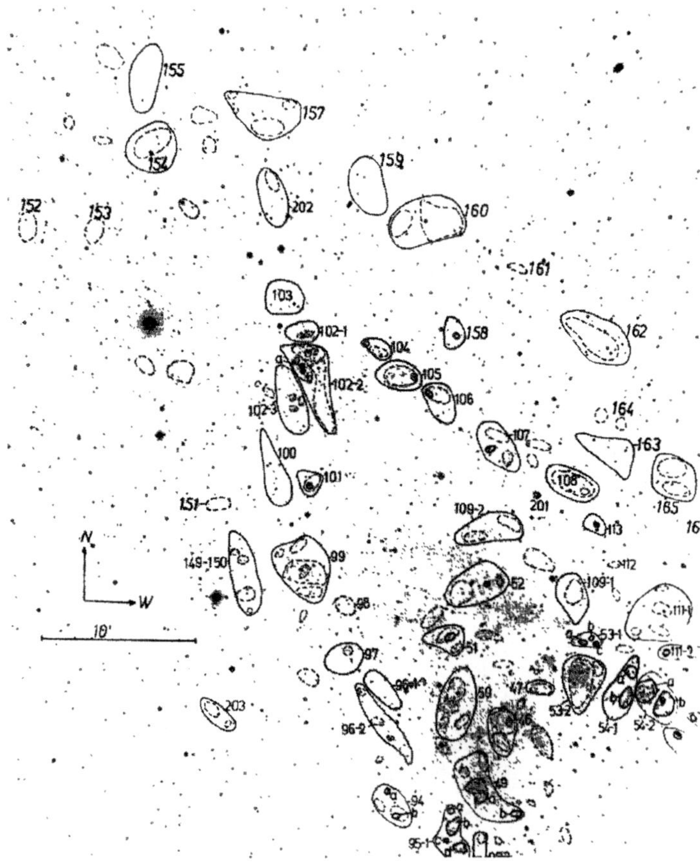

Figure 10.2 The northeast part of M31, showing van den Bergh's (1964a) associations (solid lines) and the smaller associations identified by Efremov et al. (1987).

In a comparative study of stellar associations in different galaxies, Hodge (1986) calculated that the normalized number in M31 is 0.5 associations per 10^8 galaxy solar luminosities, compared to values of 4 and 8 for the LMC and SMC, respectively. The only other Sb galaxy tested, NGC 7331, also showed an anomalously small normalized density, and it was suggested that the number of associations and, probably, the star formation rate, is smaller for these earlier Hubble type galaxies.

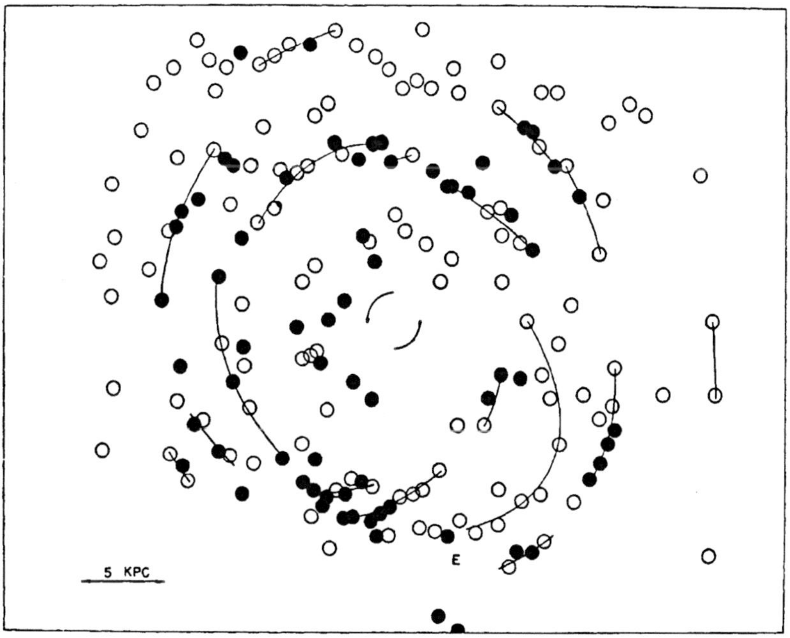

Figure 10.3 The face-on distribution of OB associations. Lines are probable arm segments. Filled circles are the younger associations (from van den Bergh, 1964a).

10.2 The Spatial Distribution

Figure 10.3 is van den Bergh's (1964a) diagram of the spatial distribution of his associations, rectified on the basis of an assumed inclination angle of 12.3°. The conspicuous features of this diagram include the dearth of associations in the central area and the apparent (rough) alignment of

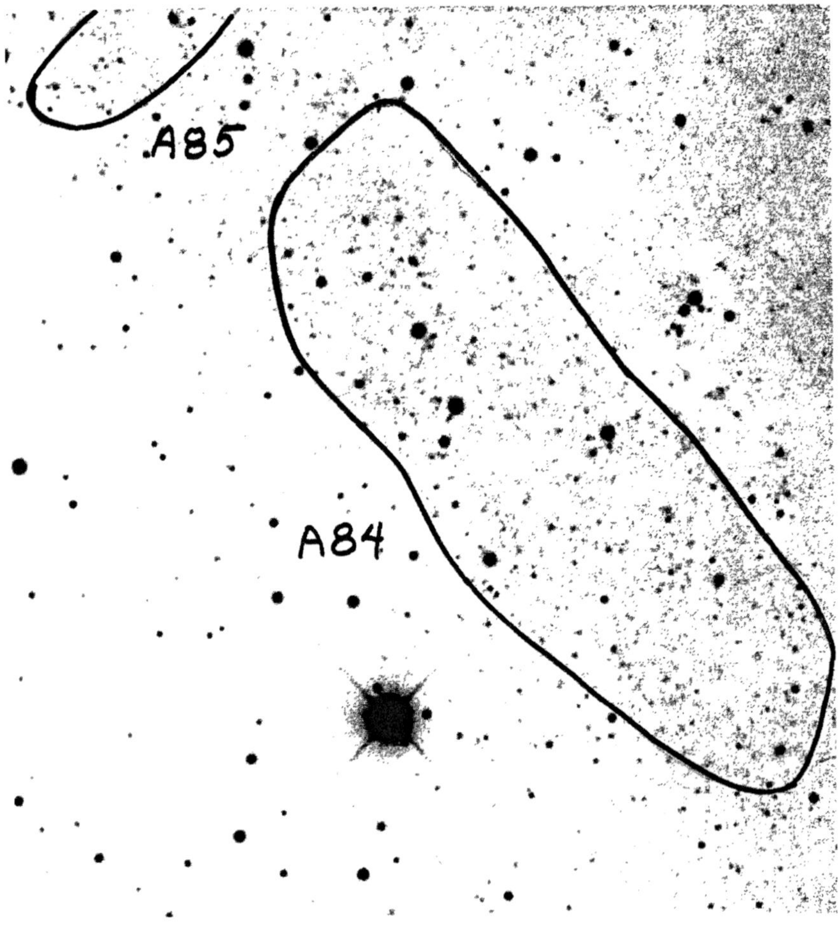

Figure 10.4 A84, a large elongated star cloud, part of the "chain" of associations running from A78 to A86 (KPNO 4-meter photograph).

some of the associations along spiral arms, indicated in the figure with lines. Van den Bergh noted an interesting arrangement of associations that extends from A78 to A86 (partially shown in Figure 10.4), crossing the major axis in the southwest quadrant. There is a tendency for this line of associations to be bordered on the inner edge by conspicuous dust lanes (D23, D43, D39, D95, D192, and D288). Subclusterings are most evident in the associations near to the dust lanes, and between A84 and A86 the inner edge includes a string of H II regions (Chapter 14). It is suggested by van den Bergh that this arrangement may indicate that star formation has most recently occurred in the inner parts of the arm, and that the outer part is relatively old.

The radial distribution of van den Bergh's associations is illustrated in Figure 10.5, which plots the number of associations as a function of true distance from the nucleus. There is, as expected, a deep minimum in the center, with no objects nearer the center than 3 kpc, and a maximum at a distance of about 12 kpc, corresponding to the maxima found in H I and other indicators of Population I. Beyond 15 kpc the density decreases rapidly, with none found in his survey beyond a distance of 25 kpc. Richter (1971) has, however, found a small number of more distant, faint associations.

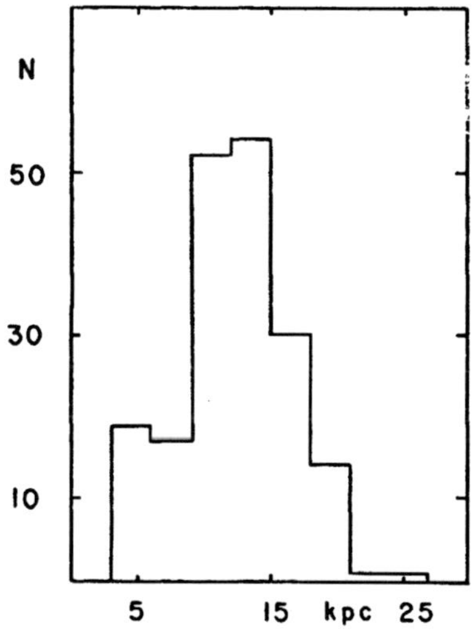

Figure 10.5 The radial distribution of OB associations (from van den Bergh, 1964a).

The surface density of associations in the 10 kpc ring is 0.2 kpc^2, which is more than a factor 10 smaller than the density in the plane of our Galaxy near the sun. No doubt much of this discrepancy is the result of the criteria used by van den Bergh; 1000 kpc star clouds in our Galaxy probably have a much smaller surface density than the OB associations. Nevertheless, there may still be a difference, as Efremov *et al.* (1987) find that most van den Bergh objects break up into two or three normal-sized OB associations. The difference may be an indication that the star formation rate in M31 is lower than in our Galaxy near the sun, a suggestion that van den Bergh (1964a) supports by arguments from star counts.

10.4 Dimensions

The sizes of the stellar associations are readily measured once their boundaries are determined. Because of his criteria, van den Bergh (1964) found very large sizes, with a mean value of 480 pc for the diameters (most published data refer to dimensions measured parallel to the major axis, to allow for projection effects). His size distribution is compared with that for the Efremov *et al.* (1987) objects in Figure 10.6, where it is clear that objects of very different scale are being sampled. On the other hand, the size distribution for the Efremov associations agrees very well with that for the LMC associations catalogued by Hodge and Lucke (1970), as shown in Figure 10.7. The mean dimension for the LMC associations is 78 pc, while that for M31 is 80 pc.

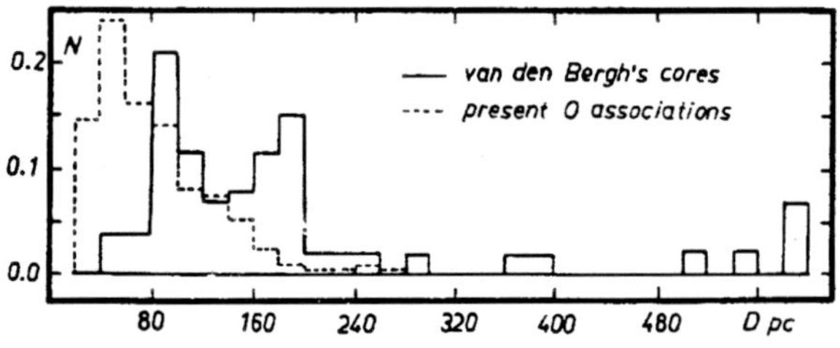

Figure 10.6 A comparison of the size distributions of the van den Bergh (1964a) sample and the Efremov et al. objects (from Efremov et al. 1987).

Very few of the associations are not distended in the direction of the major axis position angle. Much of this effect is the result of the flatness of the associations, but some may also be the result of differential rotation. The mean axial ratio for van den Bergh's associations (or star complexes) is $b/a = 0.56$. If this value is accepted at face value and is ascribed entirely to inclination effects, and if it is assumed that the associations are, on the average, circular, then their true thickness can be calculated. The result is that they are ~40 pc thick, though this result is extremely tentative because of the number of assumptions required.

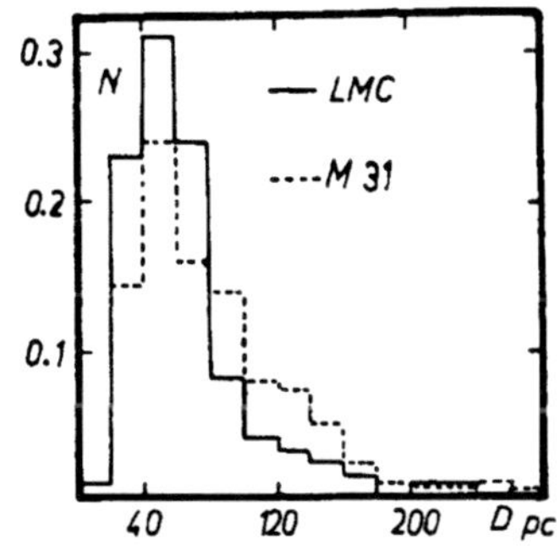

Figure 10.7 A comparison of the sizes of the Efremov et al. associations and those in the LMC (Lucke and Hodge, 1970) (from Efremov et al. 1987).

10.5 Stellar Content

Color-Magnitude Diagrams

Chapter 2 has already discussed the first color-magnitude diagram (CMD) to be obtained for one of M31's associations. This same object, NGC 206, has been the topic of more recent studies (van den Bergh 1966, Odewahn 1987). As the richest and most conspicuous star cloud in M31, it is a remarkable object (Figure 10.8). Its total integrated magnitude is $M_B = -12.8$ (Hodge 1986) and its dimensions are about 400×800 pc. It contains more than 300 stars brighter than $M_B = -3.6$ (van den Bergh 1966). The CMD that he obtained, using the Kitt Peak 84-inch telescope, is shown in Figure 10.9, where it is seen that the brightest main sequence stars have V = 17.4, corresponding to $M_V = -7.9$. There are also foreground red stars in the CMD, but the red supergiants can be distinguished, and van den Bergh estimated that the brightest has $M_V = -6.4$.

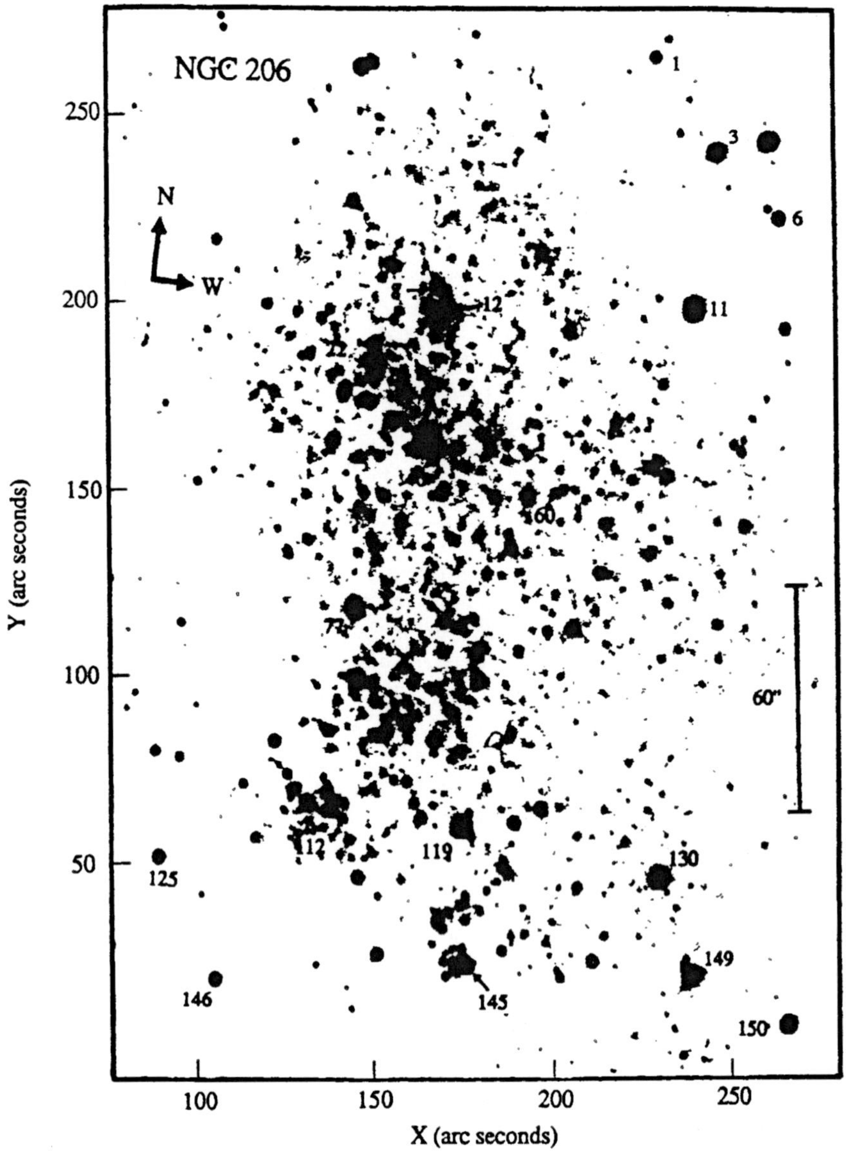

Figure 10.8 An image from the CCD photometry of NGC 206 (A78) (from Odewahn, 1987).

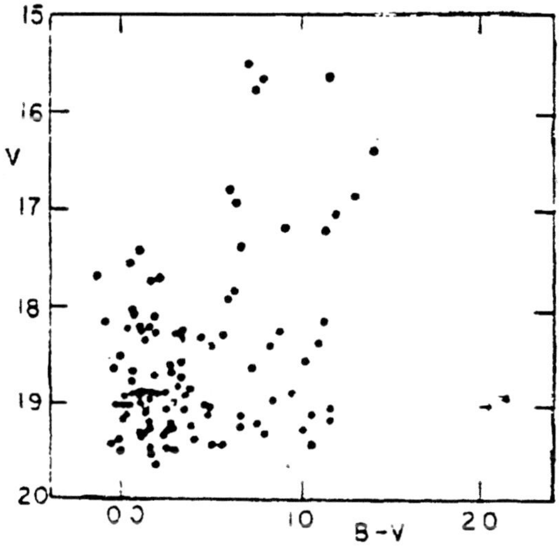

Figure 10.9 A color-magnitude diagram for NGC 206 (from van den Bergh, 1966).

Using electronic techniques, Odewahn (1987) obtained a new CMD for NGC 206, which is shown in Figure 10.10. The conclusions that can be derived are similar to those of van den Bergh. The diagram reaches to about the same limit as the previous study, but was done with a smaller telescope (the McDonald Observatory 0.9-meter). Odewahn's paper includes an extensive discussion of the star formation rate and initial mass function for the NGC 206 area.

Massey *et al.* (1986) and Hodge *et al.* (1987, 1988) have used CCD detectors and the Kitt Peak 4-meter telescope to obtain CMDs for a total of nine different fields in the disk of M31, including several of van den Bergh's and Efremov's stellar associations. Figure 10.11 gives results for van den Bergh's A99, which includes several condensations marked on Efremov *et al.*'s maps. The CMD shows a broad main sequence that extends from the limit of V = 22.5 to V = 18.5. This turn-off corresponds to an absolute magnitude of $M_V = -6$, indicating that the age of A99 must be on the order of a few times 10^6 years. Most of the redder stars in Figure 10.11 are probably foreground stars, according to the analysis of the expected foreground population. However, in their R, V−R CMD (Figure 10.12), giants belonging to A99 show up as an accumulation of stars at V − R = 0.75, R = 20.2, corresponding to $M_V = -4.0$.

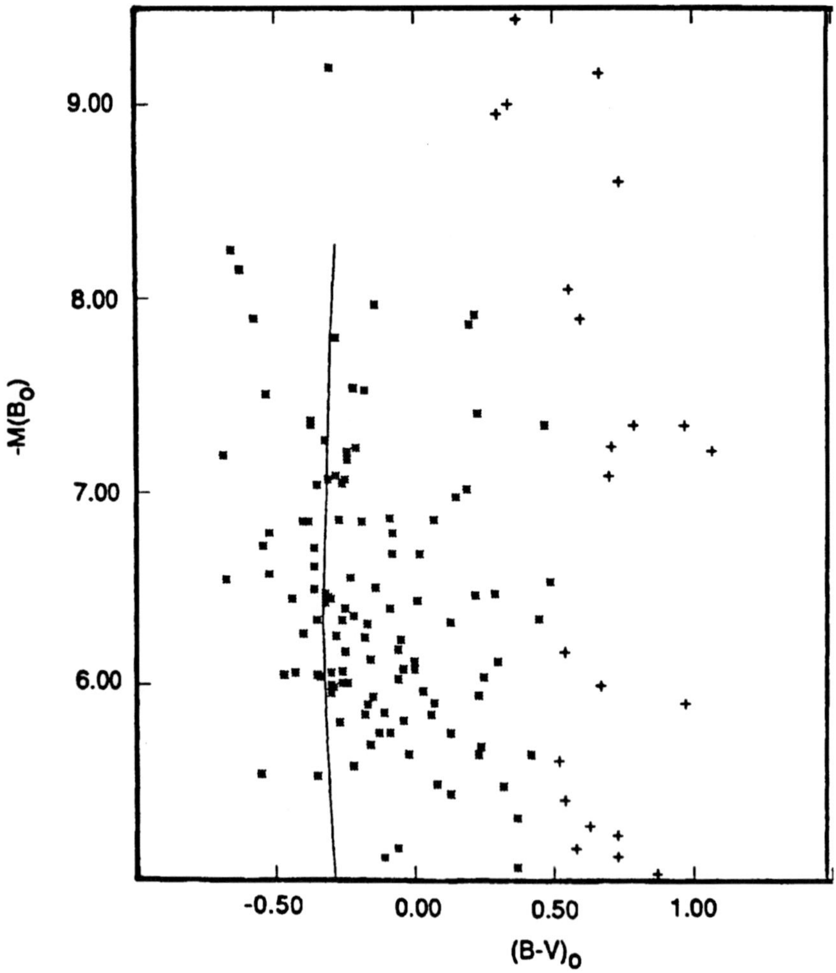

Figure 10.10 An absolute magnitude-color diagram for NGC 206 (from Odewahn, 1987). The line indicates the main sequence for a 4 million year old cluster.

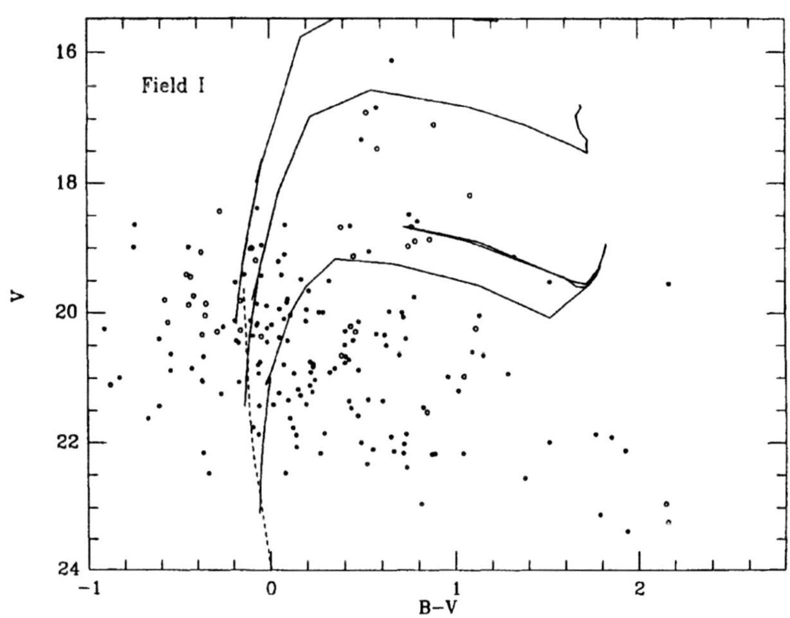

Figure 10.11 A color-magnitude diagram for an M31 field that is centered on A99. Lines are for evolutionary tracks for massive stars (from Hodge and Lee, 1988).

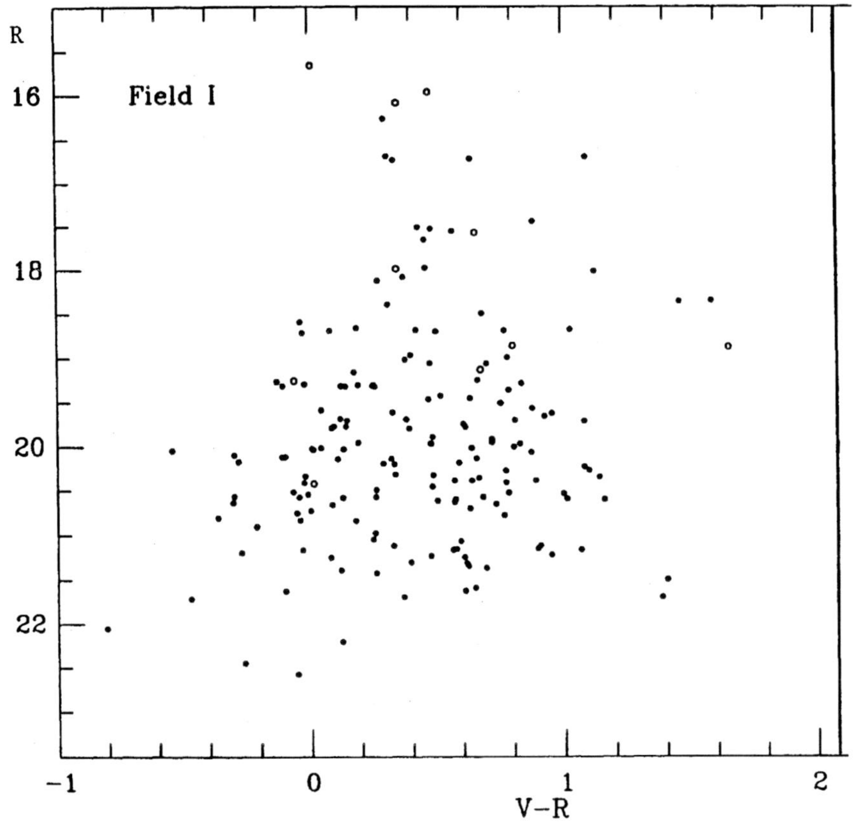

Figure 10.12 An R, V–R diagram for A99, showing red giants at R = 20, V–R = 0.7 (from Hodge and Lee, 1988).

Table 10.2 lists the 13 different van den Bergh star clouds and associations that have CMDs in the literature so far. The brightest stars on (or above) the main sequence show a large range, from V = 19.1 to V = 16.8, corresponding to $M_V = -5.6$ to -7.9 (approximately, depending on the reddening used). This is largely an age effect, though it is difficult to be certain about how much of the difference is also caused by contamination by different numbers of evolved blue supergiants and by small number statistics. Most of the red stars in all of these CMDs are probably foreground and background (M31 field) stars, with only a few likely to be member supergiants.

Table 10.2
OB Associations in M31 with CMDs

van den Bergh (1964) Number	Brightest MS Star (V)	Reference
A8	18.1	Massey *et al.* 1986
A9	19.1	Massey *et al.* 1986
A10	18.2	Massey *et al.* 1986
A24	18.4	Hodge and Lee 1988
A48	18.4	Massey *et al.* 1986
A76	19.0	Hodge and Lee 1988
A78 (NGC 206)	17.4	van den Bergh 1966
A78 (NGC 206)	17.5	Massey *et al.* 1986
A78 (NGC 206)	16.8	Odewahn 1987
A99	18.5	Hodge and Lee 1988
A102	18.6	Massey *et al.* 1986
A102	18.0	Hodge and Lee 1988
A138	18.5	Hodge and Lee 1988
A139	18.4	Hodge and Lee 1988
A185	18.5	Baade and Swope 1963
A185	18.5	Hodge *et al.* 1987

Luminosity Functions

Most of the photometric studies of associations have included data on their luminosity functions. For example, the observed luminosity functions for the Massey *et al.* (1986) fields are shown in Figure 10.13, where it is seen that the luminosity functions are fairly steep, with a flattening beyond V = 21 that is due to incompleteness. Most published data on this subject are more or less severely affected by completeness problems, even at fairly bright magnitudes, and this probably explains why some have relatively flat slopes. The only quantitative completeness correction, using the analysis of large numbers of artificial stars introduced into the digital images and recovered by the program DAOPHOT, that has been made so far is for A185 (Hodge *et al.* 1988), the most remote and thus the least affected by this problem. Even here the luminosity function slope is significantly changed when completeness corrections are made. The corrected slope is 0.53 ± 0.06 (in units expressed as log[stars/0.5 mag/arcmin^2] for main sequence stars, in excellent agreement with the global luminosity function slope found for galaxies in general by Freedman (1986).

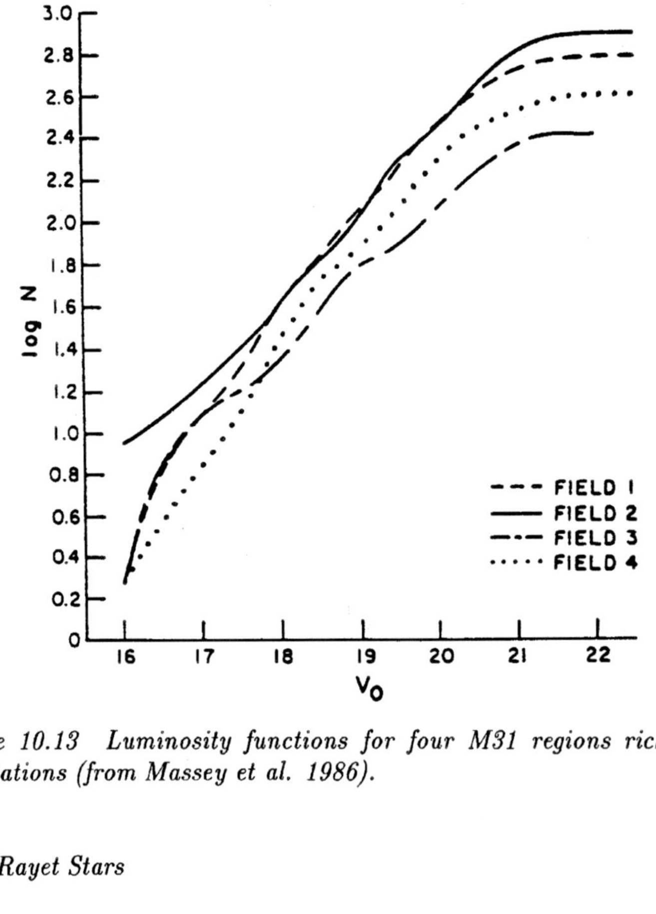

Figure 10.13 Luminosity functions for four M31 regions rich in OB associations (from Massey et al. 1986).

Wolf-Rayet Stars

Massey *et al.* (1986) greatly enhanced their study of the young stars in M31's OB associations by carrying out an interference filter search for Wolf-Rayet (WR) stars in 11 van den Bergh star clouds. They found a total of 34 WR stars and determined that these objects, which are among the most massive and thus the youngest of stars, exist preferentially in the parts of the star clouds that otherwise show evidence of the most recent star formation activity. About half of the WR stars are of the WN type, indicating that the ratio of WN to WC types is probably similar to that in the solar neighborhood. The density of WR stars in some areas is as large as it is in active regions of our Galaxy, but the average is considerably smaller. Thus, although there are some places in M31 that are forming stars at the rate found in the solar neighborhood, in general star formation is going on at a significantly reduced level there.

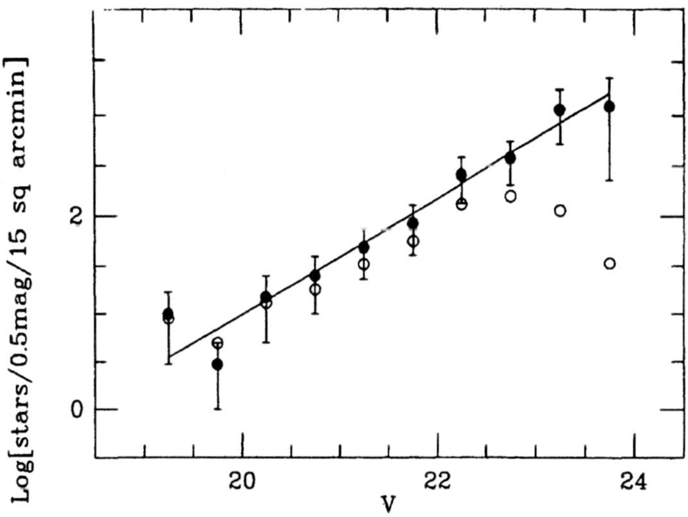

Figure 10.14 Luminosity function for an area that includes association A185. Open circles are from observed counts and filled circles are corrected for completeness (from Hodge et al. 1987).

10.6 Summary

M31 contains about 200 OB associations and star clouds, representing areas of very recent massive star formation. They are concentrated in the 10 kpc "ring", but even in this region their frequency is low compared to our Galaxy's, near the sun.

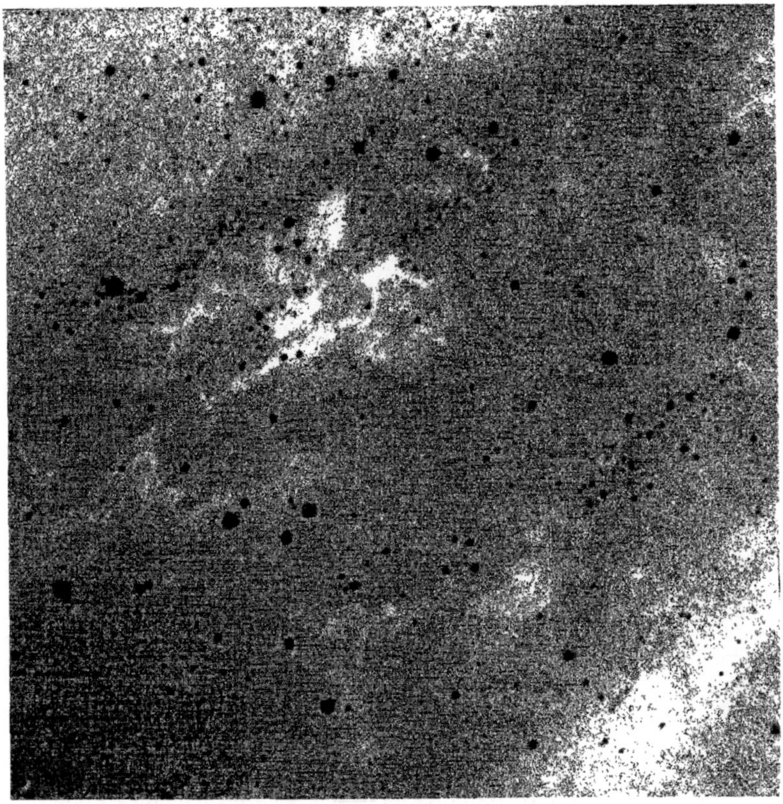

Figure 10.15 A2 is the large linear complex of bright stars at the left of this photograph and the smaller object at the upper center is A12.

CHAPTER 11

OPEN CLUSTERS

It has been realized for many years that M31 contains a number of open clusters roughly similar to those in our Galaxy. Hubble (1929), for instance, in discussing M31's stellar content, mentioned the presence of a small open cluster near his variable No. 40. (This cluster is No. 127 in Hodge's catalogue, 1979). Furthermore, he studied what he called a star cluster, NGC 206, in some detail, even publishing a color-magnitude diagram for it (see Chapter 2).

Until recently, however, a great deal more attention has been paid to the globular clusters (Chapter 9). Not only are the globular clusters generally brighter and less restricted to the complex stellar disk, but they are also more readily usable to answer certain questions about M31, such as its distance, mass, chemical abundances, and chemical history. About the only observations made recently of the open clusters were obtained in the hope or thought that the clusters in question might be globular. For example, van den Bergh (1967) mentioned that the colors of eight supposed globulars are blue, and he suggested that they are "blue globulars" of the sort found in the Magellanic Clouds. Similarly, other recent searches have found such clusters. For example, Crampton *et al.* (1985) found approximately 100 of their cluster candidates to be bluer than normal globulars and many are probably examples of populous young clusters.

11.1 Searches

The relative neglect of the open clusters is not fully justified. Although very difficult to observe, they represent an interesting component of M31 and their study potentially can shed light on questions about the recent history of star formation in M31.

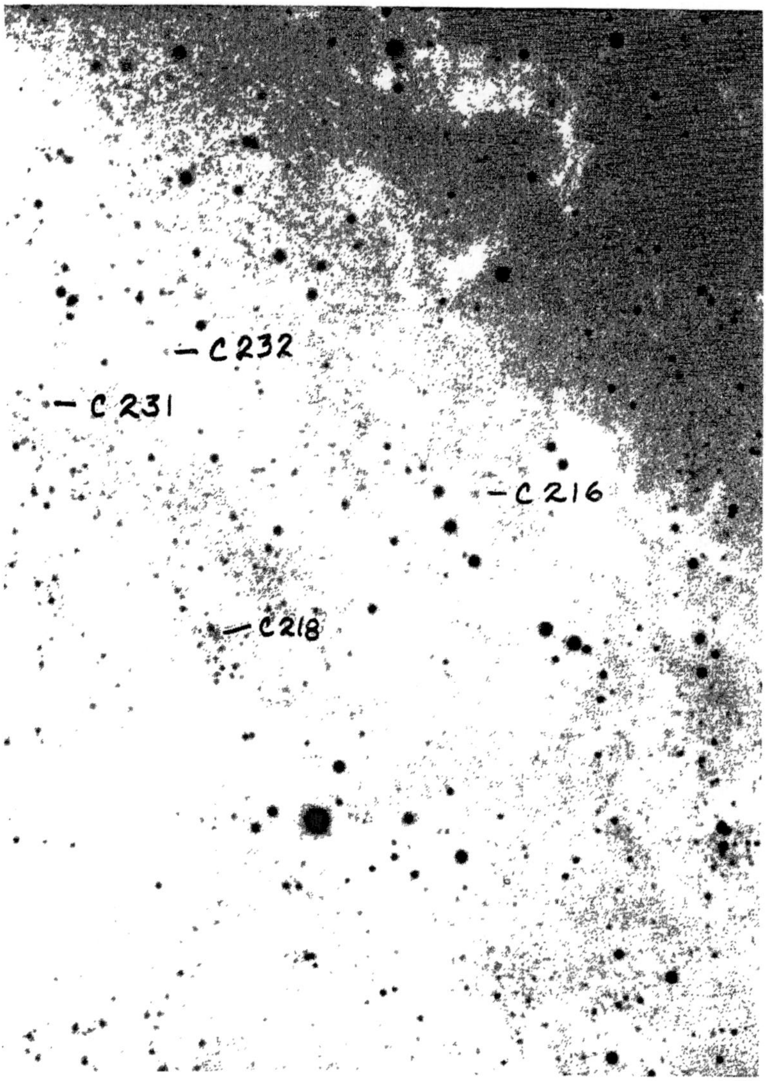

Figure 11.1 Examples of open clusters in M31 (KPNO 4-meter photograph).

A search for young open clusters, especially those with stars bright enough to be resolved, was made with the Kitt Peak 4-meter telescope (Hodge 1979). The area searched for open clusters totaled 3.4 square degrees. Judging from the distribution of discovered clusters, it is unlikely that a significant number of bright open clusters lies outside the search area. The plates were searched for clusters using a binocular microscope. Because of generous overlap in the plate series, all areas were examined on at least two plates. All central clusters were further examined on 4-meter Cassegrain plates, and several questionable clusters were checked on Palomar Schmidt plates, which were available in three colors and thus provided some clues about the integrated colors of the cluster candidates.

The greatest problem in the observational material had to do with the scale of the plates (18.6 arcsec/mm), which allows efficiency in the search but which makes identifications of very small clusters difficult. The result is that several cluster candidates are possibly spurious, either being chance groupings of unrelated stars, stars embedded in nebulosity, or stars superimposed on a distant galaxy's image.

Figure 11.1 shows a portion of a Kitt Peak 4-meter Cassegrain plate with some open clusters identified.

11.2 Dimensions

Figure 11.2 shows the size distribution for the clusters identified in the 1979 survey. Measured diameters range from $\sim$ 5 to $\sim$ 70 pc, with a mean value of 20 pc. The decrease in frequency below 15 pc is no doubt influenced by resolution effects, as this diameter corresponds to only 5 arcsec.

Also shown in Figure 11.2 are curves giving the size distribution for open clusters in our Galaxy (Allen 1974) and in the LMC (Shapley 1932). The Shapley data were used because his survey, based on Bruce 24-inch camera plates, is more likely to be similar to the M31 data in its resolution and absolute magnitude limits than more modern studies. The curves for M31 and the LMC are in fact very similar; if allowance were made for probable loss of small clusters in the M31 data, the curves would match even more perfectly, with the only conspicuous difference being the excess of very large clusters in the LMC, which form an extended tail. These are the "blue globulars," which were missed in the 1979 survey. The conspicuous difference between the above curves and the curve plotted for the Galaxy is probably due to resolution limits; the vast majority of open clusters in the Milky Way are only $\sim$ 2 pc in diameter and beyond resolution limits in M31.

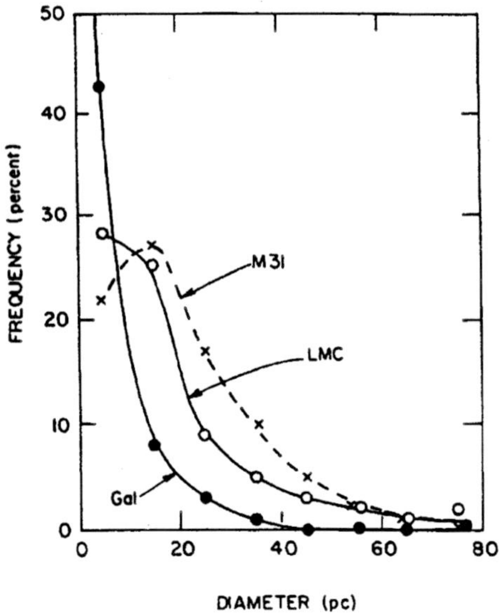

Figure 11.2 Size distribution of open cluster candidates in M31, compared to those for two other galaxies (from Hodge, 1979).

11.3 Magnitudes and Colors

Colors for young clusters that were found in the searches for globulars (*e.g.*, by van den Bergh 1969 and Crampton *et al.* 1985) range from B − V = 0 to ∼ 0.5, at which limit it is impossible to distinguish, from presently available data, intermediate-age rich open clusters from old globular clusters. Crampton *et al.* (1985) include very blue clusters with magnitudes ranging from $V_0 = 18.5$ up to $V_0 = 13.4$. The latter object, if a cluster in M31, has a remarkably high luminosity ($M_{V_0} = -10.8$) for an open cluster, and would have to be the kind of luminous blue cluster typified by the most luminous Magellanic Cloud clusters or by OB associations.

Magnitudes and UBVR colors for 33 of the young open clusters in the disk were obtained by CCD photometry with the Kitt Peak 4-meter telescope by Hodge *et al.* (1987).

The Cluster CMD

Figure 11.3 plots the values of V and B−V for the M31 young open clusters. The area occupied by M31's globular clusters (*e.g.*, van den Bergh 1969) is indicated for comparison.

The disk clusters are seen to resemble the globular clusters in their apparent magnitudes, but to be bluer in almost all cases. Figure 11.3 shows a strong resemblance to a similar diagram for M33 clusters (Christian and Schommer 1982), though there are more extremely blue objects in the M31 sample. This effect is probably a result of selection criteria; the M31 clusters include some objects that are more properly considered to be nuclei of compact OB associations than stable open clusters. The distribution in color is a good match to that of the LMC, which similarly has clusters with measured colors as blue as $B-V = -0.3$. Of course, such blue colors are somewhat unrealistic, in that even if all the light were from a single unreddened O5 star, the color would be no bluer. Probably the clusters with $B - V = -0.3$ are those dominated by early O stars and with observational errors that have moved them to the blue (1σ will change $B-V$ by 0.1). Without spectroscopy, we do not know, of course, whether emission lines may also be affecting the colors somewhat.

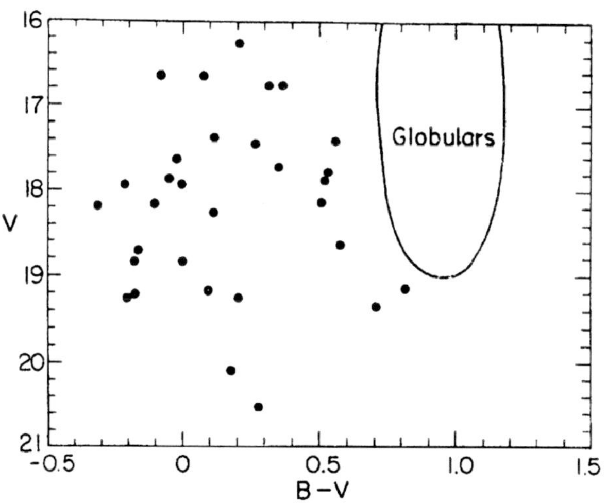

Figure 11.3 Color-magnitude diagram for young disk clusters in M31 (from Hodge et al. 1987).

The data suggest that disk clusters in M31 include a substantial population of luminous, intermediate-color objects. Figure 11.4 shows a histogram of cluster colors (uncorrected for reddening) for M31 clusters and similar diagrams for disk clusters in M33 and the LMC. M31 has an excess of very blue clusters, but this is probably a selection effect, as mentioned above. It is otherwise not unlike a composite of the M33 and LMC cluster population. It does *not* show a minimum at B − V = +0.5, as in the LMC; neither does M33. This fact may be partly the result of the larger spread in reddening values available in the two larger galaxies, but also probably reflects differences in the star-forming histories of the galaxies.

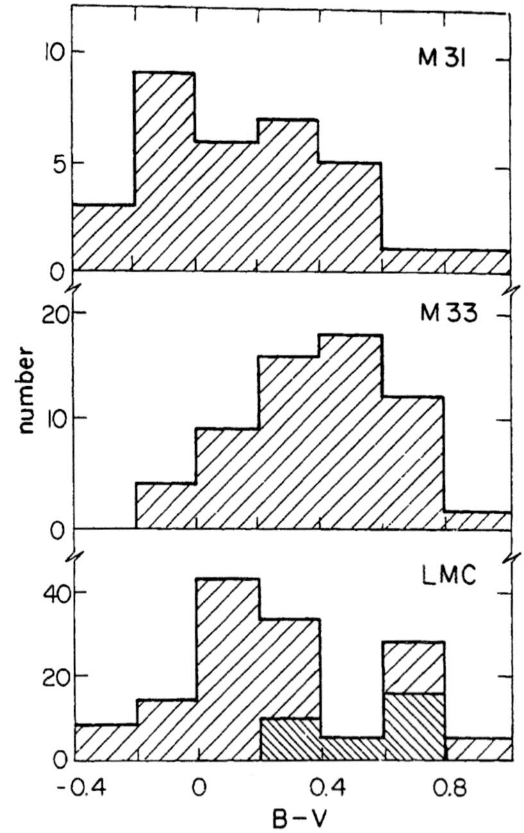

Figure 11.4 Color distributions for star clusters in three galaxies. The preponderance of blue clusters in M31 is largely a selection effect (from Hodge et al. 1987).

Differences between the populations can be the results of different cluster formation histories, different chemical abundance histories, and different stellar IMFs, as well as different selection effects. For example, the difference between the populations of bright, intermediate-color clusters probably relates to the age-mass spectrum. For M31 and M33, there are massive, luminous clusters in each age range, while for the LMC, intermediate-age clusters (those with B − V ~ 0.5) are more rare. The causes for such differences could bear upon important questions regarding galaxy evolution.

The (U−B) vs. (B−V) Diagram

Figure 11.5 shows the two-color (U−B) *vs.* (B−V) plot for the M31 disk clusters for which these three colors have been measured. The prevalence of young clusters is clearly shown by the large number with very negative U−B colors; these are clearly dominated in U by a few very hot stars.

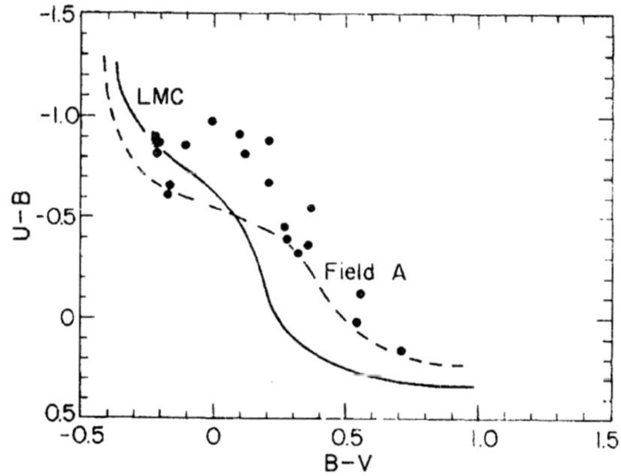

Figure 11.5 Color-color diagram for M31 disk clusters. The solid line is for LMC clusters and the dashed line is for Field A, which has only very small reddening (from Hodge et al. 1987).

Reddening has been determined for the clusters having UBV data by reference to the observed colors of LMC clusters, adopting the blue edge of van den Bergh's (1981) plot and the data of Mateo (1987) for clusters in the far northern part of the LMC, correcting both for foreground reddening. The adopted reddening-free line plotted is fit through the data for the clusters in Baade and Swope's Field IV (Baade and Swope 1963). This field, which is located in the extreme southwest part of M31, has been studied extensively (Racine 1967, Humphreys 1979, Hodge *et al.* 1986) and has a relatively small reddening of $E_{B-V} = 0.15$. Reddening for the clusters in the other fields, all of which are in areas with more obvious dust clouds present, were determined differentially from the reddening-free line. Reddenings so determined range from $E(B-V) \approx 0.6$ to 0.81 mag, in good agreement with ranges found for disk stars (Massey *et al.* 1986, Hodge *et al.* 1986).

Bohlin *et al.* (1988) obtained rocket-borne images of the open cluster C107 (also known as VdB0). This highly-luminous open cluster ($M_V = -9.75$) was detected at both $\lambda 2700$ and $\lambda 1500$, and has UV fluxes consistent with its young age, which Bohlin *et al.* (1988) estimate to be 1×10^8 yrs.

The Luminosity Function

Figure 11.6 shows the luminosity function for the M31 clusters for which integrated magnitudes are available and compares it with that for M33 (Christian and Schommer 1982). The principle difference is in the location of the peak, which might be expected to be the same if sampling effects are similar and the distances are nearly the same (the distances to both galaxies, especially to M33, are uncertain; see Madore *et al.* 1985, Sandage 1983, de Vaucouleurs 1978, and many other recent papers). However, they are somewhat different, with the peak for M31 at V = 17.5 and that for M33 at V = 18.5.

Based on the luminosity functions and the comparison with deep searches for open clusters in the Magellanic Clouds, it is highly unlikely that present surveys of M31 are as much as 10% complete. If we ignore possible intrinsic differences in either the cluster mass function or the destruction rate, it is possible to make a very rough estimate of the total cluster population for M31 by fitting the brightest 3 magnitudes to the luminosity function of LMC clusters derived by Elson and Fall (1985). Table 11.1 shows the results and suggests that present surveys are indeed very incomplete. Clearly, large numbers of disk clusters in M31 remain to be identified and studied.

Table 11.1
Total Cluster Populations

Galaxy	Number Known	Estimated Total Number
LMC	1604	6500
M33	~200	1300
M31	403	10,600

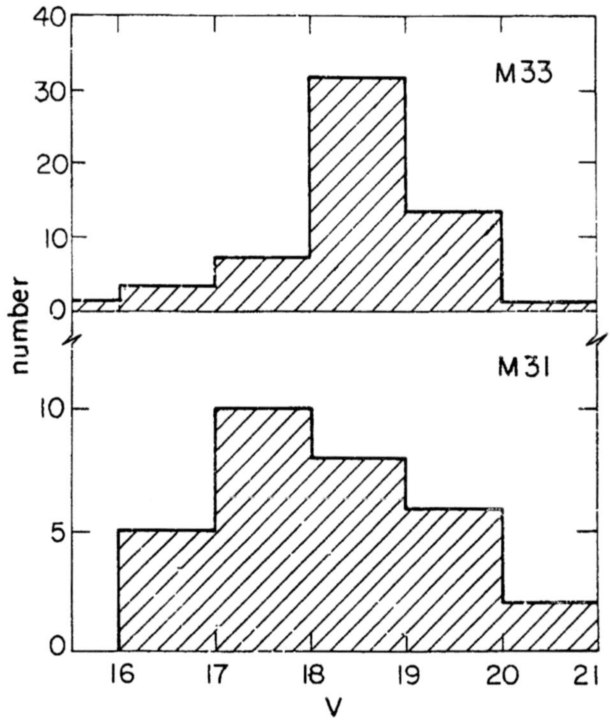

Figure 11.6 Luminosity functions for young clusters in M31 and M33 (from Hodge et al. 1987).

11.4 The Age Function

It is quite difficult to measure reliable ages of open star clusters in other galaxies, a fact that is attested to by the extensive and continuing revisions of age estimates in even the two nearest galaxies, the Magellanic Clouds (*e.g.*, see Elson and Fall 1988 for a discussion). For M31 the data are cruder; we have only the integrated photometry, a few low dispersion spectra, and estimates of the brightness and colors of the brightest stars to go on.

To find at least a crude estimate of the ages of the disk clusters, Hodge (1979) made an attempt to determine the approximate magnitude of the brightest stars. Both severe crowding effects and a highly variable background made photometry of the images completely impossible, and it was necessary to resort to eye estimates.

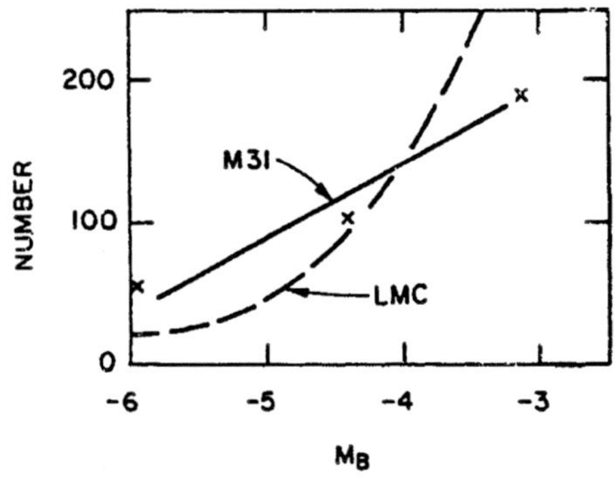

Figure 11.7 The distribution of magnitudes of brightest stars for clusters in M31, compared to similar data for the LMC (from Hodge, 1979).

Because of this, the magnitudes were divided into three rather large bins, corresponding to average absolute B magnitudes of −5.2, −4.4, and −3.1. The relative numbers of these are plotted in Figure 11.7, where the results are compared with a similar curve for the LMC. Although the data are uncertain, the general agreement is good, especially if it is realized that the faintest bin is probably affected by incompleteness because of its proximity in M31 to the plates' resolution limits. This general agreement indicates that for young clusters the formation rate history in the two galaxies cannot have been very different at least over the last $\sim 10^8$ years (that is, the relative numbers of clusters of different ages is nearly the same in the two galaxies).

Figure 11.8 shows the radial distribution of the ratio of "old" (10^8 years) to "young" (10^6 years) clusters along the major axis. The bin of clusters with brightest stars at $M_B \approx -3.1$ were taken as old clusters and the bin with brightest stars at $M_B \approx -5.2$ as young clusters. A tendency for the younger clusters to be relatively more plentiful in the inner areas (to ~ 50 arcmin) is noticeable and is probably related to the peak at ~ 50 arcmin in H I, H II regions, and bright stars, all indicating a higher than average recent rate of star formation at these positions in M31.

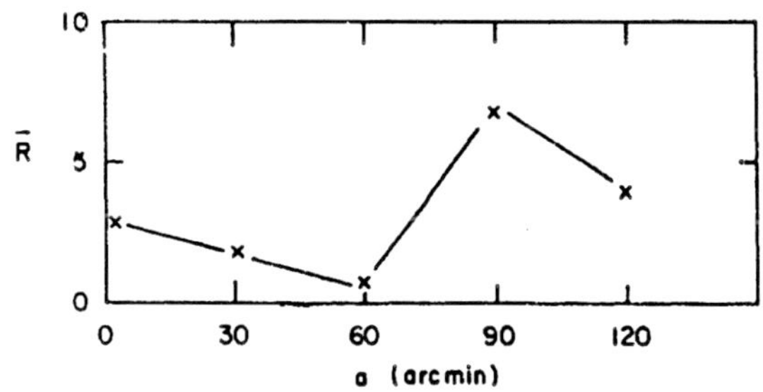

Figure 11.8 The mean ratio, $\overline{\mathrm{R}}$, *of the number of old to the number of young clusters as a function of distance from the center of M31 (from Hodge, 1979).*

An idea of the cluster ages also can be obtained from their unreddened colors. A curve of U−B *vs.* age, derived for Magellanic Cloud clusters, has been used to evaluate the integrated colors of M31 clusters. Unfortunately, it shows a degeneracy that occurs for old clusters, in which case it is impossible to distinguish between the effects of blue main sequence stars and horizontal branch stars. For clusters younger than ~ 5×10^8 years, however, the Cloud clusters can be age-dated fairly reliably from U−B colors. Ages derived by Hodge *et al.* (1987) using this technique range from ~ 10^6 years to 4×10^8 years. The great predominance of younger clusters is, of course, a consequence of M31's distance, at which old, faint clusters are very difficult to detect.

11.5 The Spatial Distribution

As Figure 11.9 shows, the distribution of the known open clusters in M31 shows no concentration toward the center of the galaxy, is exclusively limited to the disk, and follows the spiral arm pattern. Baade's

(1963) spiral arm crossings of the major axis, in the very center where they are defined by absorption lanes only, occur at positions of cluster concentration.

The most distant cluster, C1 at a distance from the nucleus of 139 arcmin, is 4 kpc beyond Baade's most distant spiral arm, which he places at 22 kpc, and may represent an eighth arm crossing too thinly populated for Baade to have detected. Baade's seventh crossing is marked by seven detected clusters.

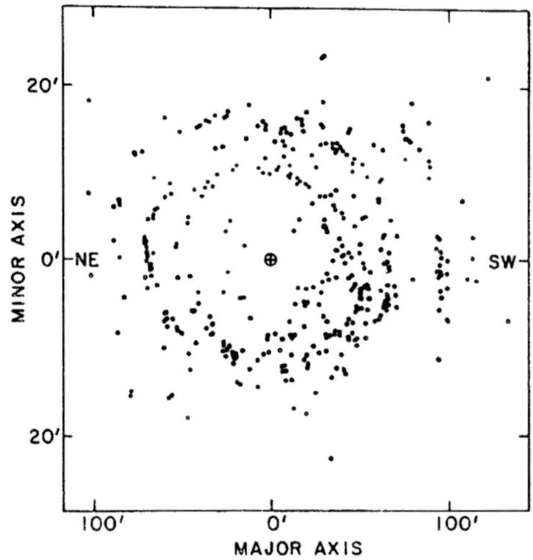

Figure 11.9 The spatial distribution of young disk clusters, corrected to face-on (from Hodge, 1979).

Figure 11.10 shows the distribution of clusters along the major axis, counted from a 5-arcmin-wide strip. The two halves of the major axis show conspicuously different amplitudes for their oscillations, but agree on the presence of peaks at 50 and 68 arcmin. The southwest axis has peaks at 35 and 100 arcmin that do not show up on the northeast profile, probably reflecting the unevenness of cluster population along the arms. The M31 clusters appear to be highly clumped in their distribution, which can explain the differences in Figure 11.10.

A comparison of the mean cluster profile and two other optical indices is given in Figure 11.11. The profile for bright stars is averaged from the counts of van den Bergh (1958) and Reddish (1962). The curve labeled "flat component" is taken from the surface photometry of de Vaucouleurs

(1958), who separated the blue intensity profile into a spheroidal and a flat component. It is interesting to see that the amplitude of the variation in density for the clusters is larger than for the bright field stars, which in turn is very much larger than for the flat disk itself. The former difference may mean that cluster formation is relatively more common in spiral arms than in interspiral regions (or that the velocity dispersion is larger for field stars than clusters). The latter difference indicates that the fainter (and largely older) stars of the disk are spread more smoothly between arms, as photometry already has shown (Chapter 4).

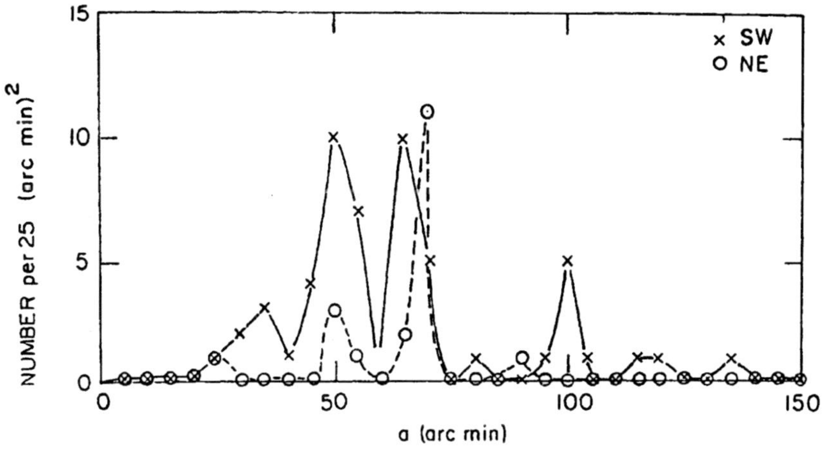

Figure 11.10 The radial distribution of young disk clusters along the two halves of the major axis (from Hodge, 1979).

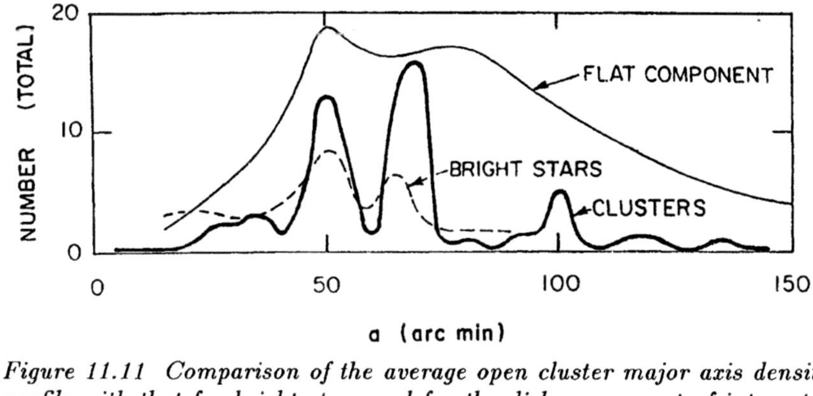

Figure 11.11 Comparison of the average open cluster major axis density profile with that for bright stars and for the disk component of integrated light (from Hodge, 1979).

11.6 Summary

Only a start has been made on studying the open clusters of M31. There are about 400 luminous, blue disk clusters known and perhaps 100 blue globular clusters. The former are young ($\sim 10^6$ to 4×10^8 years old), fairly large ($\sim$ 20 pc) objects that are concentrated to the 10 kpc "ring" and to segments of spiral arms.

Table 11.2
Open Clusters of M31

(1950 Coordinates)

No.	R.A.	Dec	desc.*	No.	R.A.	Dec	desc.*
1	$00^h33^m29^s$	39°06′27″	l,f	33	00^h04′53″	40°37′01″	s,f
2	34′04″	24′07″	l,f	34	35′04″	06′39″	s,m
3	33′50"	28′40″	l,f	35	35′03″	07′38″	s,b
4	34′03″	27′48″	s,f	36	35′03″	08′17″	m,b
5	24′19″	28′27″	l,b	37	36′04″	39°51′35″	m,f
6	33′49″	35′33″	l,f	38	36′39″	59′13″	s,m
7	35′13″	34′14″	s,m	39	36′35″	40°00′00″	s,f
8	35′15″	34′34″	l,f	40	36′33″	00′07″	s,f
9	35′16″	35′00″	s,f	41	36′06″	07′12″	l,b
10	34′58″	36′52″	s,f	42	36′32″	04′01″	m,f
11	34′53″	38′04″	s,f	43	36′25″	05′46″	l,m
12	35′12″	38′30″	s,f	44	36′34″	05′26″	s,m
13	35′11″	39′42″	s,f	45	36′34″	05′52″	l,m
14	35′04″	40′47″	l,m	46	36′37″	04′53″	l,m
15	35′03″	40′53″	l,f	47	36′51″	03′22″	l,b
16	35′01″	41′46″	l,m	48	36′55″	03′03″	l,b
17	34′54″	41′52″	l,m	49	36′50″	03′55″	l,b
18	34′47″	43′38″	s,m	50	36′46″	04′47″	l,b
19	34′47″	44′10″	s,m	51	37′03″	04′14″	s,f
20	34′49″	43′51″	l,m	52	36′38″	07′05″	l,m
21	34′48″	44′17″	s,m	53	36′58″	04′27″	vl,b
22	35′10″	41′46″	s,f	54	37′00″	04′27″	l,m
23	35′09″	42′39″	s,f	55	36′59″	04′40″	s,m
24	35′54″	36′26″	s,f	56	37′01″	04′40″	s,m
25	34′32″	52′53″	l,f	57	37′04″	04′47″	l,b
26	34′34″	53′46″	s,f	58	36′26″	09′29″	s,m
27	31′39″	52′34″	l,f	59	36′27″	09′16″	s,f
28	34‘16″	56′30″	s,f	60	36′31″	09′35″	l,b
28	35′55″	53′13″	s,f	61	35′47″	16′08″	l,b
29	35′55″	53′13″	s,f	62	36′03″	14′17″	l,m
30	34′32″	40°05′39″	l,f	63	36′06″	14′50″	l,b
31	34′56″	03′55″	s,f	64	37′14″	05′19″	l,b
32	35′01″	05′39″	s,m	65	37′12″	05′52″	s,m

Table 11.2 continued

No.	R.A.	Dec	desc.*	No.	R.A.	Dec	desc.*
66	$00^h37^m02^s$	40°07′38″	s,m	99	$00^h37'38''$	40°18′52″	l,f
67	36′18″	14′04″	vl,m	100	37′10″	23′00″	l,f
68	36′18″	14′10″	s,f	101	36′58″	25′44″	s,f
69	37′27″	07′38″	l,f	102	36′46″	27′16″	l,m
70	37′00″	20′30″	l,b	103	36′41″	27′29″	s,f
71	37′08″	13′12″	l,b	104	36′38″	27′35″	l,m
72	37′23″	12′13″	s,f	105	36′32″	31′31″	l,f
73	37′49″	10′14″	s,f	106	36′34″	31′51″	s,f
74	37′42″	11′26″	s,f	107	37′44″	19′44″	l,m
75	37′10″	12′59″	s,f	108	38′08″	15′55″	s,b
76	37′21″	13′51″	s,f	109	38′05″	17′00″	s,f
77	37′20″	13′51″	s,f	110	38′03″	17′33″	l,m
78	36′27″	20′50″	l,b	111	37′49″	20′17″	s,f
79	36′50″	19′12″	l,f	112	37′40″	21′09″	l,f
80	36′53″	20′04″	s,f	113	37′29″	22′21″	l,f
81	37′15″	16′01″	l,b	114	37′31″	22′53″	l,f
82	37′23″	14′17″	s,f	115	37′46″	22′27″	s,b
83	27′15″	17′00″	l,f	116	38′10″	19′38″	s,m
84	37′37″	14′23″	l,f	117	38′16″	18′20″	s,m
85	36′29″	23′59″	s,f	118	38′17″	18′33″	s,b
86	37′08″	19′12″	s,f	119	38′19″	18′39″	s,m
87	37′15″	19′12″	s,f	120	38′11″	19′44″	l,m
88	37′33″	16′21″	l,f	121	38′12″	20′10″	l,b
89	37′32″	16′27″	s,f	122	37′45″	24′52″	s,f
90	37′36″	15′48″	l,f	123	36′50″	31′51″	l,f
91	37′38″	16′27″	s,f	124	37′01″	33′42″	l,b
92	37′39″	16′21″	l,f	125	37′25″	29′00″	s,f
93	37′47″	16′21″	l,f	126	37′25″	29′19″	s,b
94	38′02″	14′36″	s,m	127	37′45″	25′57″	s,b
95	38′05″	14′23″	l,b	128	37′47″	26′23″	l,b
96	38′06″	14′50″	l,m	129	37′43″	27′16″	l,b
97	37′51″	16′53″	l,f	130	37′41″	28′53″	l,b
98	37′47″	16′40″	l,f	131	37′48″	26′43″	l,b

Table 11.2 continued

No.	R.A.	Dec	desc.*	No.	R.A.	Dec	desc.*
132	00^h38′43″	40°21′46″	l,b	165	00^h38′04″	40°36′43″	l,f
133	38′44″	21′40″	l,f	166	38′09″	36′54″	l,f
134	38′44″	21′51″	s,b	167	37′52″	38′48″	s,m
135	38′42″	21′38″	s,f	168	38′11″	37′24″	l,f
136	38′34″	23′26″	s,f	169	37′22″	41′25″	s,f
137	38′36″	23′26″	s,m	170	37′08″	45′36″	l,f
138	38′56″	23′01″	l,b	171	36′44″	48′39″	l,f
139	38′06″	27′29″	l,m	172	37′10″	49′38″	l,f
140	37′52″	28′53″	s,f	173	37′20″	43′42″	s,f
141	37′53″	29′19″	s,m	174	37′21″	44′09″	s,f
142	37′31″	33′32″	l,f	175	37′36″	44′07″	s,f
143	37′02″	35′00″	l,b	176	38′54″	30′59″	l,f
144	37′04″	36′07″	s,f	177	38′41″	33′54″	s,m
145	37′10″	36′20″	s,b	178	38′10″	14′23″	l,b
146	37′11″	36′20″	s,b	179			
147	37′41″	32′59″	l,f	180	38′03″	40′48″	s,m
148	37′40″	33′27″	s,f	181			
149	38′12″	30′25″	s,b	182			
150	38′43″	25′19″	l,f	183	37′55″	44′27″	s,m
151	38′46″	25,17″	l,f	184			
152	37′13″	38′41″	l,m	185			
153	37′15″	38′09″	l,f	186	38′57″	34′03″	s,m
154	37′42″	36′02″	l,b	187	39′22″	31′15″	s,f
155	39′11″	23′19″	s,f	188	39′50″	17′22″	l,f
156	38′56″	27′09″	s,m	189			
157	39′07″	27′36″	s,m	190	37′56″	47′00″	s,f
158	39′02″	28′51″	s,b	191	38′21″	43′49″	s,f
159	39′01″	29′23″	l,f	192	39′01″	38′05″	l,f
160	38′53″	29′25″	l,f	193	39′28″	33′50″	s,f
161	38′29″	31′33″	s,f	194	39′48″	29′25″	l,f
162	38′28″	32′43″	l,b	195	40′13″	32′23″	s,m
163	38′06″	35′02″	s,m	196	39′47″	35′02″	l,f
164	38′07″	35′59″	s,f	197	39′49″	35′02″	s,f

Table 11.2 continued

No.	R.A.	Dec	desc.*	No.	R.A.	Dec	desc.*
198	00ʰ39′42″	40°35′50″	s,f	233	00ʰ38′32″	40°00′52″	l,b
199	39′51″	36′59″	s,f	234	38′30″	02′02″	s,m
200	39′36″	39′01″	s,f	235	38′40″	00′57″	s,m
201	39′14″	38′48″	s,f	236	38′40″	01′49″	s,m
202	39′22″	40′15″	s,b	237	38′45″	00′32″	s,f
203	39′22″	40′35″	s,b	238	40′32″	44′27″	l,f
204	39′13″	41′02″	s,f	239	40′26″	45′09″	l,b
205	38′16″	46′26″	s,b	240	40′25″	45′20″	s,f
206	38′19″	47′57″	l,m	241	40′18″	48′29″	s,m
207	37′44″	52′33″	s,f	242	40′32″	46′40″	l,m
208	38′00″	52′49″	s,m	243	40′31″	46′49″	l,m
209	37′58″	54′07″	s,f	244			
210	37′58″	54′32″	l,m	245	38′38″	41°03′18″	s,m
211	38′05″	56′14″	s,f	246	38′40″	03′29″	l,m
212	38′31″	50′43″	s,f	247			
213	38′29″	51′18″	s,f	248	39′01″	01′46″	s,m
214	38′47″	48′34″	s,b	249	38′56″	11′11″	s,f
215	39′27″	44′54″	s,f	250	39′04″	02′27″	s,m
216	39′36″	44′39″	s,f	251	40′30″	40°52′44″	s,f
217	40′31″	37′31″	l,b	252	38′57″	41°06′00″	s,f
218	39′52″	43′08″	l,f	253	39′20″	06′07″	s,f
219	39′03″	50′23″	s,b	254	39′19″	08′39″	l,f
220	38′28″	56′43″	s,f	255	39′02″	10′40″	s,f
221	38′25″	58′40″	s,f	256	39′05″	10′42″	s,f
222				257	39′07″	12′43″	l,f
223	38′33″	51′57″	s,b	258	39′31″	08′59″	l,f
224	38′33″	51′45″	l,b	259	40′49″	40°53′32″	l,b
225	38′43″	55′33″	l,b	260	40′45″	55′42″	l,f
226	38′39″	57′14″	s,m	261	40′46″	55′40″	s,f
227	38′46″	56′52″	l,b	262	41′00″	53′07″	l,b
228				263	39′37″	39′37″	l,m
229	40′11″	43′49″	s,m	264	39′17″	14′27″	l,f
230	40′03″	44′43″	s,m	265	41′26″	56′07″	s,f
231	40′02″	45′41″	l,f	266	41′16″	58′28″	s,m
232	39′54″	46′26″	l,f	267	39′23″	41°19′40″	s,f

Table 11.2 continued

No.	R.A.	Dec	desc.*	No.	R.A.	Dec	desc.*
268	$00^h39'46''$	41°18′29″	s,f	304	$00^h42'09''$	41°14′59″	
269	41′06″	04′42″	l,m	305	41′15″	14′50″	l,m
270	41′34″	01′17″	l,b	306	42′10″	15′09″	l,b
271	40′44″	11′36″	l,f	307	40′29″	33′52″	l,m
272	41′05″	08′25″	l,b	308	40′46″	31′33″	l,f
273	41′35″	05′11″	l,f	309	41′09″	30′23″	l,b
274	41′15″	10′03″	s,f	310	42′18″	21′02″	s,b
275	40′14″	20′50″	l,f	311	42′24″	20′20″	l,b
276	41′58″	05′16″	l,f	312	42′25″	20′54″	l,b
277	42′03″	06′45″	s,f	313	42′26″	21′07″	l,b
278	41′52″	09′09″	s,f	314	42′18″	22′31″	l,f
279	41′53″	09′05″	l,m	315	42′30″	23′02″	l,b
280	41′57″	09′20″	s,f	316	41′16″	32′29″	l,f
281	42′01″	09′18″	s,f	317	40′43″	36′07″	l,b
282	41′55″	09′41″	s,f	318	40′51″	37′46″	s,f
283	41′55″	10′03″	l,m	319	41′23″	33′05″	l,f
284	41′57″	10′12″	s,f	320	41′34″	31′31″	l,m
285	41′58″	10′17″	l,m	321	41′04″	40′28″	l,b
286	41′57″	10′22″	l,m	322	41′06″	40′49″	l,b
287	41′57″	10′51″	s,f	323	41′09″	41′07″	l,b
288	40′26″	22′10″	s,f	324	42′32″	28′46″	l,m
289	40′53″	19′51″	s,m	325	42′27″	29′29″	l,f
290	40′04″	28′12″	l,m	326	41′43″	38′16″	l,m
291	40′12″	29′09″	l,m	327	42′59″	27′20″	s,f
292	40′16″	29′54″	l,b	328	42′46″	30′32″	l,f
293	40′33″	28′28″	l,m	329	42′47″	30′54″	l,f
294	41′24″	17′16″	s,f	330	42′53″	30′54″	l,f
295	41′09″	21′16″	s,m	331	42′34″	34′22″	l,b
296	41′41″	17′48″	s,f	332	42′10″	39′03″	l,b
297	40′54″	28′58″	s,f	333	42′12″	39′22″	l,b
298	42′02″	12′36″	s,b	334	42′14″	39′09″	l,b
299	42′02″	12′54″	s,m	335	43′23″	27′48″	l,f
300	42′04″	12′52″	s,m	336	43′17″	29′52″	l,b
301	42′09″	13′10″	l,m	337	43′55″	25′13″	l,b
302	42′09″	14′54″	l,b	338	43′22″	34′08″	l,m
303	42′13″	14′20″	l,b	339	41′46″	48′26″	l,f

Table 11.2 continued

No.	R.A.	Dec	desc.*	No.	R.A.	Dec	desc.*
340	$00^h42'59''$	41°39′22″	l,b	376	$00^h44'03''$	41°54′00″	l,f
341	42′46″	42′26″	l,m	377	43′50″	55′31″	s,m
342	41′28″	46′08″	s,b	378	43′50″	55′57″	l,b
343	43′28″	36′00″	l,b	379	43′44″	55′44″	l,b
344	42′30″	47′46″	s,m	380	43′26″	56′30″	l,m
345	42′28″	48′59″	s,f	381	43′45″	57′03″	l,b
346	42′09″	56′43″	s,f	382	43′46″	56′56″	l,b
347	42′46″	50′23″	l,b	383	43′48″	56′56″	l,m
348	43′32″	38′30″	l,m	384	43′07″	46′21″	l,f
349	44′12″	34′01″	l,m	385	43′21″	49′10″	l,f
350	44′19″	35′06″	m,f	386	43′22″	49′17″	s,f
351	43′46″	40′34″	s,f	387	45′27″	52′48″	s,f
352	43′43″	41′46″	s,f	388	45′30″	52′17″	s,f
353	43′42″	42′33″	s,m	389	44′51″	42°01′38″	s,f
354	43′46″	43′12″	l,m	390	44′14″	10′27″	s,f
355	43′51″	42′39″	s,m	391	44′14″	11′06″	s,m
356	44′06″	44′06″	s,f	392	44′41″	07′05″	l,f
357	43′06″	41′33″	l,b	393	45′25″	02′56″	s,f
358	42′57″	57′03″	l,b	394	44′44″	10′27″	l,f
359	43′11″	54′33″	l,b	395	44′27″	13′25″	l,f
360	43′45″	48′07″	l,b	396	45′42″	18′00″	s,f
361	43′14″	54′33″	m,b	397	45′43″	18′13″	s,f
362	43′16″	54′59″	l,m	398	44′22″	31′36″	l,f
363	43′18″	54′52″	l,b	399	45′08″	25′18″	s,f
364	43′36″	51′55″	s,f	400	34′48″	39°38′43″	s,f
365	43′45″	51′29″	s,f	401	35′26″	39′03″	s,f
366	43′52″	51′16″	s,m	402	34′55″	36′26″	s,f
367	43′47″	51′55″	l,f	403	34′59″	37′57″	s,f
368	43′48″	52′53″	s,m	404	34′43″	59′19″	s,f
369	43′52″	52′27″	l,b	405	37′15″	40°05′59″	l,b
370	43′51″	53′06″	l,b	406	42′05″	41°12′39″	l,b
371	43′48″	54′12″	s,m				
372	43′48″	54′46″	l,b				
373	43′48″	55′05″	l,b				
374	43′52″	54′20″	s,m				
375	44′05″	52′01″	s,f				

*description, (a) the first letter describes the size of the cluster:
l = large, m = medium, s = small
(b) the second letter describes the brightness of the clusters as a whole:
b = bright, m = medium, f = faint

CHAPTER 12

DUST

The dust content of M31 can be studied in a variety of ways. Optical evidence of the presence of dust has been known since the first photographic plates were examined in the 19th century. Accurate photometry has revealed the effects of dust in its reddening, as well as its absorption, of starlight in M31. Polarization by the dust provides further information on the properties of the dust. Recently, detailed infrared studies, especially by the IRAS satellite, have provided information about the temperatures in the dust component.

12.1 Dust Clouds

M31 obviously contains dust. This fact is made clear by the conspicuous obscuration by dust of the northwest side of the bulge. The clear morphology of obscuring material makes it obvious that the dark lanes are not the result of a lack of stars, but must be the result of the presence of large amounts of dust in M31. Its particular conspicuousness results from the high angle of inclination of the galaxy, which makes the dust lanes stand out starkly against the starlight of the smoothly structured bulge (Figure 12.1).

Optical photometry of the dust lanes, such as that carried out photographically by Lyngå (1959), has shown that they are deep, obscuring as much as 60% of the background starlight. The dust lanes, when examined on high resolution photographs, break up into a series of individual dust clouds, superimposed on a general, diffuse dust component.

Baade (1963) noted that the dust clouds play a significant role in delineating the inner spiral structure of M31. While the outer spiral arms are principally marked by the presence of H II regions, OB stars, and H I, the inner arms are best traced optically by the dust clouds and lanes.

The only comprehensive optical survey of the dust was carried out on plates taken with the Kitt Peak National Observatory's 4-meter telescope (Hodge 1980). A total of 730 different dust clouds were identified and catalogued. An example is shown in Figure 12.2, which compares a photograph of D307 with a photometric tracing through it.

Figure 12.1 Dust clouds along the southeast minor axis of M31, photographed with the Kitt Peak 4-m telescope.

The distribution of the 730 catalogued dust clouds is shown in Figure 12.3, which is a plot of the centers of the clouds. This diagram shows the prevalence of the dust clouds on the northwest side of the center. It also shows the disk shape of the dust, which corresponds closely to the position of the inner disk of stars. The spiral arms show up only roughly; the inner arms in particular, however, would be better defined if the elongations of each cloud were also shown, as there is a tendency for the larger clouds to be elongated in the direction of the optical spiral arms.

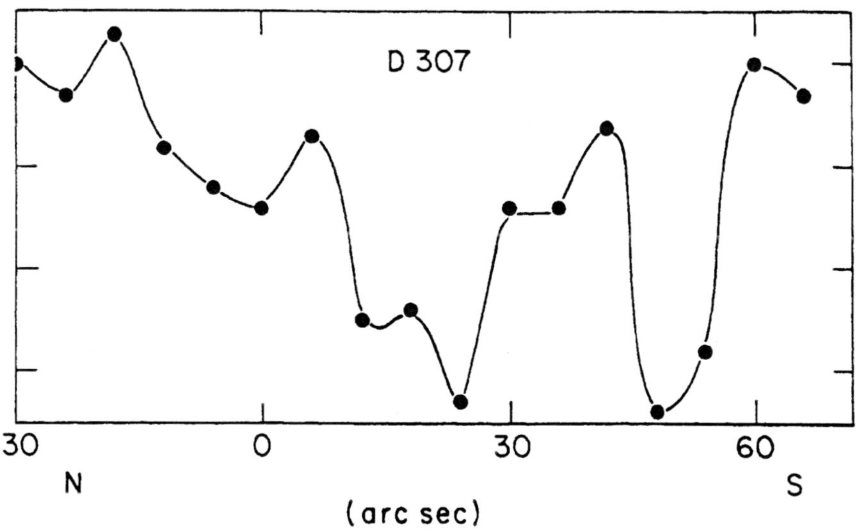

Figure 12.2 An east-west photometric tracing of the intensity of blue light from the area of dust cloud D307. (from Hodge, 1980).

Dust clouds very near the nucleus were first detected by Johnson and Hanna (1972) and have been further studied by Gallagher and Hunter (1981) and McElroy (1981). Johnson and Hanna's objects were traced to within 6 arcsec from the nucleus, and a spiral shape for them could be detected up to approximately 30 arcsec from the nucleus. Figure 12.4 shows McElroy's remarkable photograph of the central dust clouds. Whether the small clouds even nearer the nucleus can be said to form a spiral structure, however, is not clear; they do not fit a logarithmic spiral in the plane of the galaxy. Johnson and Hanna (1972) suggest that the inner dust clouds are not confined to the plane of the 25-pc thick inner gas disk, and not even to the 100-pc thick disk of the Rubin and Ford

(1971) model. Instead, there is the suggestion that the inner dust lanes and clouds extend out into the bulge in three dimensions, much as the faint wisps of ionized gas are found to do (Chapter 14).

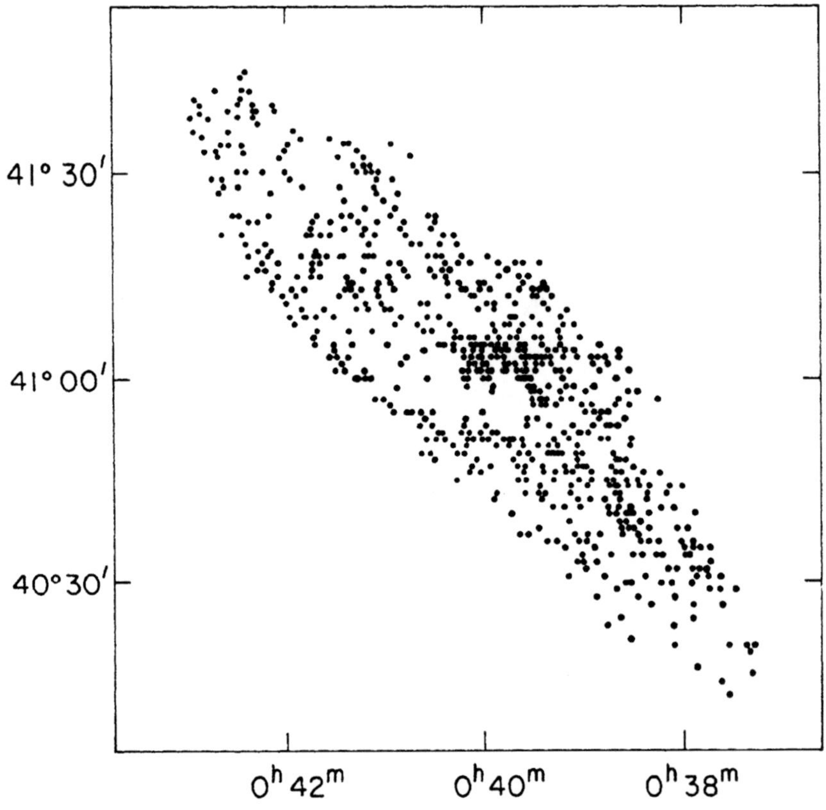

Figure 12.3 The distribution on the sky of the centers of catalogued optical dust clouds in M31 (from Hodge, 1980).

In general, however, the dust clouds are found to lie in the main disk of M31, as their large-scale distribution (Figure 12.3) illustrates. When the positions of the dust clouds are corrected to face-on positions, their radial distribution is found to fall off nearly linearly with distance from the center (Figure 12.5). This is in contrast to the exponential decrease

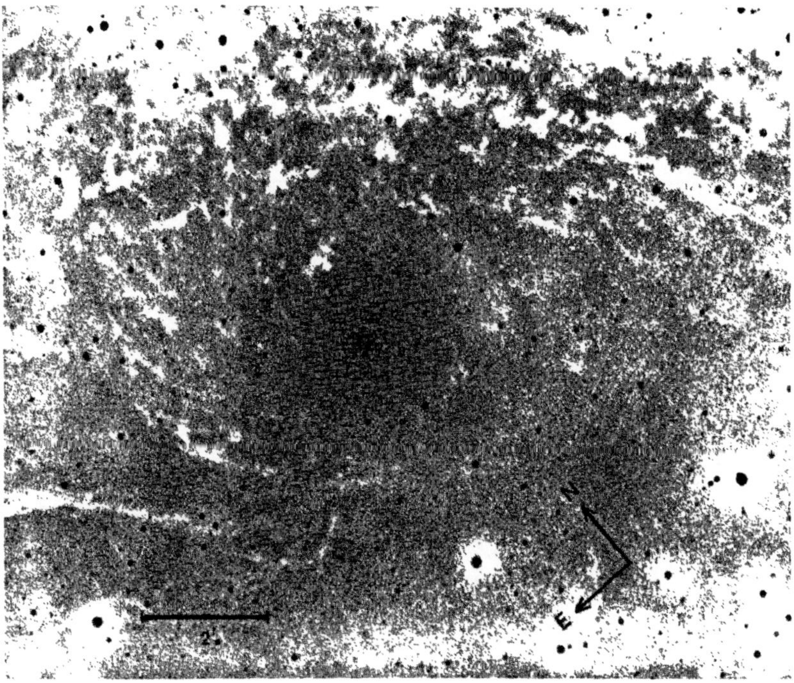

Figure 12.4 The central bulge of M31 after masking out the smooth component. The dust clouds near the nucleus show up well in this image, in which the major axis is aligned vertically, with the northwest minor axis to the right (courtesy D. McElroy, Kitt Peak National Observatory photo).

of the starlight in the disk, and illustrates the very different spatial characteristics of the dust from other components, graphically illustrated in Figure 12.6, which compares the major axis distributions of the dust clouds, the OB stars, the open clusters, and the general reddening.

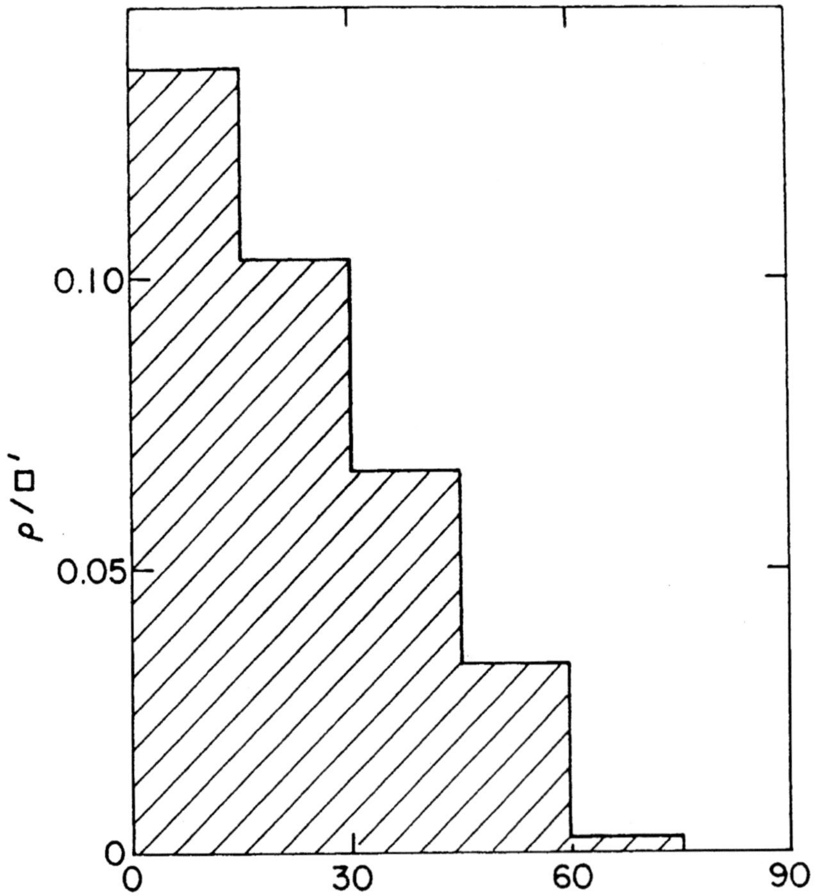

Figure 12.5 The radial distribution of positions of dust clouds in M31 (from Hodge, 1980).

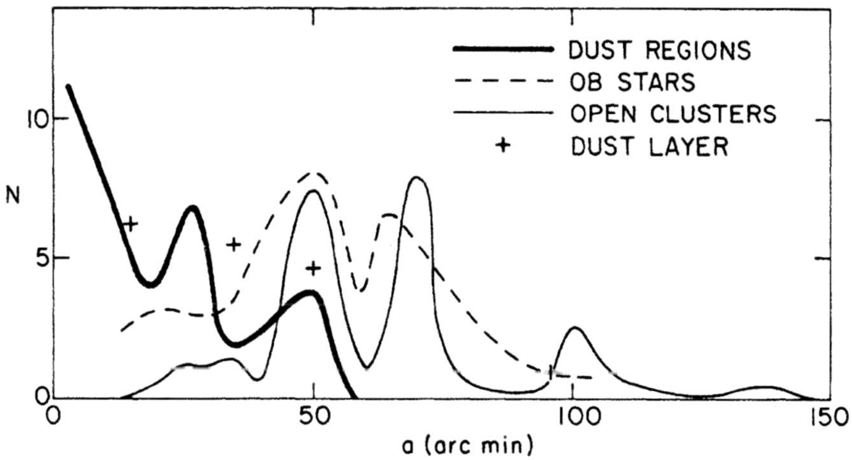

Figure 12.6 A comparison of the radial distributions of dust clouds, OB stars, open clusters, and the dust layer in M31 (from Hodge, 1980).

The catalogued dust clouds range in size from the discernible limit of about 5 pc up to nearly 2700 pc (Figure 12.7). These data are probably largely determined by observational selection; at higher resolution, the clouds might be broken into smaller components (*e.g.*, as found by Gallagher and Hunter 1981 in the inner regions), and for a galaxy at a greater distance, the discernible dust clouds would probably be fewer and larger. Figure 12.8 compares the size distribution of the M31 dust clouds with those of other galaxies, showing a wide variety. If M31's dust clouds are truly larger in scale than those in either the LMC or the Galaxy, then we should look for some medium-scale hydrodynamical effect. However, at this time the data are too likely to be affected by selection problems to warrant a search for a theoretical model.

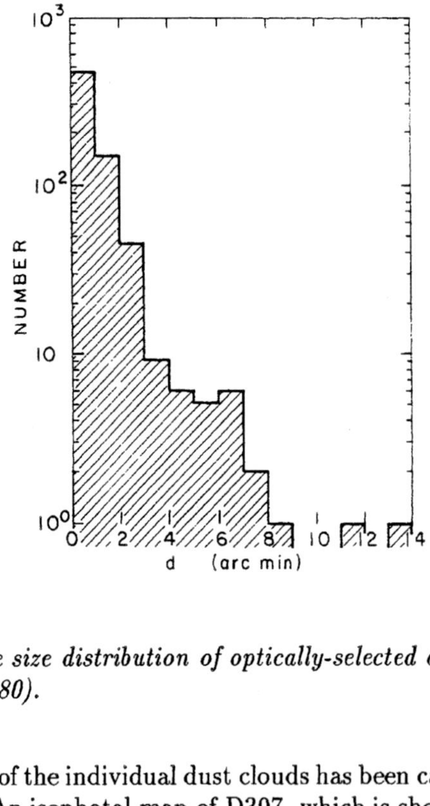

Figure 12.7 The size distribution of optically-selected dust clouds in M31 (from Hodge, 1980).

Photometry of the individual dust clouds has been carried out to a very limited extent. An isophotal map of D307, which is shown in Figure 12.2, was constructed from Palomar Schmidt plates by Hodge and Kennicutt (1982), who found a maximum extinction of 0.60 mag at the core of the cloud and an average extinction of 0.35 mag over the 220 arcsec^2 area of the cloud. Dust cloud D441 (Figure 12.9), a relatively conspicuous, isolated example, showed a maximum extinction of 0.76 mag and a mean over the cloud of 0.43 mag. Most of M31's dust clouds, however, show smaller amounts of extinction, and are, in the mean, much less optically thick than the typical dust clouds that are found in star-forming regions in our Galaxy.

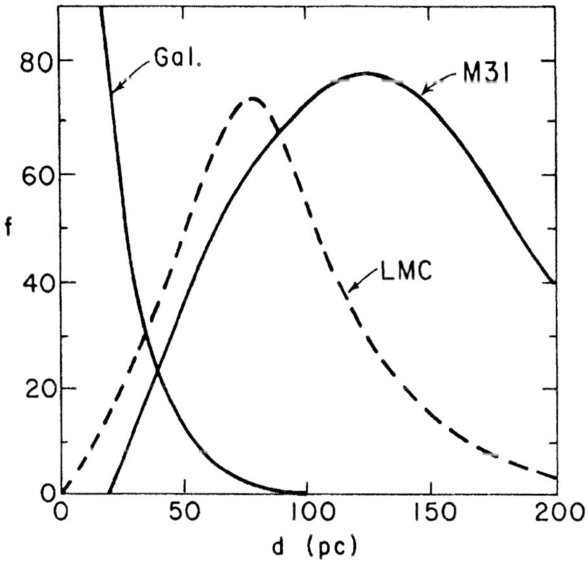

Figure 12.8 Dust cloud sizes compared for the Galaxy, M31 and the LMC. Differences may be influenced by observational selection (from Hodge, 1980).

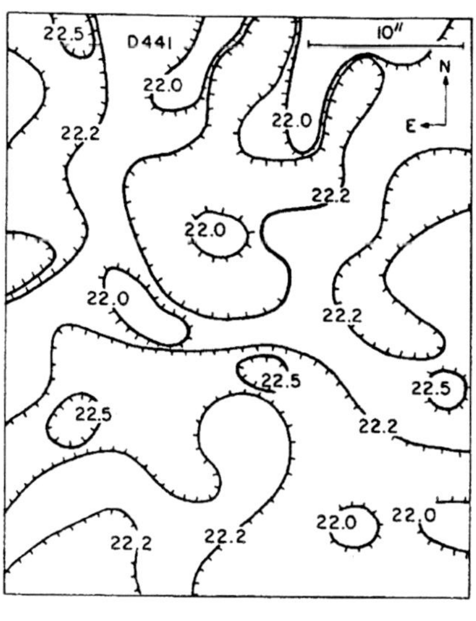

Figure 12.9 Isophotes in B for the dust cloud D441 (from Hodge and Kennicutt, 1982).

Are there any dense, optically thick dust clouds in M31? Gallagher and Hunter (1981) suggest that only shallow, diffuse clouds exist in M31's bulge, and correlate this fact with the lack of large mass star formation there. Presumably denser, smaller-scale dust clouds do, however, exist in the outer parts of the disk, where there are many large-mass stars being formed, and where there are molecular clouds (Section 12.5 and Chapter 16). Present studies are probably not able to detect any very dense cores or components that may exist in the catalogued dust clouds; this is an area that needs further optical high-resolution study.

12.3 Reddening and Extinction

The reddening of various kinds of objects is the prime optical source of information on the dust in M31. In the discussion below the results from each different kind of object are given separately, with a comparison and general conclusions given at the end.

From Foreground Stars

Measurements of the reddening of foreground stars in the direction of M31 can give the effects of Galactic dust that is unrelated to M31. Using the standard cosecant law of reddening, for instance, predicts a foreground reddening of E(B−V) = 0.15 mag (Arp 1962), while the use of H I data predicts a value of 0.08 mag (Burstein and Heiles 1984).

A specific measurement of the foreground reddening in front of M31 was made by McClure and Racine (1969). They used intermediate- and wide-band photometry of late-type field stars, mostly K giants with V = 7 to 12, to determine the color excess as a function of distance. Their analysis showed that Galactic dust could be detected towards M31 up to distances that indicated that the absorbing layer extends to -200 ± 50 pc from the Galactic plane. The total reddening was found to be $E(B - V) = 0.11 \pm 0.02$ mag. This value agrees with other, similar results for foreground objects, which range from 0.09 to 0.20 (Taylor and Kellman 1978).

From Globular Clusters

Multicolor photometry of M31's globular clusters can be used to gauge the reddening in different parts of the galaxy. For the outer halo, these data indicate primarily the overall foreground reddening, while for those globulars located within the disk and bulge, they act as probes of the inner dust.

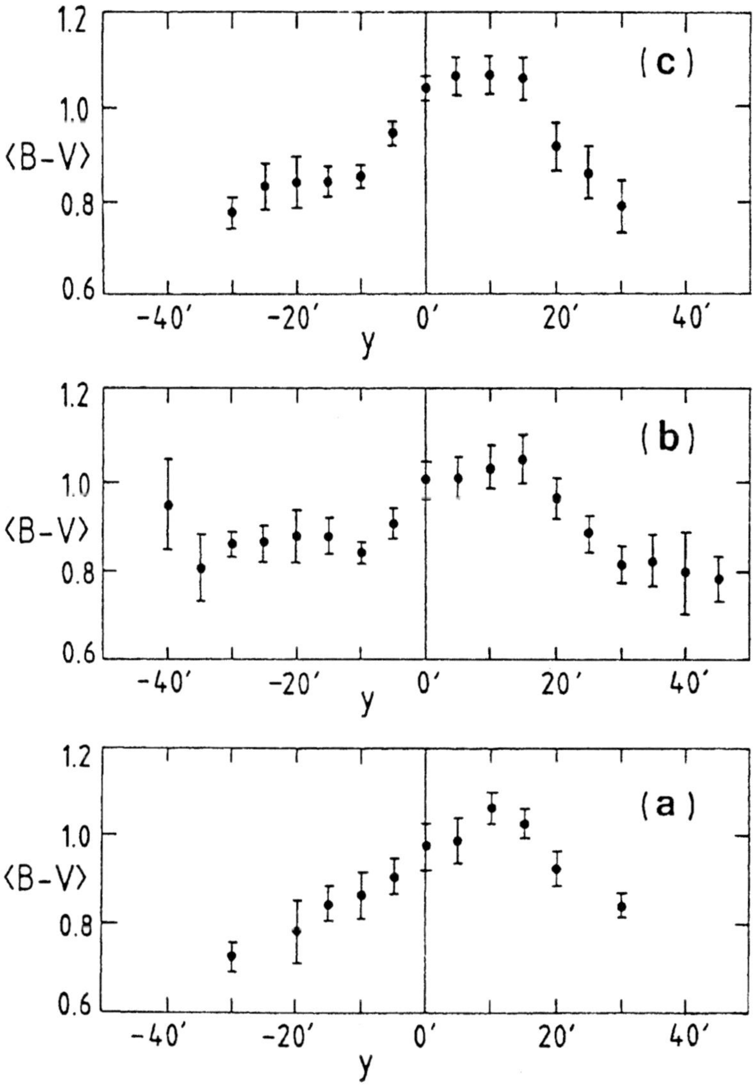

Figure 12.10 The variation of mean B-V color of globular clusters along the direction of Y (perpendicular to the major axis). Samples used are: (a) van den Bergh (1969), (b) Sharov and Lyutyi (1983), and (c) Buonanno et al. (1982) (from Iye and Richter, 1985).

Assuming that the intrinsic colors of M31's globulars are the same as those for Galactic globulars, van den Bergh (1963) determined that the outer "halo" globulars suffered a total reddening of $E(B-V) = 0.20 \pm 0.03$ (the globular cluster photometry used, published by Kron and Mayall 1960, was on the old P, V system). He later redetermined the reddening from additional data, obtaining E(B-V) = 0.09 mag (van den Bergh 1969). As it is presumed that the halo globulars suffer no intrinsic reddening in M31, these data give foreground reddening information. When combined with the data discussed in the previous paragraphs, it can be concluded that the best value for this datum is $E(B-V) = 0.11 \pm 0.04$ mag.

Globular clusters closer to the center of M31 have been used to determine the reddening as a function of position and to address the nature of the extinction law. Vetesńik (1962) was one of the first to use globular cluster colors to characterize the extinction at various positions in the galaxy. More recently, Iye and Richter (1985) have used a carefully selected set of homogenous color data to more fully analyze the problem. They find that there is an asymmetry with respect to the minor axis in the reddenings for globulars in the inner part of M31, with the clusters seen in projection onto the southeastern half being less reddened than on the opposite side (Figure 12.10). The difference at 20 arcmin is approximately 0.17 magnitudes, which corresponds to a mean total extinction due to the effect of the disk (assuming a Galactic extinction law) of V = 0.56 mag.

Many early studies of the colors of globular clusters in M31 (*e.g.*, Kron and Mayall 1960 and van den Bergh 1969) showed that the ratio of total to selective extinction, R, might be smaller than that for the solar neighborhood. The best value for R is approximately 2.5 ± 0.1, rather than the common value of 3.2 found locally. This difference probably reflects a difference in the size distribution of the dust particles. As described below (Section 12.3), the polarization also indicates such a difference and suggests that the dust particles are approximately 20% smaller in the areas of M31 sampled.

Iye and Richter (1985) also examined the reddening law in the M31 disk. Figure 12.11 shows one of their three-color diagrams. They concluded that the ratio $E(U-B)/E(B-V) = 1.01 \pm 0.11$, a much larger value than the Galactic ratio of 0.72. A similar conclusion was reached by Sharov and Lyutyi (1989 and numerous previous papers), who found that the ratio is larger than in our Galaxy and that it depends on the cluster's spectral type. Their value for R is 2.7.

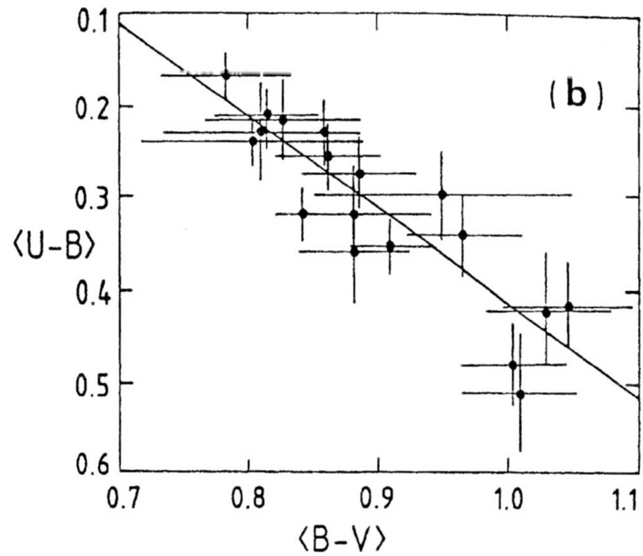

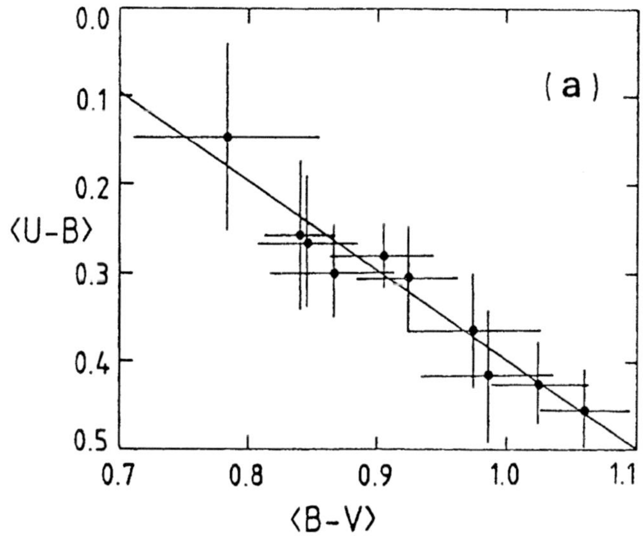

Figure 12.11 Two-color diagrams for the globular clusters of M31, using the samples of (a) van den Bergh (1969) and (b) Sharov and Lyutyi (1983). The fitted curves give a slope of $E_{U-B}/E_{B-V} = 1.02 \pm 0.12$ *(from Iye and Richter 1985).*

Iye and Richter (1985) also found a color gradient in the mean colors of globulars with distance from the center of M31, either indicating a systematic intrinsic color gradient or a gradient in the reddening ratio. If the latter, then they find that the anomalous ratio in the central areas gradually becomes approximately the same as the "normal" (Galactic) value near the edge of the disk.

From Open Clusters

Although there has been much less photometry done on open clusters than on globulars, some three-color data are available and they have been used to investigate (in a preliminary way) the reddening of these inner disk objects. The first attempts to use this method were made by Schmidt-Kaler (1967), who derived a mean value for the color excess in the disk of 0.12. Using open clusters is made complicated, of course, by their widely ranging intrinsic color differences, which are primarily age-dependent, but which also can be influenced by small number statistics for the smaller clusters. Figure 12.12 shows a three-color diagram for open clusters in M31, compared to colors in the LMC and the Galaxy. While there does seem to be reddening affecting the M31 clusters, it is difficult to be very quantitative from these preliminary data.

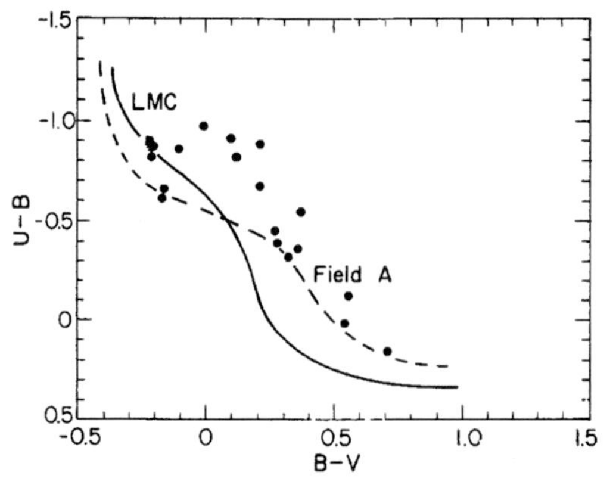

Figure 12.12 The two-color diagram for a star field in M31. The solid line is the position of MS stars and the dashed line that for supergiants. A sample reddening line is indicated (from Hodge and Lee 1988).

From Disk Colors

Multicolor surface photometry of the M31 disk has shown the results of reddening, especially in terms of an asymmetry between the two directions along the minor axis. As described in Chapter 4, the pioneering photometry of Lyngå (1959) and de Vaucouleurs (1958) clearly showed these effects. In B – V, for example, the disk along the NW minor axis is approximately 0.15 mag redder than the SE minor axis. As shown also in the comparison of the brightness of the two halves of the minor axes in Hodge and Kennicutt's (1982) photometry, there is approximately 0.70 mag (in B) more extinction in and in front of the NW axis than the SE. This can be interpreted in physical terms, depending on how the distribution of dust and stars is modeled (*e.g.*, see Lyngå's 1959 model based on his photographic photometry).

From Individual Main Sequence Stars

The large number of early-type stars in the disk of M31 makes it possible to map out the reddening in the disk in detail. The first use of this method, which, of course, is the standard method in our Galaxy, was made by van den Bergh (1964b), who derived a mean color excess of 0.06. A much larger value was obtained shortly afterwards by Racine (1967), who used a clever scheme for enhancing the reliability of photographic photometry in order to produce an accurate CMD for Baade's Field IV. He found that the reddening of early-type stars in this remote part of the M31 disk was E(B – V) = 0.15, and that the range in reddening must be as large as 0.20 mag.

An even larger range in reddening was reported by Ivanov and Golev (1984), who measured the UBV colors of stars in two large OB associations, A 78 and A 184. For the latter they found little reddening beyond the foreground value, while for A 78 the mean reddening was E(B – V) = 0.45 ± 0.15 mag. The more extensive studies of field stars in the disk of M31 have been based on CCD photometry with the KPNO 4-m telescope. The first (Massey *et al.* 1986) examined the CMDs for eight CCD fields in various places in M31, mostly in areas rich in OB associations, and derived reddening for them. A second, similar study was by Hodge and Lee (1988), who observed six fields, one of which largely overlapped one of Massey *et al.*'s fields, giving a rare opportunity to compare two independent studies of the same objects and the same questions. Agreement between the studies was quite good, except for the derived reddening, which was very different. For example, Massey *et al.* (1986) obtained a reddening of E(B – V) = 0.08 for their Field 4, while Hodge and Lee (1988) derived a value for their Field J of 0.24. The disagreement was traced, at least largely, to different methods of fitting the stellar data to the unreddened main sequence line; that is, the different approaches separated the observational spread from the reddening spread in the data differently, and they apparently defined "mean reddening" differently. Both studies did demonstrate that the reddening in a field varies greatly from place to place and star to star, which is not surprising, considering the highly variable dust density that is indicated by the dust clouds.

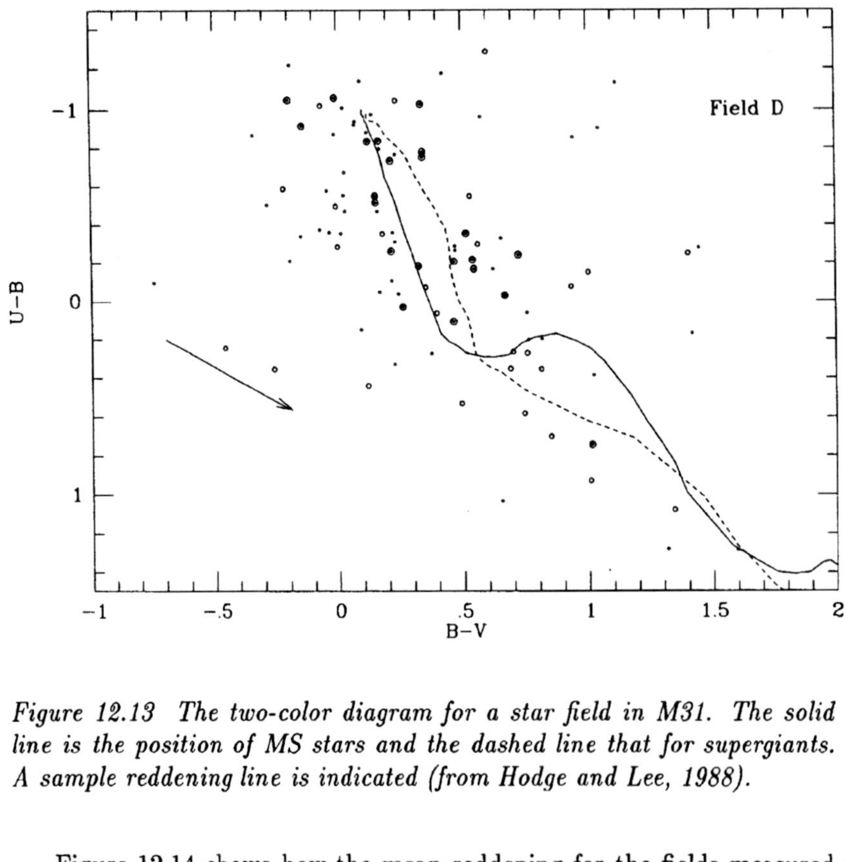

Figure 12.13 The two-color diagram for a star field in M31. The solid line is the position of MS stars and the dashed line that for supergiants. A sample reddening line is indicated (from Hodge and Lee, 1988).

Figure 12.14 shows how the mean reddening for the fields measured by Hodge and Lee (1988) varied with distance from the center of M31 projected onto the disk. There is a clear tendency for the reddening to increase to a maximum at about 10 kpc, where so much material and activity is concentrated, and then to decrease outwards. The maximum color excess (at 9 kpc) is E(B − V) = 0.45.

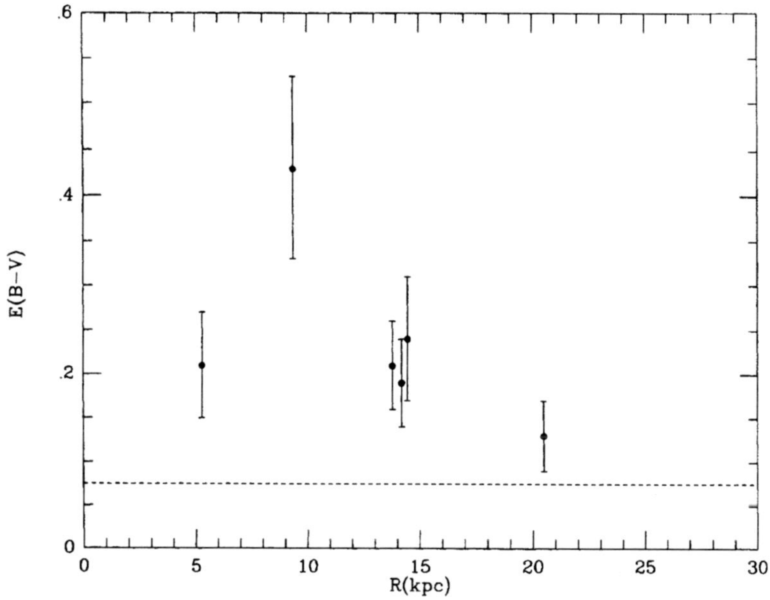

Figure 12.14 Color excess as a function of distance from the center of M31 (projected onto the plane) for six star fields (from Hodge and Lee, 1988).

From Cepheids

Baade and Swope's (1963) survey of Cepheids in M31 (Chapter 13) led to a determination of the mean reddening in Field IV, which they found to be E(B-V) = 0.16 mag. This low value agrees very well with Racine's almost identical value based on blue star colors. In more central parts of the disk, however, the reddening of Cepheids is greater. An example is found in van Genderen's (1973) discussion of the distribution of absorbing material detected from a study of Cepheids, using the relation between the amplitude of the light curve and the deviation from the mean period-luminosity relation. Rather than finding some vaguely defined "mean" reddening, van Genderen modeled the dust layer, explored its geometry, and derived optical half-thicknesses for the different Baade variable star fields. Positions ranged from 3 kpc for the inner field (Field I) to 20 kpc for the outer one (Field IV). The mean absorptions for the fields ranged from a high value of 0.98 (blue magnitudes) for Field II (at 35 arcmin from the nucleus) to a low value of 0.31 mag for Field IV (these values do not include the effects of foreground extinction). The latter can be compared to values for the reddening for Field IV, such as Racine's (1967) and Hodge *et al.*'s (1988), which, when corrected for foreground and converted to A_B using the Galactic relationship, give blue absorptions of 0.36 and 0.20, respectively.

From H II Regions

Intensities of the Balmer lines in H II regions can be compared to give a value for the reddening within these regions. Kumar (1979) used this method, basing his data on spectra of 22 H II regions at different distances from the center of M31, and concluded that the extinction, A_V, is highest in the region of 4–8 kpc, decreasing sharply outwards (see Figure 12.14).

Conclusions

The reddening in front of M31 is $E(B - V) = 0.10 \pm 0.03$. Within M31 itself, the additional reddening varies from approximately 0.10 in the inner disk (at $R = 5$ kpc) to a maximum value of approximately 0.45 at $R = 10$ kpc, decreasing outwards to a value of about 0.02 at the position of the outermost measurements available ($R = 21$ kpc). In individual regions, such as dust clouds and star-forming areas, much higher reddenings are detected.

12.4 Polarization

The first photoelectric study of the interstellar polarization of light from objects in M31 was that of Hiltner (1958), who examined 21 globular clusters, finding polarization ranging from 1.5% to 2.6% (in the blue) for six of them. Elvius and Hall (1964) determined the foreground Galactic polarization to be $0.8\% \pm 0.1\%$ with position angle $86° \pm 6°$. Martin and Shawl (1979) observed three reddened globular clusters and found polarizations of about 2% in the green ($\lambda = 5300$ Å) for two of them (G78 and G150) and a value of 6% in the yellow ($\lambda \sim 5800$ Å) for the third (Baade 327, which is very heavily reddened). Figure 12.15 shows G78, which lies at the edge of a large dust cloud.

The wavelength dependence of polarization, which is dependent on the physical characteristics of the polarizing dust particles, was examined for G78 by Martin and Shawl (1979) at five wavelengths, from 3700 Å to 8100 Å. The result was significantly different from the relation commonly found in our Galaxy; the value of λ_{max} was found to be 4300 Å $\pm$ 600 Å, implying that, if the dust particles are otherwise similar, they must be on the average about 20% smaller than in our Galaxy. Such a situation would be consistent with the observed difference found for the ratio R of total to selective extinction (Section 12.2).

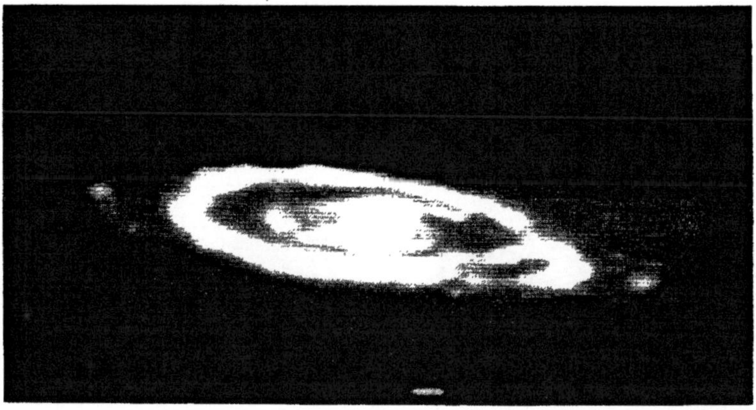

Figure 12.15 An IRAS image of the infrared radiation from M31 (courtesy C. Beichman).

12.5 Infrared Radiation

Particularly important information about the dust in M31 came when it was observed by the infrared astronomical satellite IRAS (Habing *et al.* 1984). A detailed study of the nuclear bulge was published by Soifer *et al.* (1986), who concluded that the short wavelength radiation (at 12 and 25 microns) comes mostly from stars and circumstellar dust, while the 60- and 100-micron emission is from interstellar dust. They estimated that the far infrared emission flux implies a total mass in the bulge of approximately 1500 solar masses of dust at a temperature of 45 K.

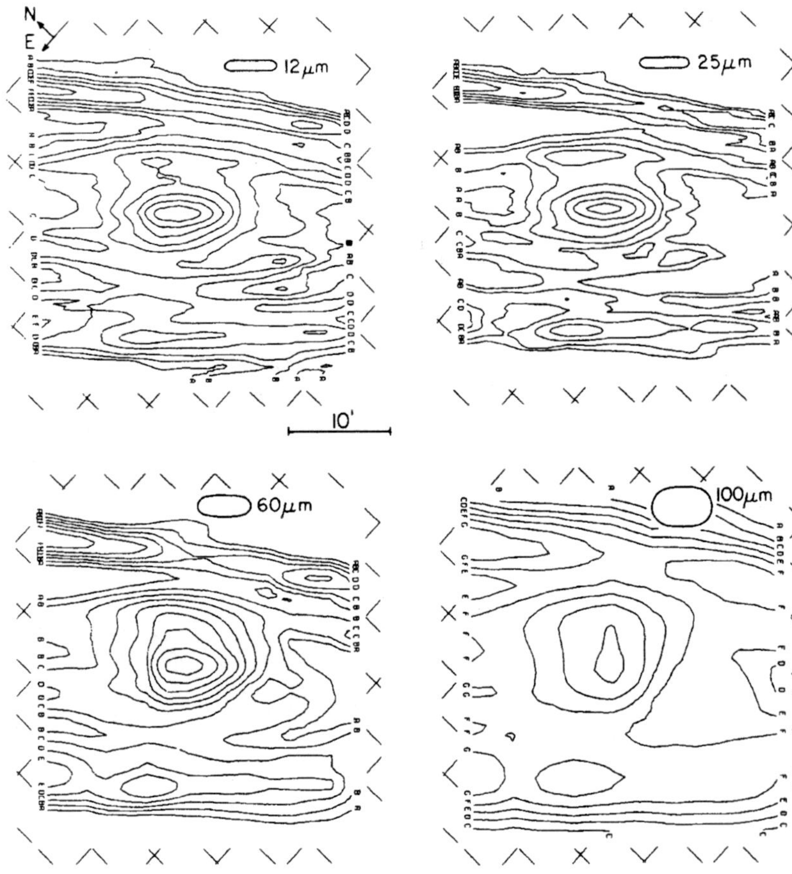

Figure 12.16 Contour plots of the infrared radiation from the central bulge of M31 from IRAS observations. Wavelengths and beam sizes are indicated (from Soifer et al. 1986).

In the disk, the situation is a little different. An exhaustive study of the IRAS data by Walterbos and Schwering (1987) concluded that the very strong disk radiation at 12 and 25 microns implies a component of hot (380 K) dust, as well as the more normal, widespread cool (20 K) dust. Near star-forming regions, a warmer (40 K) dust component is found, and this is attributed to warming of the general interstellar dust by the nearby hot stars. However, there is a smaller fraction of this warmer dust in the M31 disk than in our Galaxy's, which may be related to the fact that M31's star-formation rate is smaller than ours. A comparison of the spatial distributions of the cool and warm dust in M31 is shown in Figure 12.17.

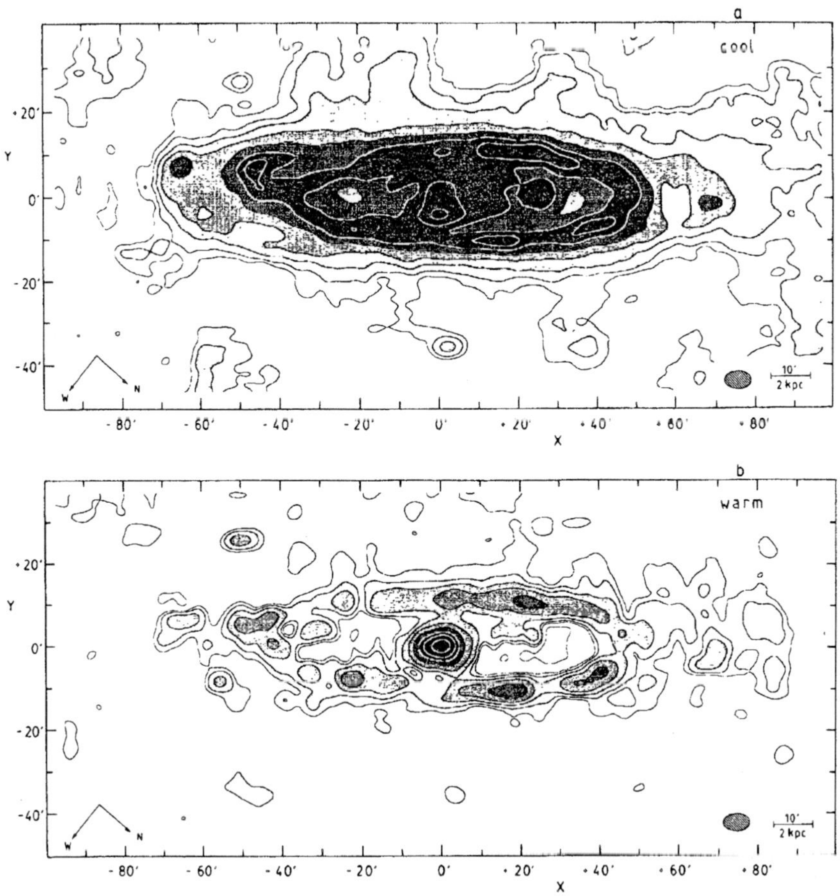

Figure 12.17 Cool and warm dust distributions in M31, based on IRAS observations (from Walterbos and Schwering, 1987).

When Walterbos and Schwering compared the dust distribution inferred from the IRAS data with the H I distribution in M31, they found a strong radial decrease outwards in the ratio of dust to atomic gas. There are complications introduced by the question of the influence of very small grains, but a qualitative conclusion seems clear that the dust density decreases more rapidly outwards than the atomic gas density; a similar conclusion was found for the dust clouds, as exemplified by Figure 12.6.

Table 12.2
Infrared Luminosities of M31 and
of the Galaxy
(from Walterbos and Schwering 1987; the M31 luminosities
are extrapolated from appropriate IRAS bands
and assumed emissivity laws.)

Component (heat source)	Galaxy luminosity $L_\odot$	M31 luminosity $L_\odot$	T_d K
total	1.5×10^{10}	2.6×10^9	
cool dust (ISRF)	5.7×10^9	1.4×10^9 ± 0.2	21
warm dust (O, B stars)	7.6×10^9	2.0×10^8 ± 0.5	40
hot dust (ISRF, M giants with dust shells)	2.0×10^9	8.0×10^8 ± 2.0	380

12.6 The Dust-to-Gas Ratio

The overall dust-to-gas ratio for M31, without regard to any radial variations, is similar to that for our Galaxy. Schmidt (1964) was one of the first to take advantage of the available H I maps to determine the dust-to-gas ratio, using low-resolution H I maps and photographic colors of globular clusters. More accurate data was used by van den Bergh (1975), who compared moderate-resolution H I maps and photoelectric colors for globulars. The result was a value of the parameter a, defined by

$$N_H = aE(B - V) \qquad (eq.\ 12.1)$$

of $a = 5 \times 10^{21}$ at cm^{-1} mag^{-1}. This is almost exactly equal to the Galactic value of 5.2×10^{21} at cm^{-1} mag^{-1} (Knapp *et al.* 1973). No radial dependence was found. In a comparison that used even better resolution H I data, Bajaja and Gergeley (1977) found a mean value of 3.3×10^{21} at cm^{-1} mag^{-1}, and noted a radial dependence that showed a peak at about 10 kpc and a decrease inwards and outwards from that position. Subsequent close looks at this parameter agree that there is a radial effect, and that the general value of the dust-to-gas ratio, measured in the spiral arms, is similar to that in our Galaxy.

12.7 Summary

M31 is rich in dust. More than 700 discrete dust clouds have been mapped and the photometric effects of light extinction by dust are well-determined. Aside from foreground reddening of ~0.1 mag., amounts of reddening within the disk of M31 have been found to range from near 0.0 at the edge of the disk to a maximum average value of 0.45 at R=10 kpc. Higher values are found in dust clouds and star forming areas. The reddening law is significantly different from that in the solar neighborhood.

IRAS data show that there is ~1500 solar masses of dust in the bulge. The infrared emission from the disk includes radiation from three components: hot (380 K) dust, cool (20 K) dust, and intermediate-temperature (40 K) dust, associated with star forming regions.

CHAPTER 13

VARIABLE STARS AND NOVAE

The variable stars of M31 have been the subject of a number of rather limited studies. The very brightest are fairly well understood, but only a few areas have been searched very deeply. There is still an immense amount of information about this component of the Andromeda Galaxy that remains untouched and unexplored.

13.1 Surveys

Following the systematic search for variables carried out by Hubble (1929) described in Chapter 2, most of the early work in this area was the result of Walter Baade's study using the Palomar 200-inch telescope during the years 1950 through 1956. He selected four areas at different positions along the SW major axis, in order to sample different populations in the galaxy (Figure 13.1). The innermost includes a portion of the bright inner bulge, while the outermost is beyond the easily visible optical disk. Because these areas are at present the only parts of M31 that have been thoroughly surveyed for variables, a summary of the properties of the variable star population of each is given in the following paragraphs.

Baade had taken between 80 and 100 200-inch plates of each field. He and Henriette Swope used between 12 and 32 pairs of these plates in the Mt. Wilson blink microscope to discover the variable stars, finding on the order of 100 per field.

The study of the variables was carried out, mostly by Swope, with a comparison microscope, using a "fly spanker", a portion of a 200-inch plate on which graduated star images were exposed. The magnitudes for the images were established for Field IV by a photoelectric sequence measured by Arp (Baade and Swope 1963), while the magnitudes for the other three fields were established by photographic transfer from Selected Area 68, using magnitudes determined photoelectrically by Baum and by Stebbins *et al.* (1950). Intercomparisons were made between the SA 68 magnitudes and the Field IV sequence, with the result that no systematic differences larger than the errors of measurement appear to be present; this claim was made also with respect to the results (the period-luminosity relations) of the photometry of Fields I and III, when compared to Field IV (Baade and Swope 1965). These studies were all made on blue sensitive plates, mostly taken without a filter, and thus correspond to pg magnitudes. However, Baade took a small number of yellow-sensitive plates that gave photovisual magnitudes, and they were used to estimate the approximate colors of the variables.

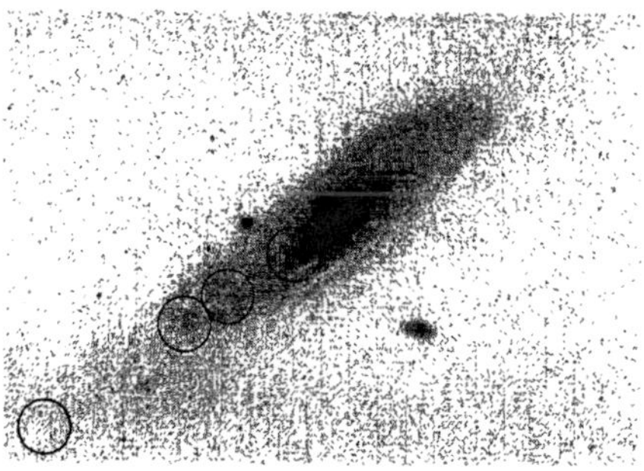

Figure 13.1 The locations of the Palomar surveys for variable stars in M31. Except for a few very small areas in the halo, the rest of M31 remains unsearched.

Field I

Only 15 arcmin from the nucleus, Field I is located in the area of overlap between the spiral arms and the central, amorphous bulge. The arms are primarily defined by conspicuous dust clouds. A total of 109 new variables and 7 novae were found. Of the variables, only 31 were found to be Cepheids. The rest were either irregular, eclipsing, or indeterminate (most of which are probably short period Cepheids too faint or too crowded for photometry). The spatial distribution of the variables is interesting; most are located near the periphery of the field, where the beginning of Population I is apparent (Figure 13.2). Only two Cepheids lie near the center, in the amorphous bulge area, V106 (period = 5 days) and V125 (period = 12 days).

Field II

Located in the thick of the inner arms, 36 arcmin (about 7 kpc) from the nucleus, this field includes 223 variable stars (Gaposchkin 1962). Of these, 120 were found to be Cepheids with well-established periods, a few more were thought to be probable Cepheids, and the remainder were red irregular and eclipsing variables. The periods of the Cepheids ranged from 2 to 42 days.

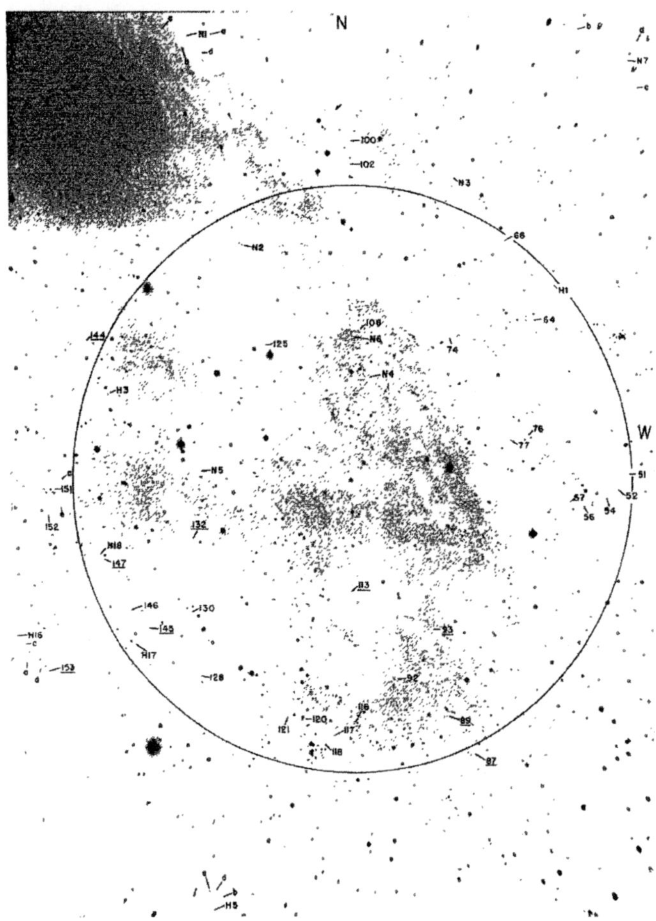

Figure 13.2 Identification of the variable stars in Field I (from Baade and Swope, 1965).

Field III

This field is in a richly populated section of the galaxy, just south of NGC 206. It is centered on Baade's arm 4, where it crosses the major axis. There are five van den Bergh OB associations (A75, A79, A80, A81, and A82) that are spread across the field and most of the variables are found in these giant star complexes. Baade and Swope (1965) found 334 variables, of which 232 could be recognized as Cepheids. They are found primarily in the densest parts of the spiral arm, clearly avoiding the areas between arms (Figure 13.3). The search also turned up 36 certain and probable eclipsing systems, 2 novae, 3 R Corona Borealis stars, and 2 long-period variables.

Field IV

The most distant field, this area was added to the program after the plates for the other three fields had been taken. It presented the opportunity to examine a remote area where the effects of dust and crowding were expected to be at a minimum. It is centered on the major axis crossing of Baade's arm 6, about 96 arcmin from the nucleus. Bright stars are sparse in the area, which includes OB association A185, plus a few young open clusters and faint H II regions. Twenty Cepheids were found in the field, and it is this group of relatively-unreddened objects that led to the best available (until recently) distance modulus for M31 (Baade and Swope 1963, who quote $(m - M)_0 = 24.20$). The remaining of the 54 well-studied variables are eclipsing binaries, irregular red variables, and long-period variables. Baade, in addition, found about 100 more variable stars by blinking photovisual plate pairs; these were too red to be studied on the blue plates and nothing further has been learned about them.

13.2 The Cepheids

Light Curves and Periods

The 403 Cepheids in M31 studied by Baade and Swope (1963, 1965) and Gaposchkin (1962) were found to be quite similar to local Galactic Cepheids in the shape of their light curves (Figure 13.4). Similarly, the range in periods was determined to be normal; the shortest periods were about 2 days and the longest over 100 days, with most variables having periods between 5 and 125 days.

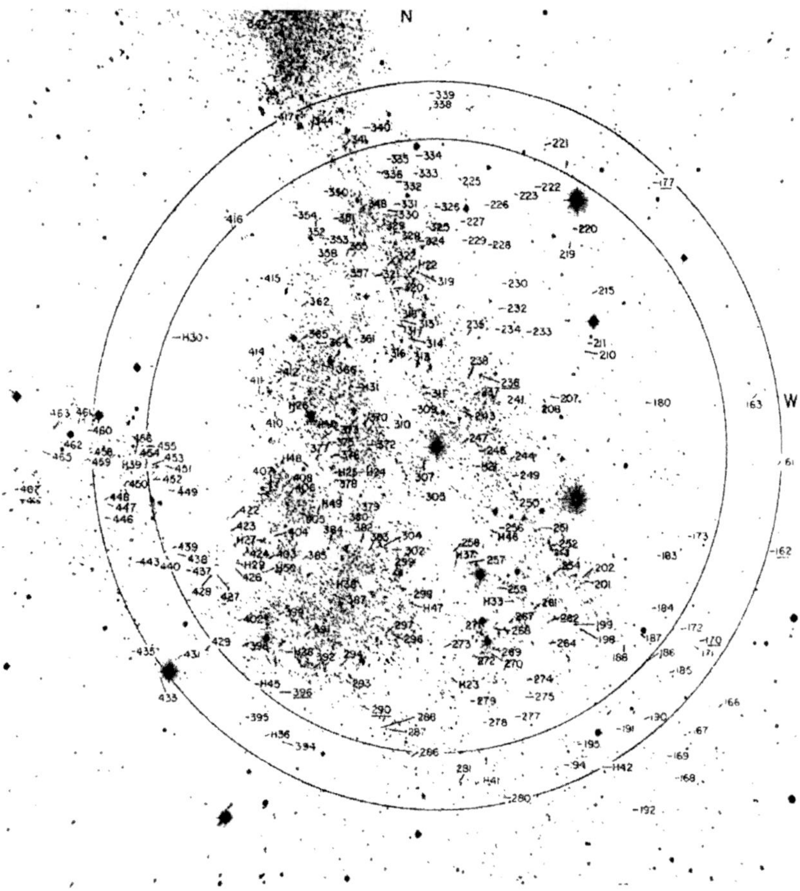

Figure 13.3 Variable stars in Field III (from Baade and Swope, 1965).

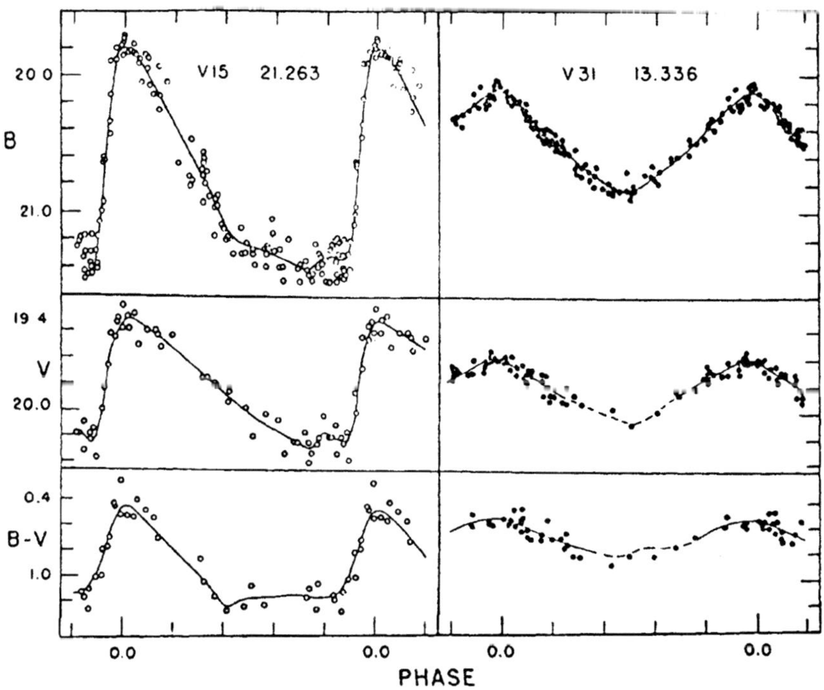

Figure 13.4 Light curves in two colors for two of the M31 Cepheids studied by Baade and Swope (1965).

The Period-Luminosity Relationship

Baade and Swope (1963, 1965) published period-luminosity diagrams for the Cepheids in the three Baade fields that they studied. The inner fields were found to have a large spread in the relation, which was interpreted to be the result of differential reddening (Figure 13.5). Nevertheless, the derived period-luminosity relationships were consistent with those found for the Galaxy and the Magellanic Clouds.

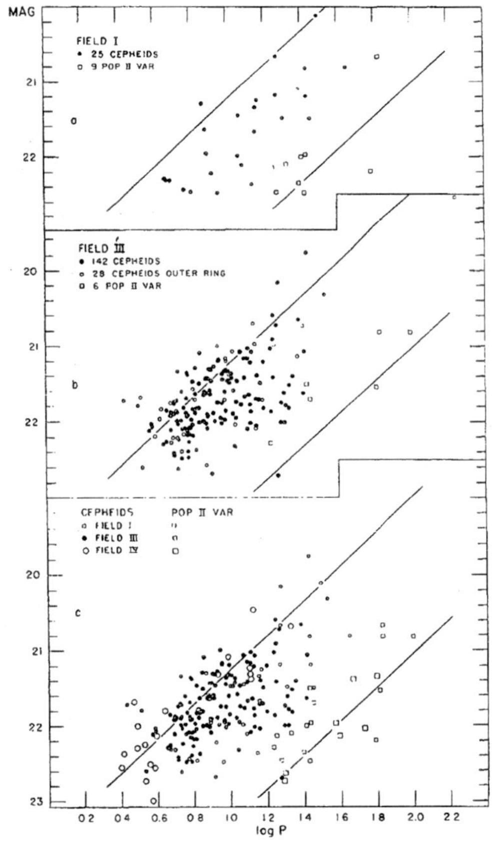

Figure 13.5 Period-luminosity diagrams for Cepheids in Palomar Fields I and III, together with a composite diagram for these fields and Field IV (from Baade and Swope, 1965).

Field IV, on the other hand, has a small amount of reddening, determined by Baade and Swope to be E(B − V) = 0.16, and subsequently found to be 0.14 by Humphreys (1979) and 0.13 by Hodge *et al.* (1988). Its period-luminosity diagram (Figure 13.6) shows a narrow relationship, with only small errors attributable to differential reddening, most of the width being caused by observational errors and the intrinsic dispersion of the relationship. A reasonably good fit was found for the blue and visual P-L diagrams when compared with those for Galactic Cepheids, and it is this comparison that led to the distance determination described above.

The appropriate equation for the blue P-L relation is

$$M_B = 23.51 - 2.25 \log P, \qquad (eq.13.1)$$

uncorrected for reddening.

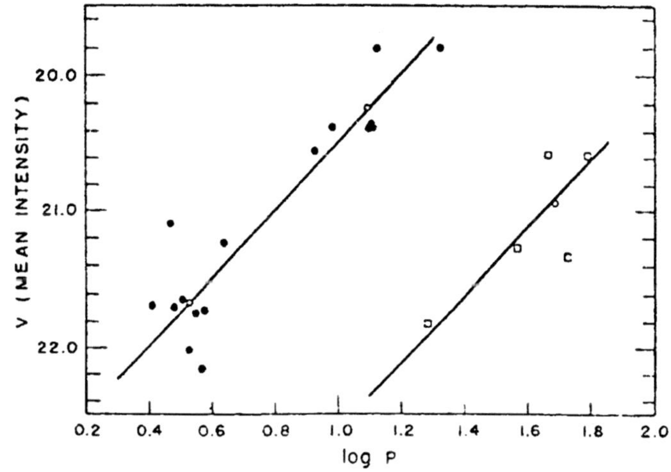

Figure 13.6 The period diagram for the Cepheids in Palomar Field IV, in which Population II Cepheids represent a significant fraction (from Baade and Swope, 1963).

The Distance from Cepheids

Baade and Swope's (1963) distance modulus from the Field IV Cepheid P-L relationship was

$$(m - M)_0 = 24.20 \pm 0.14 mag. \qquad (eq.13.2)$$

This value remained a good one, in agreement with several more recent studies. For example, Sandage and Tammann (1974) obtained a true modulus of 24.12, van den Bergh (1977) obtained a value of 24.10 mag, de Vaucouleurs (1978) reanalyzed the data from a number of criteria and obtained a value of 24.07, and Cohen (1985) used novae to obtain a value for the apparent modulus of 24.35. The value of the distance adopted in this book (p. ii), 725 kpc, corresponds to a distance modulus of 24.3.

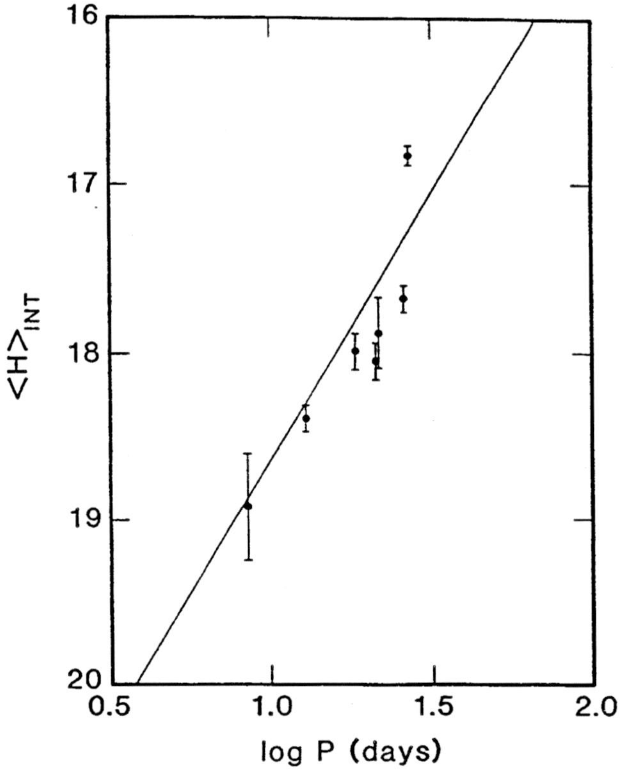

Figure 13.7 The near-infrared period-luminosity diagram for seven Cepheids in M31 (from Welch et al. 1986).

To avoid the problem of reddening even further, Welch *et al.* (1986) observed seven Cepheids in Fields III and IV using infrared (H) photometry. Rather than measure light curves, they found that they could obtain adequate precision from the intensity-weighted mean magnitudes of the randomly timed observations and found a true distance modulus of

$$(m - M)_0 = 24.26 \pm 0.08 mag., \qquad (eq.13.3)$$

in very good agreement with the 1963 Baade and Swope value. This determination corresponds to a linear distance of 710 ± 30 kpc.

The Period-Amplitude Relationship

The main reason for examining the period-amplitude relationship for M31 Cepheids is to look for any possible significant differences from the Galactic sample that might suggest physical differences, which, in turn, might argue against the reliability of using the Galactic Cepheids for calibration of the M31 P-L relationship. Baade and Swope (1965) found no evidence for any such difference, either between the data for M31 Cepheids and the Galactic ones, or between the data for the three different M31 fields tested (Figure 13.8).

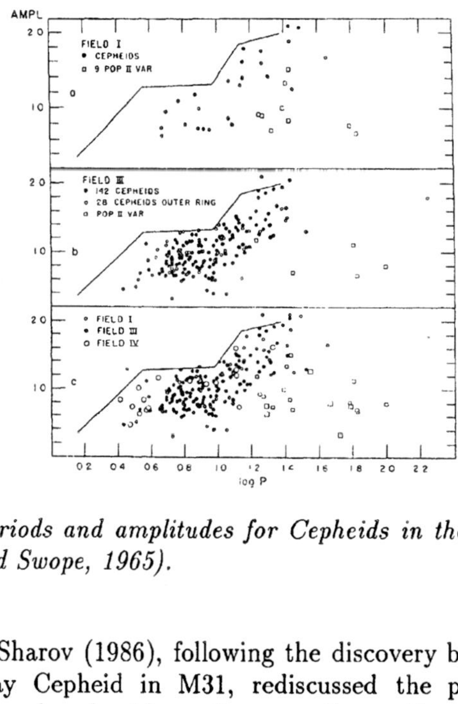

Figure 13.8 Periods and amplitudes for Cepheids in the Palomar fields (from Baade and Swope, 1965).

Ivanov and Sharov (1986), following the discovery by Ivanov (1985) of a new 75-day Cepheid in M31, rediscussed the period-amplitude relationship. They found evidence for generally small amplitudes for the longest-period Cepheids. Sharov and Kholopov (1979) and Welch *et al.* (1986) have noted the peculiar nature of the star V42 (sometimes called H42), discovered by Hubble, but apparently not a normal Cepheid, based on considerations of the variability of its light curve.

The Period-Frequency Diagram

Differences in the period-frequency diagrams for different galaxies, for instance, for the Magellanic Clouds, are thought primarily to result from different detailed histories of star formation rates. For M31, the data suffer from different efficiencies of detection (different galaxy background levels, different amounts of crowding, and different reddenings), but the results are probably fairly reliable for periods greater than about five days. Baade and Swope's (1965) result is shown in Figure 13.9, where it is compared to similar data for the Galaxy and the SMC. The Galactic and M31 relationships are very similar, but both disagree with the SMC diagram, which displays a proportionally much larger number of short-period Cepheids.

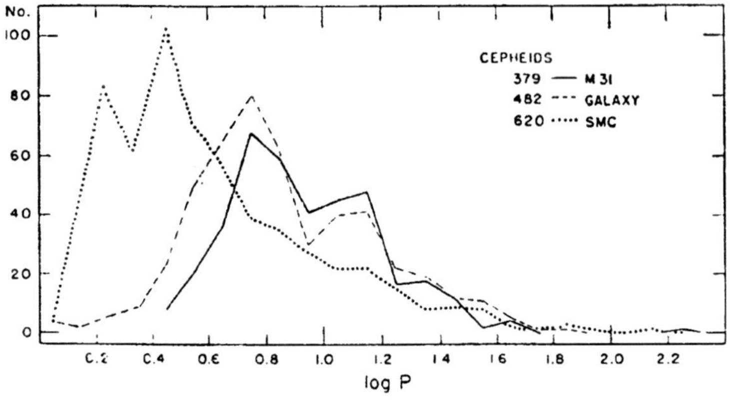

Figure 13.9 A comparison of the period frequency diagrams for M31, the Milky Way Galaxy and the SMC (from Baade and Swope, 1965).

13.3 The Hubble-Sandage Variables

As described in Chapter 2, Hubble discovered an extremely luminous variable star in M31 that was among the brightest stars in the galaxy. This kind of variable star has since become recognized in other galaxies (Hubble and Sandage 1953) and are generally called either "Hubble-Sandage variables" or "luminous blue variables". While at first their importance was because of their possible use as distance indicators, extensive recent work on them has shown that they span too large a range in mean absolute magnitude (from −8 to −11) to be very useful in that enterprise. Instead, they are interesting as examples of extremely luminous, somewhat unstable stars, occupying the very top of the stellar luminosity function.

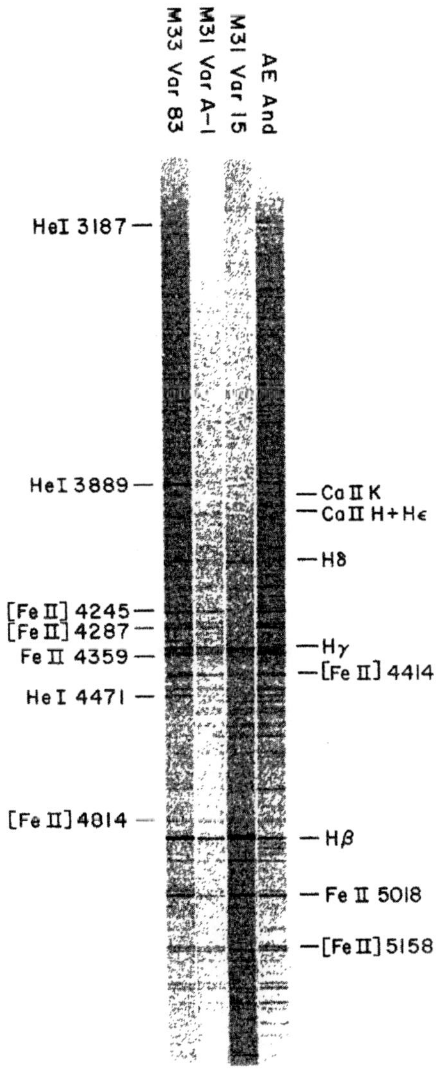

Figure 13.10 Spectra of Hubble-Sandage variables in M31 and M33 (courtesy R. Humphreys).

The Hubble-Sandage variables are characterized by a strong, hot continuum and emission lines, especially of H, He, Fe II and [Fe II] (Humphreys 1975, 1978, Humphreys and Warner 1978, Gallagher *et al.* 1981, Humphreys *et al.* 1984). Sometimes referred to as "S Doradus variables" after an example in the LMC, they may be related to the Galactic object η Carinae. The first to be identified in M31 was Hubble's V19, now known as AF And. Gallagher *et al.*'s model for AF And is a hot star with a surface temperature of $> 25,000$ K with a surrounding envelope of gas and dust that extends out to $\sim$ 1 AU, with electron densities of the order 10^{10} cm^{-3}. The total luminosities of these objects are of the order 10^6 $L_\odot$. The luminosities vary considerably. Sharov (1990) has shown that four HS variables in M31 (AE And, AF And, Hubble V 15 and VA 1) show a variety of long-term photometric characteristics. Blue amplitudes usually range from two to three magnitudes, but larger variations occasionally occur. Maximum magnitudes for AE And, AF And and VA 1 have approached values of 15 in the blue, corresponding to absolute magnitudes of nearly -10. There is a tendency for the light variations to be accompanied by corresponding color changes, and Sharov (1990) believes that the data indicate a long-term systematic decrease in brightness over the 80–100 years of observation. Figure 13.11 shows Sharov's (1990) compilation of light curves for these four M31 H-S variables.

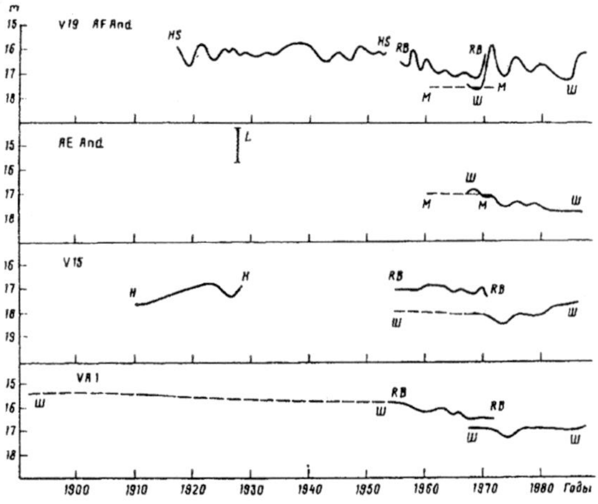

Figure 13.11 Light curves for four Hubble-Sandage variables in M31 (from Sharov, 1990).

13.4 Novae

M31's proximity and its large mass mean that it has many detectable novae per year. As related in Chapter 2, novae were the first clues to its extragalactic nature and they still are being discovered and used to further our knowledge, both of M31 and of the nature of novae in general.

Rate of Occurrence

The first novae in M31 were found by accident and did not allow a reliable estimate of the true rate of occurrence in the galaxy. The first attempt, and a highly successful one, to determine this datum was made by Arp (1956), who systematically surveyed the galaxy over a period of eighteen months, obtaining plates at Mt. Wilson on a large fraction of the clear nights. A total of 30 novae were detected and, allowing for completeness, he concluded that the annual rate of occurrence in M31 is 26 ± 4. Hubble (1929, see Chapter 2) had estimated a value of 30 per year.

Other systematic searches for novae include that of the Asiago Observatory (*e.g.*, Rosino 1973, Rosino *et al.* 1989). In the interval 1958 to 1986, a total of 119 novae were discovered at Asiago. Currently other searches are taking place, for example, at observatories in Latvia and Crimea (Sharov and Alksnis, 1991).

Taking all of these searches into account, positions, magnitudes and at least partial light curves currently exist for nearly 300 novae in M31.

Distribution of Novae

The spatial distribution, initially examined from limited data by Hubble (Chapter 2), is now well established. Hubble found that the novae were preferentially found in the central bulge of the galaxy and that the radial distribution mimicked that of the total galaxian light. This conclusion was confirmed by Arp's (1956) study, which further showed that the novae are distributed in three-dimensional space like an intermediate population, not in as flat a system as the Cepheids, but also not in as spherical a system as the globular clusters. Rosino's (1973) plots show a similar distribution.

All three of these searches had very few nova discoveries in the inner 2 arcmin of the nucleus of M31. Hubble expressed the opinion that this represented a real effect, rather than being the result of the difficulty of discovering them against a bright background. Arp (1956) seems to have agreed, though with less conviction. Against such a proposition is the more recent result of Ciardullo *et al.* (1983), who overcame the difficulty with the background by searching at the wavelength of Hα. The bright emission lines of novae at outburst made it possible for them to discover novae right up to the nucleus, and their statistics indicated that the distribution of novae continues to follow the light distribution at small radial distances.

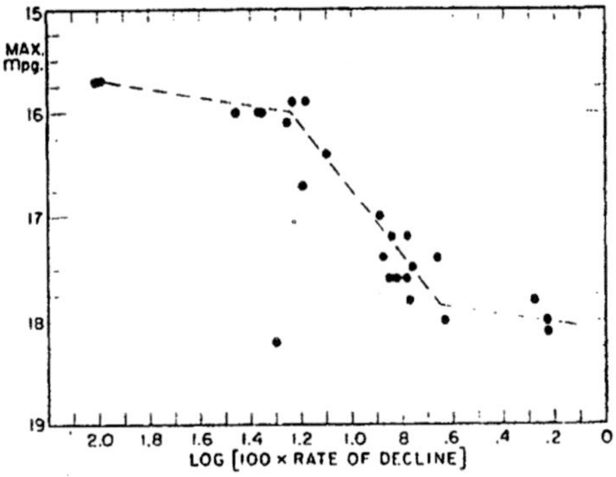

Figure 13.12 The relationship between the maximum magnitude and the rate of decline for novae in M31 (from Arp, 1956).

Light Curves

Light curves of Hubble's novae were discussed in Chapter 2. The main contribution to this subject made possible by Arp's and Rosino's systematic surveys was the establishment of the relationship between a nova's maximum brightness and its rate of decline (Arp 1956, see Figure 13.12). Very bright novae, with blue magnitudes of $\sim$ 15.5, faded very rapidly, having dropped below magnitude 20 within about 10 days. Those with maximum blue magnitudes of only $\sim$ 18, on the other hand, remained above this limit for as long as 150 days. The significance of this relationship is twofold; it provides insight into the nature of the physical phenomenon (that is, into the time scale for the energy dissipation at the surface of the presumed white dwarf), and it provides important information for the use of novae as extragalactic distance indicators, if the relationship found for M31 is a universal one (Cohen 1985).

The emission-line light curves, on the other hand, do not show a similar relationship. Ciardullo *et al.* (1990) have analyzed Hα light curves of 11 M31 novae and conclude that the emission line maximum flux occurs many days after the blue light flux maximum, and that the Hα flux becomes a major contributor to the object's total flux after the time of its maximum (Figure 13.13).

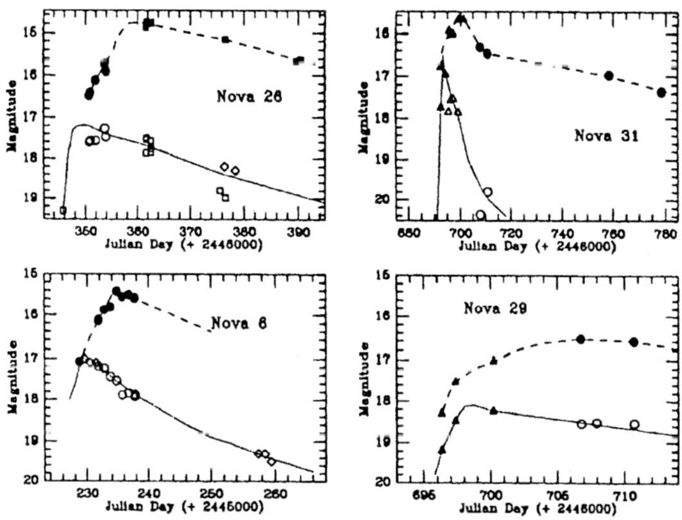

Figure 13.13 Light curves for four novae in M31. The open circles are for blue light and the filled circles are Hα emission line fluxes (from Ciardullo et al. 1990).

Spectra

Many of the M31 novae are bright enough to allow spectroscopic study. For example, Ciardullo *et al.* (1983) obtained spectra of four of the M31 novae that they discovered near the nucleus and found typical nova post-maximum spectral features: strong Hα and Hβ that is 15–20% of the Hα flux. Other, weaker emission lines were of Fe II. The Hα line showed structure, with the two components separated by ~ 1000 km/sec, with the blue component the brighter.

A particularly well-studied nova is 1986 No. 32, for which a spectrum was obtained by Cowley and Starrfield (1987, see Figure 13.14). It resembles a normal Galactic nova spectrum as it appears after maximum, when it is approximately 1 or 2 magnitudes fainter than maximum brightness. No continuum was detected. There were very strong hydrogen emission lines and weaker lines of Fe II, Ti II, Ca II, Na I, and V I, all in emission. The Balmer decrement was higher than found for Ciardullo *et al.*'s (1983) novae, possibly because of greater reddening. The linewidth of the hydrogen lines was much smaller than for the other novae as well: 665 km/sec, compared to ~ 2000 km/sec. This is possibly related to the fact that 1986 No. 32 was a relatively slow nova, which have lower velocities of ejection than fast novae (Payne-Gaposchkin 1957).

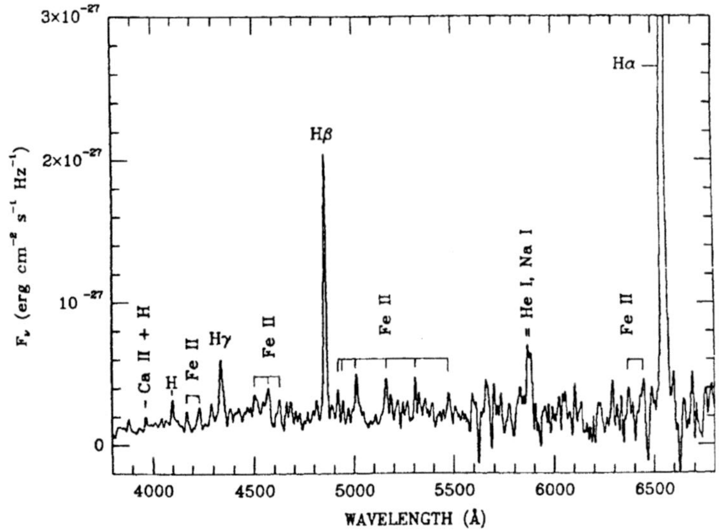

Figure 13.14 The spectrum of an M31 nova (1986 No. 32). Strong emission lines are identified (from Cowley and Starrfield 1987).

13.5 RR Lyrae Variables

The hunt for the RR Lyrae variables of M31 has a long and interesting history. Baade (1963) attempted to find them in 1950 with some of the first plates to be obtained with the Palomar Observatory 200-inch telescope. They should have been visible in large numbers, but he found none. This was one of the facts that led him to conclude that the distance scale then in use was grossly in error. Not until the combination of a superior site (Mauna Kea) and superior detector (a CCD), were these elusive objects finally found. In 1985 Pritchet and van den Bergh (1987) used the Canada-France-Hawaii 3.6m telescope to survey a small area in the M31 halo for RR Lyrae variables and found them. The luminosities were less than 1/10th the brightness that Baade had expected back in 1950.

Pritchet and van den Bergh (1987) found 28 RR Lyraes for which light curves and periods could be obtained. All were of Oosterhoff Type I, with a mean period of 0.548 days. Their mean apparent magnitude was $< B >= 25.68 \pm 0.06$, which leads to a distance of 740 kpc, using standard

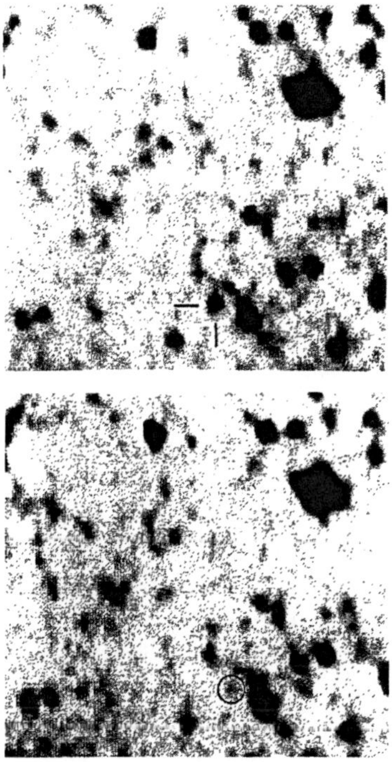

Figure 13.15 An RR Lyrae variable in M31, shown near maximum and minimum light on two CCD frames obtained with the CFHT (from Pritchet and van den Bergh, 1987).

values for the reddening and the absolute magnitudes of RR Lyraes. This value is in good agreement with the value from Cepheids (and novae). The intrinsic number of RR Lyrae variables was found to be surprisingly large, with the specific frequency being very like that for variable-star-rich Galactic globulars. Pritchet and van den Bergh (1987) emphasized that this fact, together with the possibility that the globulars of M31 have unusually strong hydrogen lines, indicates that M31's halo population may be different from the Galaxy's in having very strong horizontal branches. The field of their variable No. 1 is shown in Figure 13.15.

13.6 Other Types of Variables

The various general surveys of variable stars in M31, especially those of Baade and Swope (1963, 1965), turned up several other types of variable stars, including Population II Cepheids, long-period variables, R CrB stars, irregular red variables and eclipsing variables. Most of these are similar to typical objects of their class in our Galaxy.

Population II Cepheids

In Baade's various fields in M31 Swope found 22 Population II Cepheids, all of which fit a period-luminosity diagram approximately two magnitudes fainter than that of the classical Cepheids (Figure 13.5). Nine were in Field I, the innermost one, six were in Field III and seven were in Field IV. The ratio of Pop I to Pop II Cepheids is dramatically different for the three fields (Figure 13.16). The difference can be understood in terms of the different radial distributions of the two populations; the Population I objects (the classical Cepheids) are concentrated to intermediate distances along the disk, while the Population II objects are in a halo distribution that shows only a moderate radial decrease in density outwards over this interval.

Long-Period Variables

The long-period variables found by Swope were detected only near maximum and therefore we have rather little information about them. Most had maximum blue magnitudes near 22.

R CrB Variables

The R CrB variables (three were found in Field III) showed photometric behavior similar to typical stars of that type in our Galaxy. They were normally at blue magnitudes of ~ 20 until their rapid drops in luminosity, which took them below the plate limits at ~ 23.

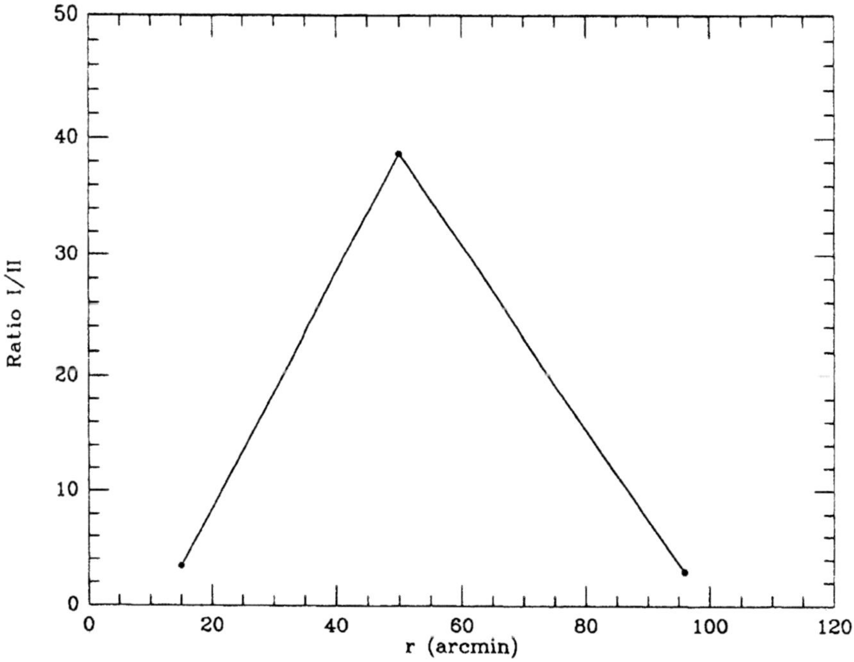

Figure 13.16 The ratio of Population I to Population II Cepheids in three fields in M31 at different radial distances, calculated from Baade and Swope's (1965) survey.

Irregular Variables

The irregular variables were both numerous and varied. Swope identified 67 such stars, some of which were red and some blue. Many showed cyclic, quasi-periodic behavior, much like those in our Galaxy. These objects showed a strong concentration toward the inner part of M31.

Eclipsing Variables

Eclipsing variables have been found by Baade and Swope (1963, 1965), as well as by others. Figure 13.17 shows typical light curves, these being from

Baade's Field III. Almost all the discovered eclipsing systems are either Algol types or Beta Lyrae types. Periods range from 511 days down to 2 days.

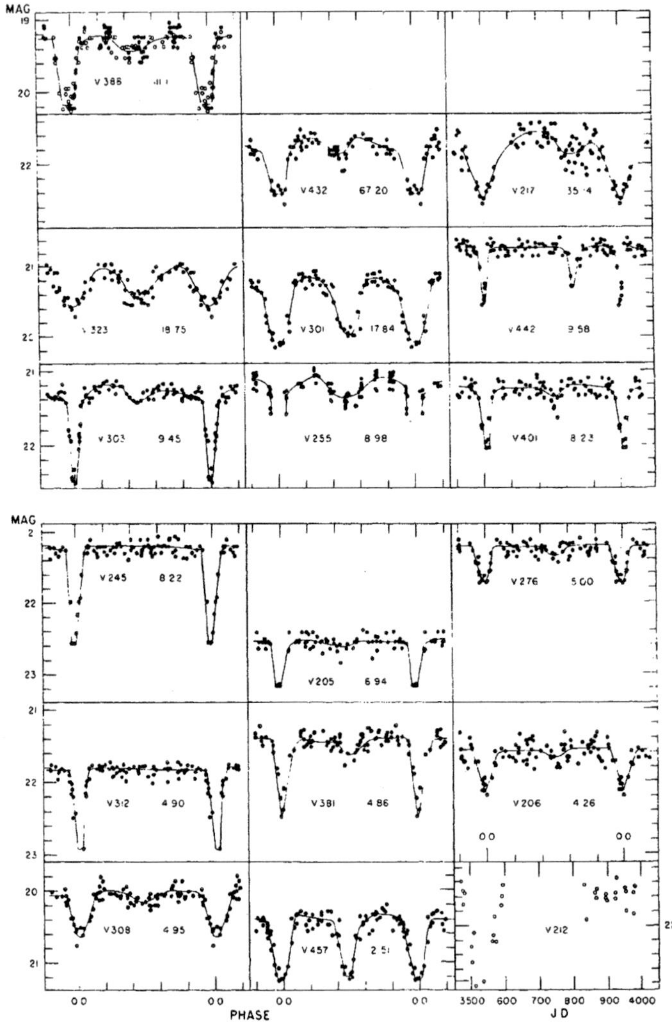

Figure 13.17 Light curves for several eclipsing variables in M31 (from Baade and Swope, 1965).

Peculiar Variable

Finally, a very peculiar variable will be described. It was discovered by Rich *et al.* (1989), who were studying the M supergiants of M31. The star, which they designated as RV (for red variable), lies 3 arcmin from the nucleus of the galaxy and when originally discovered had a J (green) magnitude of 12.8. Its computed bolometric magnitude at that time was −10, making it one of the most luminous stars in M31. However, its luminosity faded rapidly after the discovery date, falling by a factor of 13 in 70 days and by a factor of 80 in 290 days. Subsequently, it dropped entirely from view. Its spectral type at the time of discovery was M0 Ie, and the type became later as the star faded. The Doppler shifts of the lines showed a blueshift of $\sim$ 150 km/sec with respect to M31. Mould *et al.* (1990) suggest that M31 RV is an unusual type of nova, caught by chance near maximum luminosity. Its properties are inconsistent with the idea that it is a normal white dwarf nova outburst; its remarkable properties need further exploration. It is hoped that astronomers who might have images of the area obtained before the discovery may be able to more narrowly define the properties of the rise in light, about which nothing at all is known at this time.

13.7 Summary

The variable stars of M31 are still a relatively untouched resource. Only a few fields have been searched thoroughly.

The variables discovered so far are mostly quite similar to variable stars in our Galaxy. Their numbers and spatial distribution are informative about the stellar populations in M31.

The Cepheids have a P-L relation that indicates a distance to M31 of 710 ± 30 kpc.

There are many other types of variables in M31, including H-S variables, novae, RR Lyraes, Population II Cepheids, long-period variables, R CrB stars, irregular variables and eclipsing systems.

CHAPTER 14

H II REGIONS AND SUPERNOVA REMNANTS

There has been relatively little research done so far on the H II regions of M31, beyond the task of identifying them and cataloging the brightest several hundred. At the time of this writing, however, at least two groups of scientists are working on detailed studies of the H II regions and so this chapter is likely to be outdated to some extent rather soon. For supernova remnants, on the other hand, more has been completed, especially at radio wavelengths.

14.1 H II Region Identifications and Catalogs

Rather surprisingly, Hubble (1929) found no emission nebulae in M31, in spite of his rather complete census of most of the galaxy. He remarked that the lack of emission nebulae might be associated with the relative lack of resolution into stars of M31 and speculated about the relationship of such facts with the meaning of what became the "Hubble sequence" of galaxy types. The H II regions of M31 are inconspicuous and Hubble's failure to find them is evidence that they differ from H II regions in other nearby galaxies, such as M33 and the Magellanic Clouds. This fact has been confirmed quantitatively by Kennicutt *et al.* (1989).

The first H II regions in M31 to be detected were those used by Babcock (1939) to derive the first rotation curve for the galaxy. Later, Baade (1963) examined the entire image to locate H II regions and study their spatial relationship to other population components. The positions of 688 emission nebulae were published by Baade and Arp (1964). A comparison of Figure 14.1, which is a Schmidt photograph in Hα with Figure 14.2, which is Baade and Arp's map, shows that their listing included essentially all of the easily recognizable emission regions in M31.

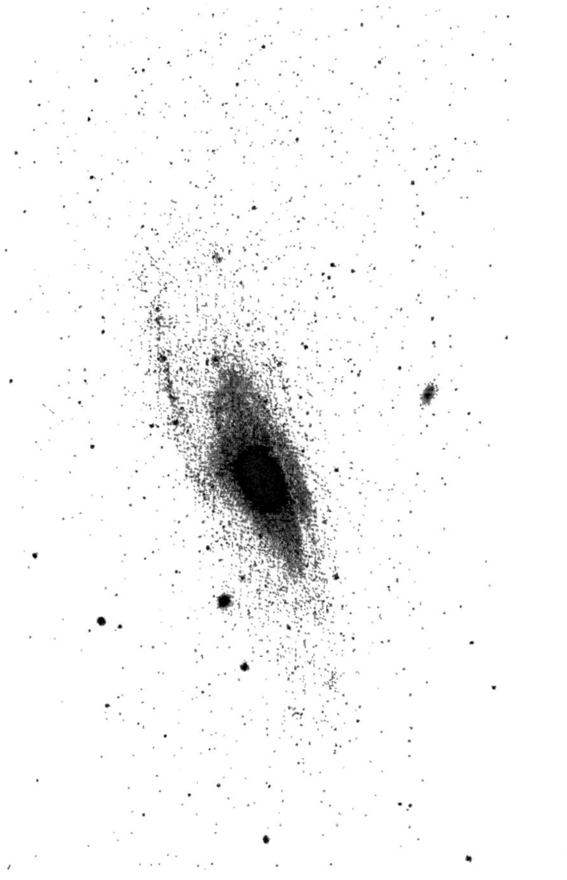

Figure 14.1 An Hα photograph of M31 taken with the Burrell Schmidt on Kitt Peak. The bright HII regions in the northeast and northwest are especially conspicuous.

The most complete survey compiled so far is that of Pellet *et al.* (1978). Their search used a narrowband interference filter with a focal-reducer on the 1.9m telescope of the Haute Provence Observatory and resulted in a catalog of 981 emission regions. Many of them are individual peaks that are involved in complexes of emission of much larger size (see Arp and

Brueckel (1973) for a discussion of this problem). As this catalog is widely used for identifications, owing largely to the fact that Pellet *et al.* (1978) included a map of the H II regions, many detailed investigations refer to the Pellet numbers, rather than the Baade and Arp numbers, of objects studied. Figure 14.3 reproduces one of their photographs; comparison with the previous two figures shows that their use of narrow filters and a short focal length allowed them to reach fainter surface brightness H II regions than in previous surveys. Their study also included the properties of many of the emission regions, as described below.

A more recent survey was carried out by Kennicutt *et al.* (1989), who used the Kitt Peak Observatory's Burrell Schmidt to measure emission line luminosities of the brighter H II regions and to determine the Hα luminosity function (Section 14.2).

In a search for planetary nebulae in M32, Ford and Jenner (1975) reported the discovery of seven faint diffuse nebulae. Subsequently-obtained spectra of two of them showed that they possessed the wrong velocities to be members of M32, but were rather faint H II regions in M31, in an outer arm behind M32.

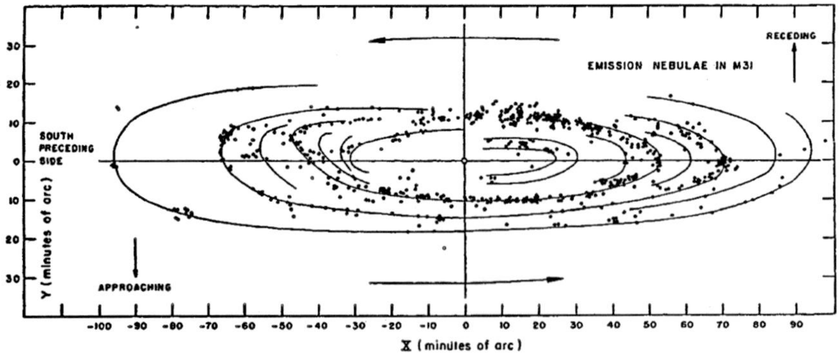

Figure 14.2 A map of Baade and Arp's HII regions in M31, together with Arp's interpretation of the spiral arm structure (from Baade and Arp, 1964).

Figure 14.3 A portion of the Pellet et al. (1978) survey of HII regions in M31, showing the northern half of the main part of the disk (from Pellet et al. 1978).

The most distant H II regions from M31's nucleus in Pellet *et al.*'s (1978) survey are at a distance of 15 kpc. Baade and Arp, however, found more distant objects. Lee (1988) has studied three very faint H II regions in Baade's remote Field IV, at a distance of 20 kpc from the center.

Two very recent studies based on CCD large-scale imaging have been completed by Walterbos and Braun (1991) and by Hodge and Lee (1992). An example of a field studied in the latter reference is shown in Figure 14.4.

The spatial distribution of M31's H II regions was discussed in Chapters 4 and 5. A comparison of the radial distribution with those of other galaxies has been published by Hodge and Kennicutt (1983), in which it is shown that the M31 distribution has a deep minimum in density in the central area, a maximum density about 0.3 radii out, and an exponential fall-off outwards from that. This is typical for type Sb galaxies, but is quite different from the case, for example, of Sc galaxies, which tend not to have a central deep minimum.

Figure 14.4 Pellet 550 and surrounding regions of M31, taken in Hα light with the Kitt Peak 0.9-m telescope (from Hodge and Lee, 1992).

14.2 Luminosities and the Luminosity Function

Luminosities for 356 different H II regions in M31 were measured by Kennicutt *et al.* (1989). The brightest single H II region had an Hα luminosity of 4×10^{38} erg/sec, while the faintest measured had a luminosity of just under 10^{37} erg/sec. Fainter H II regions do exist in M31, as Lee (1988) has shown; his faintest, HKL 3, had a luminosity of only 2×10^{36} erg/sec. No doubt many fainter emission nebulae exist, but are below current detection limits.

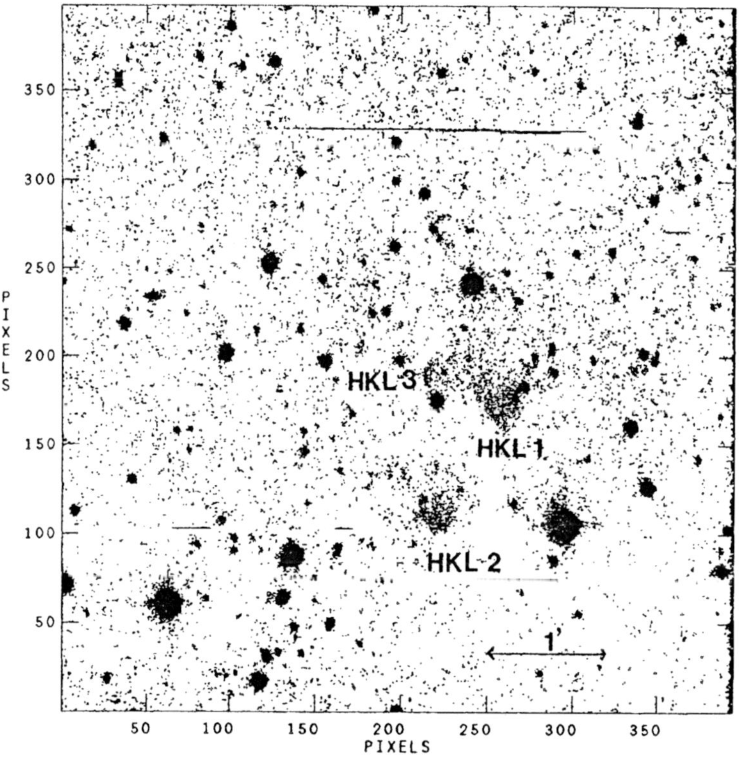

Figure 14.5 Three very faint HII regions in the outermost parts of the optical disk of M31 (from Lee, 1988).

The luminosity of M31's most luminous H II region is surprisingly modest. Even the semi-dwarf SMC has a more luminous one, and M33 and the LMC both have H II regions that are far more luminous, even though both galaxies are intrinsically much fainter than M31; 30 Doradus in the LMC has a luminosity of 6×10^{39} ergs/sec, while that for NGC 604 in M33 is 2×10^{39}. The lack of highly-luminous H II regions explains why Hubble failed to detect them and it illustrates one of the main differences in H II region populations in galaxies of different Hubble types. Most Sb galaxies, like M31, have no giant H II regions. Reasons are not yet known, but it may be related to the kinematics of the gas in the disk (Waller and Hodge 1990).

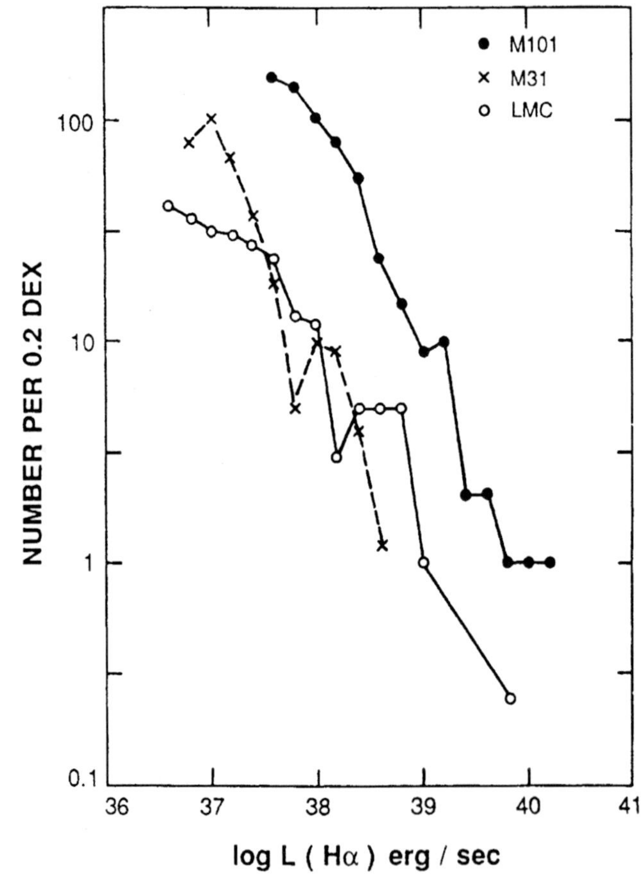

Figure 14.6 The Hα luminosity function of the HII regions of M31, compared to that for the giant Sc spiral M101 and the Irr galaxy, the LMC (from Kennicutt et al. 1989).

The point is clearly made by the full luminosity function of H II regions in M31 (Kennicutt *et al.* 1989), which has a sharp cut-off at the high-luminosity end (Figure 14.6). The figure compares the luminosity function with those of the LMC (a fainter Irr galaxy) and M101 (a giant Sc galaxy), showing that M31's H II regions not only define a lower high-luminosity cut-off than both, but also seeming to show a steeper slope than do the LMC H II regions. As Kennicutt *et al.* (1989) demonstrate, these two features are general properties of the H II region populations in an Sb Hubble type; M31 is completely normal for its class.

14.3 H II Region Sizes and Morphologies

Pellet *et al.* (1978) measured the sizes of most of their H II region identifications. Although the data are not measured to a well-established isophote level, the use of a fairly uniform set of plates suggests that their data probably reach a nearly uniform level and are therefore useful for at least preliminary quantitative comparisons. In particular, it is of interest to establish whether the sizes (and morphologies) of the H II regions of M31 differ in a significant and physically understandable way from those for other types of galaxies. Little discussion of these data has yet been published, but Hodge and Lee (1992) have looked at the problem. A brief discussion follows.

Figure 14.7 shows a plot of the size distribution found for Pellet *et al.*'s H II regions in M31. As found for other galaxies (van den Bergh 1981), the distribution has a nearly exponential shape, though, in this case, the fit is worse than for most galaxies. A visual fit gives a size parameter of 84 ± 20 pc. It has been shown (Hodge 1987) that the slope of this relation depends on both the Hubble type (for Irr and Sc galaxies, the only types tested) and absolute magnitude of the galaxy. From his data, which did not include M31, it is seen that the size parameter for an Sc galaxy with the same luminosity as M31 should be in the range 65–130 pc. Thus, the size scale of M31's H II regions is perfectly normal for a galaxy of its luminosity, and it provides no significant evidence that sizes in Sb galaxies are systematically smaller than those in Sc galaxies, though M31 and other Sb galaxies seem to have few large H II regions. There may be a cut-off at large sizes, as there is for high luminosities.

Pellet *et al.* (1978) commented to a limited extent about the morphology of M31's H II regions, especially with respect to the "bubble-like" regions. Their concern was primarily with the possible use of these objects for distance determinations. They compared the 28 bubble-like regions with those of M33 and found that the latter are on the average 50% larger than those of M31.

Hodge and Lee (1992) have made a preliminary survey of H II region morphologies, comparing the distribution of different types of H II regions with those found in other nearby galaxies, finding that M31 has a completely normal distribution of types, virtually identical to that of nearby irregular galaxies. They classified H II region morphologies into the following types, for purposes of statistical comparisons:

1 = composite
2 = bright compact
3 = faint compact
4 = loops, rings, bubble-like
5 = diffuse emission

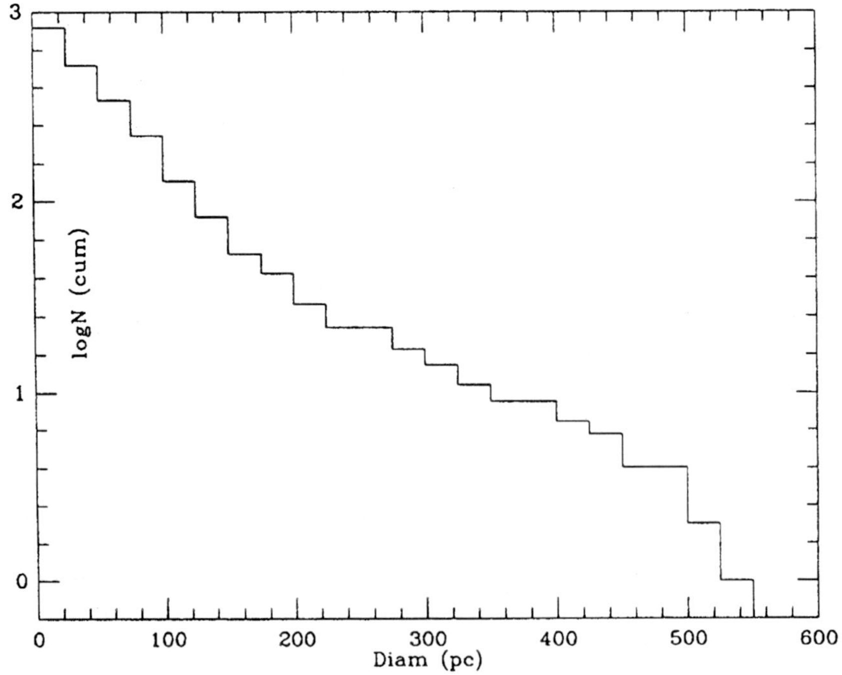

Figure 14.7 The size distribution for the HII regions measured by Pellet et al. (1978).

Figure 14.8 compares the percentage distributions of H II regions with these different morphologies with those for the nearby galaxy NGC 6822, showing the remarkable similarity. It had been thought that there would be large differences in morphology because of the very different host galaxy structure, gas density and kinematics. Apparently, the mechanism that determines the numbers of different kinds of H II regions is the same regardless of Hubble class, however. Figure 14.9 shows an image near a section of M31's minor axis, on which are seen several different examples of types of H II region morphology.

M31 is a near enough galaxy that current technology allows detailed looks at the H II regions at many wavelengths. High-resolution optical and radio H I and continuum imaging provide information useful for investigating the physical properties of the gas within the gas cloud. Very little has been published on this problem, though progress is anticipated soon (*e.g.*, Braun and Walterbos 1991). In areas so far investigated, there is a strong anticorrelation between the density of excited gas and that of the neutral gas and a tendency for the peak H I brightness temperatures to be located where there is little excited hydrogen.

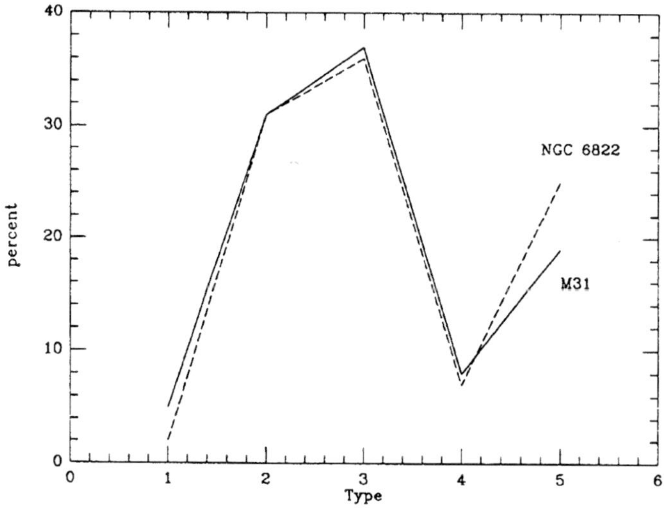

Figure 14.8 A comparison of the distribution of morphologies for HII regions in M31 and in the LMC. See text for definitions (from Hodge and Lee, 1992).

Brinks *et al.* (1990) have investigated the H II region Pellet 722, using a scanning Fabray-Perot interferometer for optical data and the VLA for radio. They found that this "bubble-like" emission region is a large (330 pc) spherical cavity, apparently blown out by stellar winds from its embedded stellar association, which has an age of 20 million years. Outside of the cavity is a gas shell that contains very young objects (OB stars, WR stars, optical H II regions), possibly formed as a result of the density enhancement caused by the expanding gas.

14.4 Spectra of H II Regions: Abundances and Velocities

Until recently, most spectra of H II regions in M31 were taken primarily for the purpose of measuring its rotation curve. For instance, the various papers reviewed in Chapter 7 included several that dealt with optical spectra, but few that discussed more than the velocities of the emission regions. More recently, detailed velocity mapping has been possible, using such instruments as Fabray-Perot interferometers (*e.g.*, Deharveng and Pellet 1969).

Figure 14.9 A portion of M31 (including stellar associations A29 and A30) that contains HII regions with different morphologies (from Hodge and Lee, 1992).

The first comprehensive look at the differences in spectra for H II regions at different localities in M31 was that of Rubin, Kumar and Ford (1972), who measured line intensity ratios for 53 H II regions at distances from the nucleus ranging from 3 to 24 kpc. They found that, compared to the nuclear values, the [N II]/Hα ratio is small in the disk, with a trend that decreases outwards at least to R = 15 kpc. While this ratio is 2 in the nucleus, it decreases to a value of about 0.2 at 15 kpc. The [O III]/Hβ ratio, on the other hand, increases from the inner disk outwards. They interpreted these data as indications for an abundance trend with distance, in which the ratios N/H and O/H are decreasing outwards by factors of 2 and 1.2, respectively. Beyond R = 15 kpc, there is no clear trend.

Dennefeld and Kunth (1981) measured abundances in eight H II regions, confirming the abundance variation trend with distance. Figure 14.10 shows the derived oxygen abundance variation (relative to hydrogen) for distances in the disk ranging from 3 to 20 kpc.

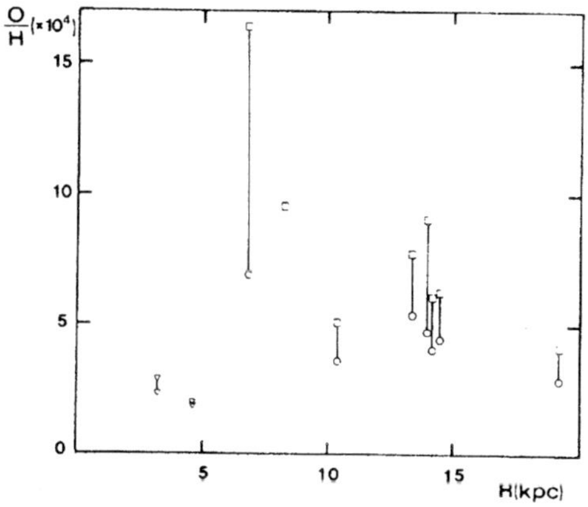

Figure 14.10 The oxygen abundance (relative to hydrogen) of HII regions at different distances from the center of M31 (from Dennefeld and Kunth, 1981).

14.5 Hα in the Bulge

Although not strictly H II regions, the emission regions in the central bulge of M31 will be discussed here. Emission lines are readily detected in the bulge. Rubin and Ford (1971) used them to establish the rotation curve in the central regions and Rubin *et al.* (1972) showed that the abundances across the nuclear disk (for the detected elements, H, N, O, and S) were quite uniform.

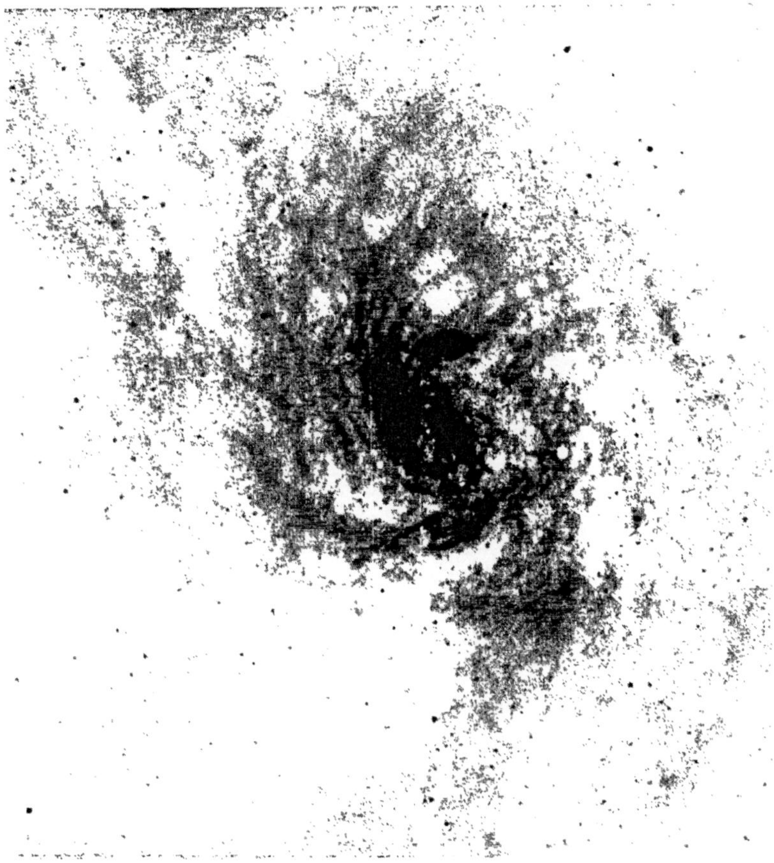

Figure 14.11 An Hα image of M31's bulge, showing diffuse hydrogen emission (from Ciardullo et al. 1988).

An image of the ionized gas in the bulge was obtained by Ciardullo *et al.* (1988), who combined hundreds of CCD frames taken over a period of five years. They found diffuse emission from gas that is spread out over the bulge in a distribution that appears to be unrelated to the M31 disk. A long circular arm and several spiraling loops and threads are spread out over the area (Figure 14.11). The distribution of the gas was interpreted as indicating that the gas in the bulge has collapsed to a rotating disk that occupies a plane significantly different from that of the main M31 disk; further, the gas is asymmetrical, suggesting that the portion of the disk south of the nucleus is more face-on even than the rest of the inner disk. In the inner 200 pc the gas density is two orders of magnitude smaller than in the surrounding regions, suggesting the effects of a galactic wind.

14.6 Supernova Remnants

Kumar (1976) suggested that some of the "anomalous H II regions" identified by Rubin *et al.* (1972) might have unusual line strengths because of being supernova remnants (SNRs) instead of normal H II regions. He listed 7 of the 52 objects for which spectra had been obtained. Some of these corresponded in position with non-thermal radio sources, a strong indication that they are SNRs.

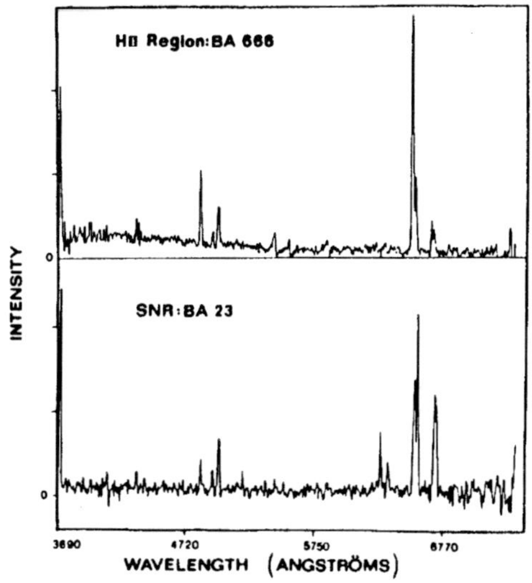

Figure 14.12 A spectrum of the M31 supernova remnant BA 23, compared to the spectrum of the normal HII region, BA 666 (from Dennefeld and Kunth, 1981).

An optical search specifically for supernova remnants was carried out by D'Odorico, Dopita and Benvenuti (1980), using a criterion that takes advantage of the fact that SNRs have high ratios of [S II] to Hα line strengths. Their search resulted in the identification of 18 SNRs in M31, five of which were in common with Kumar's list.

Dennefeld and Kunth (1981) followed up the search with spectrophotometry of candidate SNRs, confirming 12 of them. Figure 14.12 shows representative spectra of an SNR (BA 23) and a normal H II region (BA 666), illustrating the very different emission line intensity ratios.

Using the D'Odorico *et al.* (1980) census, Dennefeld and Kunth (1981) calculated that the supernova rate in M31 is approximately one in 200 years, a much lower rate than expected from statistics for galaxies in general.

Figure 14.13 An image of the bulge of M31, made by dividing an Fe I 3908λ image of an off-band 3476λ image. The Fe-rich cloud of the remnant of the 1885 supernova (S And) shows up as a small, round dark spot to the right and slightly below the nucleus (courtesy R. Fesen).

Only one supernova has been observed in M31, the object SN 1885A, observations of which have been analyzed in detail by de Vaucouleurs and Corwin (1985) and discussed by Graham (1988) and Chevalier and Plait (1988). The remnant of that supernova was detected optically in 1988 in a remarkable image published by Fesen, Hamilton and Saken (1989). They used a CCD detector on the Kitt Peak National Observatory's 4-m telescope and employed two filters, one centered on the Fe I resonance line at 3860 Å and one at an adjacent wavelength. The former showed an unresolved absorption feature located within 1 arcsec of the reported position of the supernova; it is most certainly the SNR from SN 1885. It can be concluded that the SNR is rich in iron, implying that it was a Type I supernova. Also, the ejecta must still be expanding and are relatively cold.

Walterbos and Braun (1991) carried out a deeper search for SNRs in M31, using both optical data and high-resolution radio maps. Initially, 14 SNRs were identified in a few selected areas, but many more tens (if not hundreds) of candidates were indicated. Their preliminary results indicated that the SNRs had optical and radio properties very similar to those in our Galaxy.

14.7 Summary

Nearly 1,000 H II regions have been detected in M31. These represent only the most luminous objects; large numbers of fainter ones are indicated in a few recent deep surveys.

Although individually normal, the ensemble of H II regions in M31 is anomalous in its lack of many high-luminosity objects. This may be a result of the galaxy's Hubble type. The sizes and morphologies of the H II regions are similar to those of other galaxies of most Hubble types and luminosities.

Spectra of the H II regions show a modest radial dependence of abundances, in which N/H and O/H decrease outwards by factors of 2 and 1.2, respectively.

There are several supernova remnants, recognized by their optical and radio spectra.

It is not always a simple matter to separate planetary nebulae from small H II regions. Ford and Jacoby argued that their objects are likely to be planetaries because they are entirely unresolved, the continuum image is completely absent, and they do not seem to be spatially arranged in the spiral pattern shown by M31's bright stars and dust.

A more sensitive survey of the central area of M31 was carried out using velocity modulated photographs by Lawrie and Ford (1982). They used a temperature- tuned 2 Å λ5007 filter to search the area within 250 parsecs of the M31 nucleus and found 42 planetaries there. This allowed them to estimate the total number for the bulge and for the galaxy as a whole, assuming that the luminosity function is constant and that the density follows the galaxy's light distribution. More recent estimates, using more complete data, are described below.

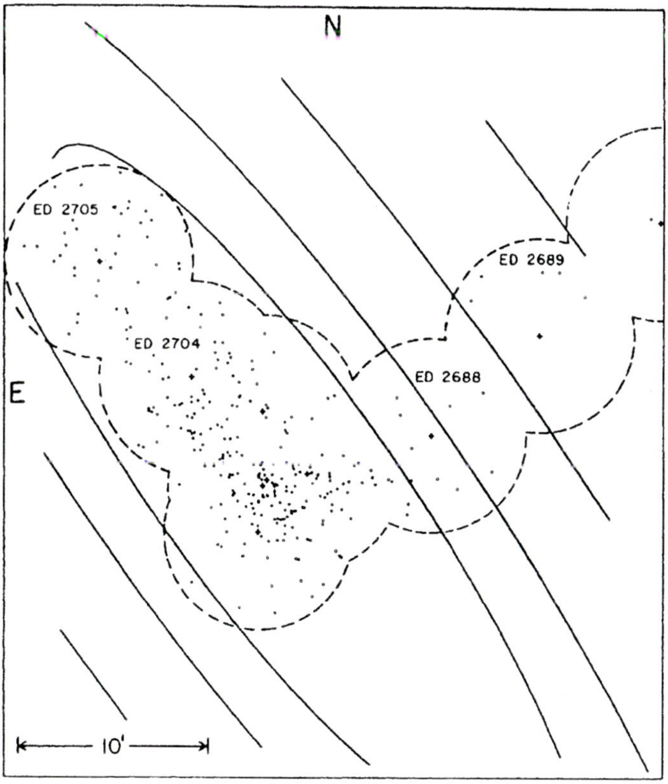

Figure 15.1 Locations of the fields in M31 searched for planetary nebulae in the Lick Observatory 3-m telescope survey (from Ford and Jacoby, 1978).

In a more recent survey, Ciardullo, Jacoby, Ford and Neill (1989) used a CCD detector to search for fainter planetaries and to examine the luminosity function more accurately and completely than had been possible with photographic techniques. Thirteen positions, with considerable overlap, provided them with coverage of much of the inner disk and bulge. A total of 429 planetary nebulae were detected and 17 of the previously reported nebulae were determined to be continuum sources or otherwise unlikely candidates.

It is clearly true that so far we know of only a small fraction of the planetary nebulae of M31. In their 1978 paper, Ford and Jacoby estimated that the total number of planetaries might be as large as 27,000, basing this estimate on a comparison of the luminosity function of the detected objects (which ranged over three magnitudes) with the more complete luminosity function for planetaries in the Galaxy. In a subsequent re-evaluation of this figure, Nolthenius and Ford (1987) made separate estimates of the disk and the bulge populations and concluded that the best estimate of the total is 12,000 planetary nebulae, with a lower limit of 6,000 and an upper limit of possibly 20,000.

One of the reasons that these searches were carried out so thoroughly and that the results were given a careful statistical treatment is that a reliable census of planetary nebulae can give important astrophysical information about stellar lifetimes. According to some theoretical models (for example, that of Renzini and Buzzoni 1986), the stellar death rate is relatively insensitive to the characteristics of the stellar population – specifically, it does not depend very much on either the age of a population or the initial mass function. Because most stars end their lives as planetary nebulae, a determination of the specific density of planetaries in different environments can test this theoretical result. The number of planetaries per magnitude in different Local Group galaxies was found to be nearly the same for all galaxies by Jacoby (1980). Therefore, it was important that Ciardullo *et al.* (1989) were able to calculate this number for M31's central bulge and to compare it with the figure found for the disk by Nolthenius and Ford (1982). The former concluded that the specific density in the bulge is

$$\text{No. (PN)}/L_B = 3.6 \times 10^{-8}, \qquad (eq.\ 15.1)$$

while in the disk, at projected distances between 15 and 20 kpc, it is

$$\text{No. (PN)}/L_B = 3.8 \times 10^{-8}, \qquad (eq.\ 15.2)$$

with L_B in solar units.

Thus, at least for M31, the stellar death rate seems to be independent of the stellar population, as predicted by evolutionary theory.

15.2 Spatial Distribution

Although no complete survey of the planetaries in M31 has been published, the partial surveys so far carried out provide good coverage of much of the galaxy, enough to determine the spatial distribution reasonably well. The Ford and Jacoby survey (Figure 15.1) covered the central areas and the inner parts of the major and minor axes. The Lawrie and Ford (1982) and Ciardullo *et al.* surveys covered the central areas more completely. The Nolthenius and Ford (1987) analysis provided information on the distribution of planetaries in sampled portions of the outer disk (Figure 15.2). Together, these have provided a good idea of the spatial distribution for both the halo and the disk planetaries.

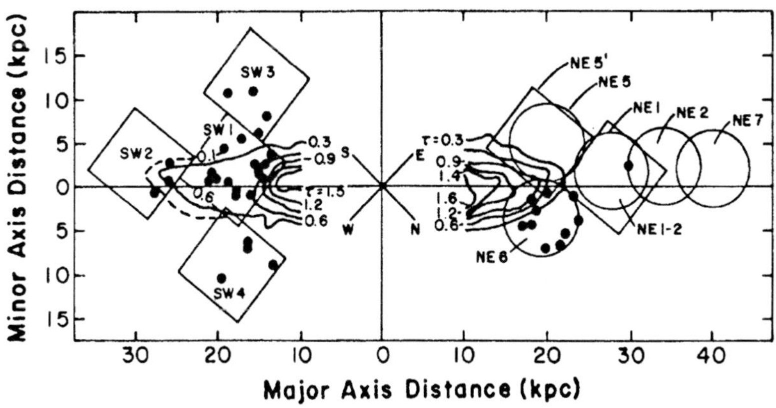

Figure 15.2 Planetary nebulae in the outer disk of M31 (from Nolthenius and Ford, 1987).

For the disk planetaries, Nolthenius and Ford (1987) found that the distribution favored the model of the warped disk of M31 that was derived by Newton and Emerson (1977). Using that model, they found that the planetaries showed an exponential decrease in density outwards, with a scale length of 4.6 kpc, which can be compared with the scale length of ~5.2 kpc found for the starlight (Chapter 4). They found no evidence for any variation of number density per stellar background luminosity in the disk.

Ciardullo *et al.*'s (1989) CCD survey found similar results. Figure 15.3 shows the number of planetaries found, compared to the photometric galaxy profile, and also illustrates the near constancy of the specific density.

In the bulge, the planetaries also follow the background light distribution (Lawrie and Ford 1982), although the numbers are small enough to make detailed agreement an open question. As Figure 15.4 shows, there does seem to be a dearth of planetaries in the northwest quadrant of the bulge, which Lawrie and Ford ascribe to the effect of a thin dust sheet, which partially obscures the far side planetaries.

15.3 Kinematics

The first indication of the velocity distribution and kinematics of the M31 planetary nebula population came from Lawrie and Ford's (1982) velocity-tuned images of the central area. Of the 45 detected planetaries, they were

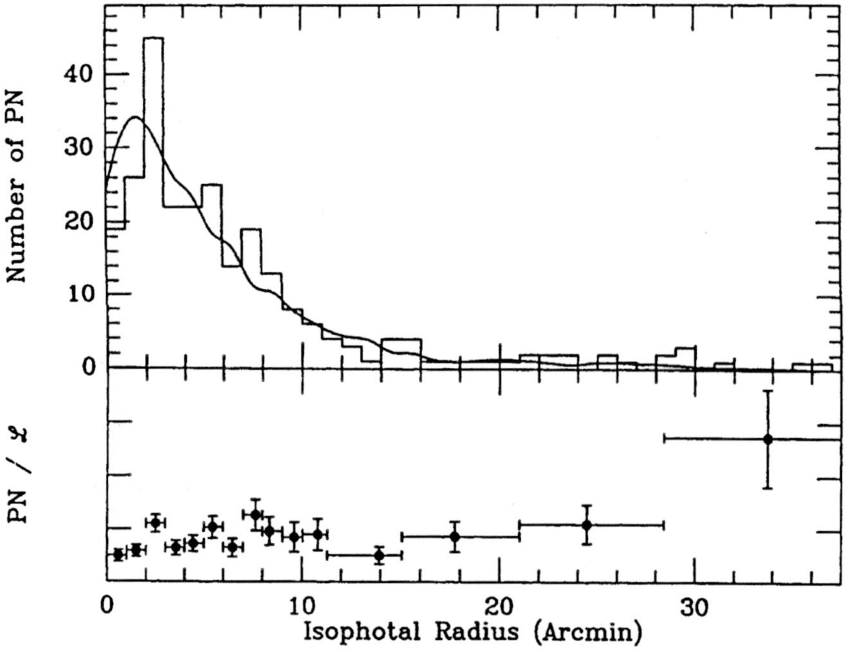

Figure 15.3 The upper panel shows the number density of planetaries at different isophotal radii vs. the amount of galaxy light contained in the isophote. The bottom panel plots the specific density of planetaries (from Ciardullo et al. 1989).

able to determine the radial velocities of 33. The mean velocity was found to be -309 ± 25 km/sec, in good agreement with the systemic velocity of the galaxy. The velocity dispersion was calculated to be

$$\sigma = 156 \pm 23 \text{ km/sec}, \qquad (eq.\ 15.3)$$

This value agrees excellently with the stellar velocity dispersion for the central area, which is (Chapter 8)

$$\sigma = 151 \pm 16/\text{km/sec}, \qquad (eq.\ 15.4)$$

according to Whitmore (1980).

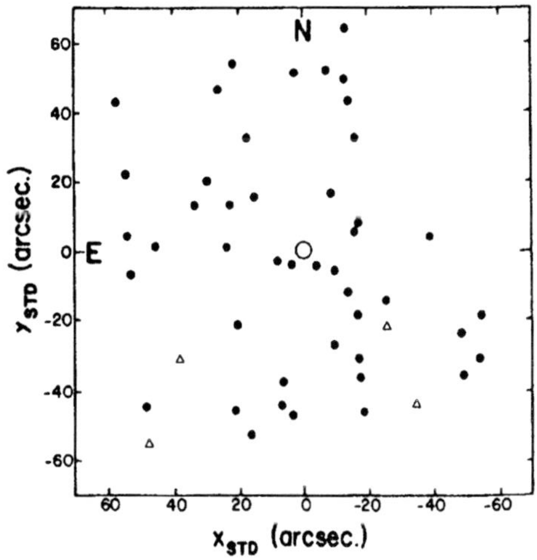

Figure 15.4 The distribution of planetary nebulae in the M31 bulge (from Lawrie and Ford 1982).

For the outer disk and the halo, the best data are those of Nolthenius and Ford (1987), who obtained radial velocities of 34 planetaries found in their fields, with radial distances from the nucleus ranging from 15 to 30 kpc. They separated the planetaries into disk and halo objects, ignoring the possibility that any might belong to some intermediate population, which was felt to be demonstrated as unlikely, according to studies of edge-on galaxies.

Depending on which warping model was adopted, it was found that approximately 10 of the planetaries are probably halo objects, judging from their fit to the H I rotation curve (Figure 15.5). The halo planetaries were found to show a rotation that has the same sense as the disk rotation, but much slower, with a velocity dispersion of

$$V = 92 \pm 43 \text{ km/sec}, \qquad (eq.\ 15.5)$$

instead of the 232 km/sec value of the disk. The velocity dispersion for the halo was calculated to be

$$\sigma = 116 \pm 48 \text{ km/sec}, \qquad (eq.\ 15.6)$$

smaller than, but not significantly so, the velocity dispersion of the bulge. Agreement is reasonable with the velocity dispersion for halo globular clusters (Chapter 9), which have

$$\sigma = 80 \pm 28 \text{ km/sec}. \qquad (eq.\ 15.7)$$

The disk planetaries in the Nolthenius and Ford sample fit the H I disk rotation curve very well (Figure 15.5). The rotational velocity, taken to be constant at the relevant distances, is 218 ± 12 km/sec, comparable to the H I velocity of 232 km/sec. There is a marginally significant detection of an asymmetrical drift, with a velocity of approximately 14 km/sec, similar to values for the Galaxy's planetaries. If true, this drift indicates that the orbits of the planetaries are not perfectly circular.

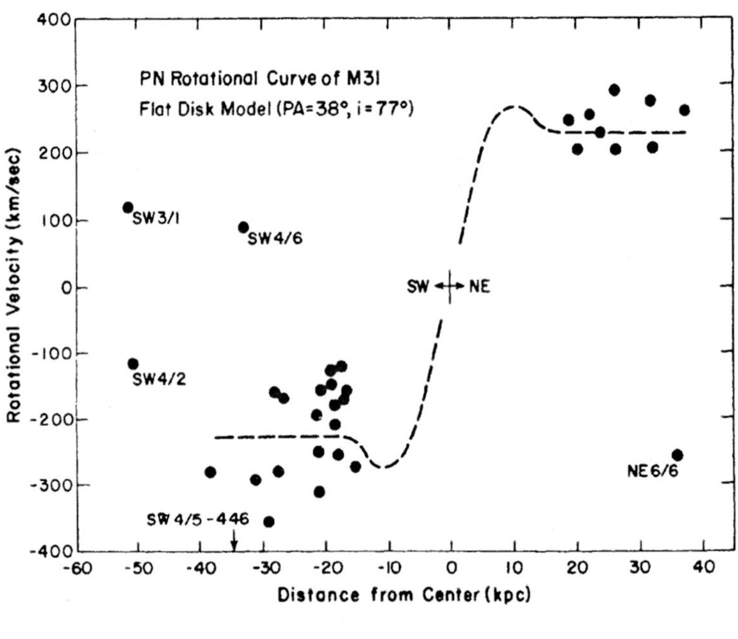

Figure 15.5 Rotational velocities of the planetaries, compared to the flat disk rotation curve based on HI and HII velocities. The halo planetaries are labeled with their names (from Nolthenius and Ford, 1987).

15.4 Abundances

Because emission-line spectra of planetary nebulae are comparatively easy to obtain, the abundances of the elements in planetary nebulae can provide important information about the abundances of stars in other galaxies, where it is still difficult to obtain absorption-line spectra of individual stars. The power of the method can be appreciated by noting that the only discussion of spectra of M31's planetary nebulae published so far (Jacoby and Ford 1986) was able to use spectra of only three planetaries to deduce an amazing amount of information about the history of element enrichment in the disk and halo.

The three planetary nebulae studied are at widely different projected distances from the center of M31; the nearest is at r = 3.5 kpc, the next at r = 18 kpc and the farthest at the large distance of 33 kpc, in the distant halo and beyond the visible disk of stars. The same equipment and technique was used to obtain spectra of a small, high-surface brightness H II region, BA 685, in the disk at a projected distance of r = 17 kpc; this provided a comparison with abundances in material involved in current star formation activity in the disk.

Table 15.1
Elemental Abundances in Three Planetaries
and an H II Region in M31
(from Jacoby and Ford 1986)

Object	[12 + log (N/N_H)]			
	He	O	N	Ne
M31-290	11.20	8.55	7.81	7.79
M31-363	11.21	8.67	7.88	8.11
M31-372	–	8.06	<7.71	7.49
BA 685	10.93	8.51	7.39	7.66

Figure 15.6 and Figure 15.7 show the spectra of the three M31 planetaries and the H II region. Lines of H, He I, [N II], [O II], [O III], [Ne III], and [S II] were detected and measured. Elemental abundances are given in Table 15.1. It is of interest to note that the abundances in the H II region at 17 kpc are quite comparable to those in the Orion nebula. This is somewhat of a surprise, because H II regions in the outer parts of our Galaxy tend to have lower heavy-element abundances than this. At 17 kpc, neither galaxy has very much stellar material and it is

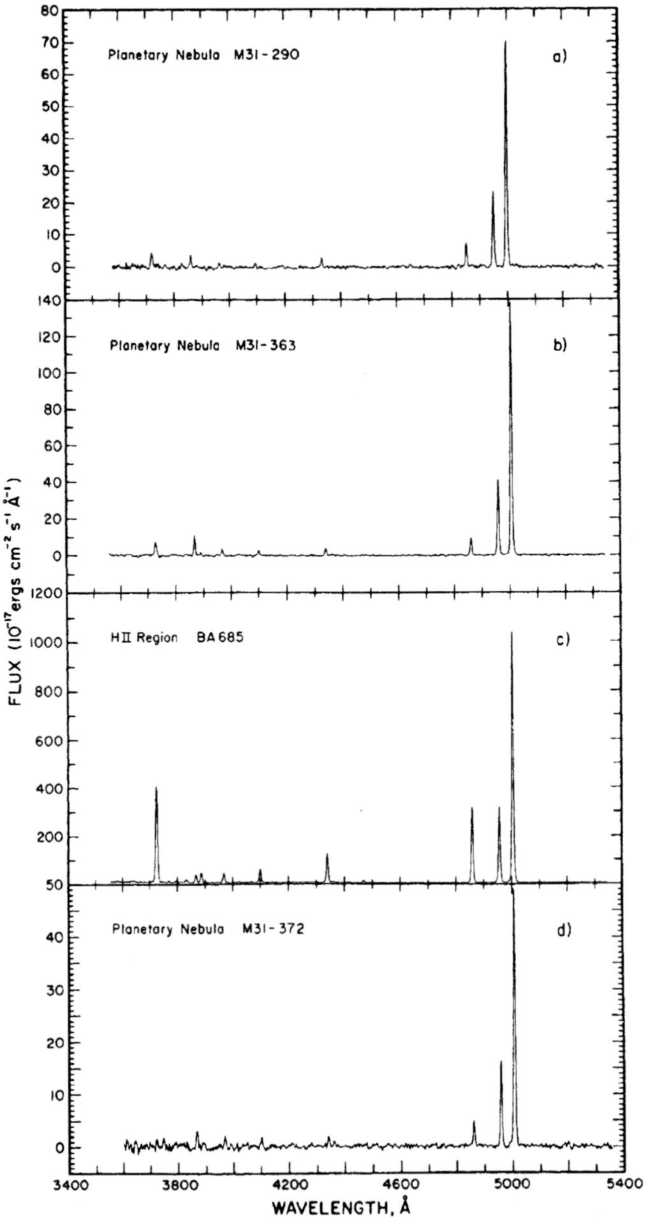

Figure 15.6 Blue spectra of three planetary nebulae and an HII region in M31 (from Jacoby and Ford, 1986).

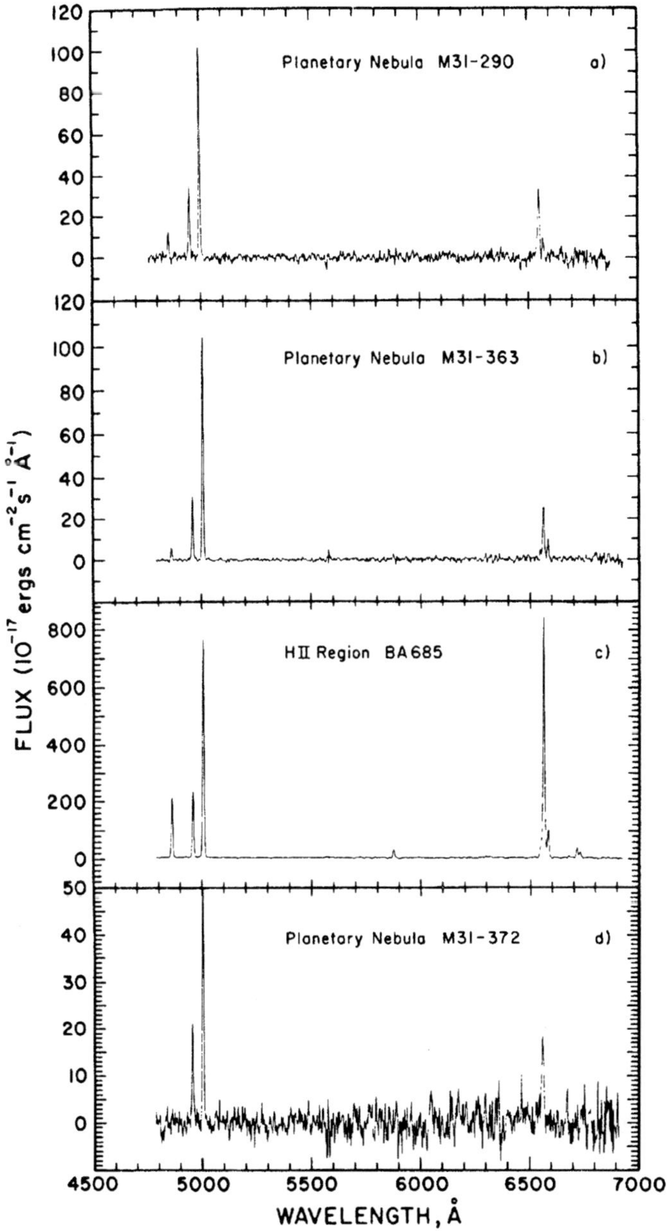

Figure 15.7 Red spectra of the nebulae shown in Figure 15.6 (from Jacoby and Ford, 1986).

interesting that M31, which presently is forming stars at a lower rate than the Galaxy, should have evidence of a history of a formerly higher rate of stellar processing of material in its outer disk.

Two of the planetaries, M31-290 at r = 3.5 kpc and M31-372 at r = 33 kpc, are halo members and the third, M31-363 at r = 18 kpc, is a disk object. Mindful that the oxygen abundances suffer little change in the stellar evolutionary processing, compared to that of C and N, Jacoby and Ford (1986) based their conclusions about abundance differences primarily on the O abundances. The fact that the two halo planetaries have very different values indicates that the M31 halo is (and was) chemically inhomogeneous. Combining this fact with similar conclusions based on spectra of halo globular clusters (Chapter 9) leads to the conclusion that there was a large range in heavy element enrichment taking place when M31's halo collapsed.

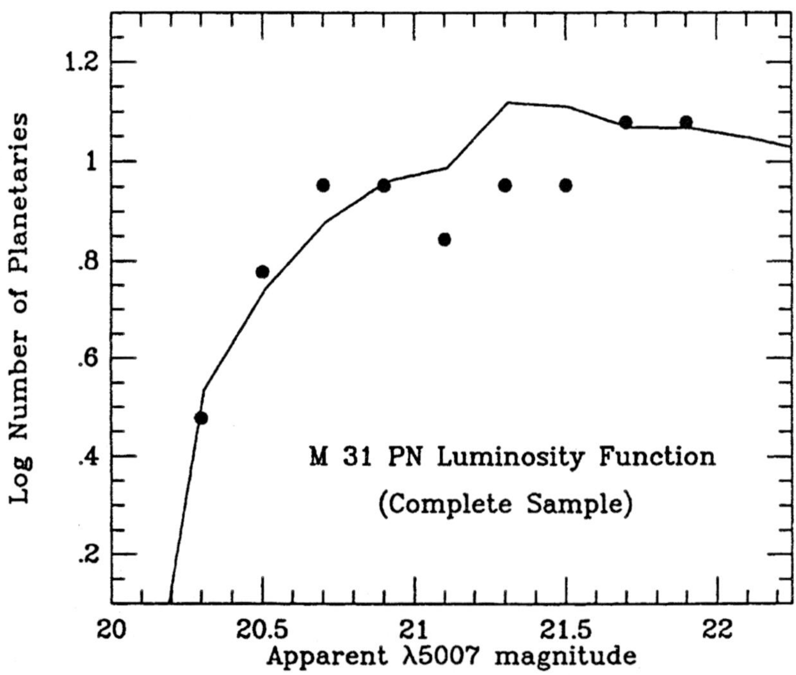

Figure 15.8 The luminosity function of the M31 planetary nebulae compared to a curve generated on theoretical grounds (from Ciardullo et al. 1989).

The disk planetary, M31-363, has an oxygen abundance that is equal (within the uncertainties) to that of the nearby H II region. This is of importance because the planetary is the remnant of a star that is probably at least 2×10^9 years old. The surprisingly high heavy element abundance of the present disk gas, as exemplified by the H II region, is therefore not a recent product of stellar evolution; the enrichment of the gas must have occurred several billion years ago.

15.5 The Planetary Nebula Luminosity Function

Because of the relatively large number of planetaries found in M31, the sample has been used to establish the characteristics of the planetary nebula luminosity function, which is of importance for the extragalactic distance scale. Jacoby (1980) showed that the [O III] luminosity function of planetaries in the Magellanic Clouds showed two characteristics that made them useful distance indicators: they appeared to have a sharp upper luminosity limit and this limit appeared to be independent of galaxy type.

Ciardullo *et al.* (1989), following the earlier work on planetary nebulae in M31 and its elliptical companions of Ford and his collaborators, established detailed statistical properties of the M31 luminosity function from a survey of the bulge. Their goal was a solid calibration of the luminosity function for application to more distant galaxies. They found that the shape of the function is favorable for such use, as it is decidedly not a power law. Instead it shows a sharp cutoff at an absolute magnitude (at λ5007 [O III]) of -4.48. The data (Figure 15.8) show the beginning of a turndown at $M = -3.8$. The curve plotted in the figure is from a theoretical luminosity function generated by Jacoby (1980). Agreement is excellent.

15.6 Summary

Nearly 500 planetary nebulae have been found so far in M31, but the total number present in the galaxy is estimated to be on the order of 12,000. The planetaries provide useful information on the distribution of the intermediate-age population in the disk and halo. The velocity dispersion in the bulge agrees with that for bulge field stars and that in the halo agrees with that of the halo globular clusters.

Abundances of two halo planetaries differ significantly, suggesting that there is a large range in the chemical abundances within the halo. One disk planetary, at a distance from the center of 18 kpc, has been studied spectrographically. It has a surprisingly large heavy element abundance, considering its position.

CHAPTER 16

MOLECULES

The discovery of emission from molecules in M31 is a recent development. As it involves very faint sources and requires high resolution, the field is still in an early phase of its history.

16.1 CO Observations

Following experience with our Galaxy, the most easily-studied molecular species is CO. From the intensity of the CO emission, the density of the more-abundant molecules, such as molecular hydrogen, can be calculated on the basis of simple assumptions, which, in our Galaxy, can be checked. Thus most of the remainder of this chapter discusses CO emission, but the conclusions can be accepted as applying in general to other molecules, especially H_2.

The initial discovery of CO emission in M31 was a somewhat controversial issue. Rickard (1975), for instance, looked at the nucleus of M31 and found none, whereas Solomon and Zafra (1975), using the same equipment, found a rather strong, broad emission line of CO from the nucleus. The latter result, however, has remained puzzling, especially because the velocity indicated by the line had a peak at −125 km/sec instead of at the systemic velocity of M31, −300km/sec. Whatever was being detected apparently was not the nucleus.

The earliest surveys of molecular emission in M31 occurred as recently as 1977, when Combes *et al.* (1977a) published the result of a moderate-resolution (64 arcsec) search for CO emission along a strip coinciding with the major axis of M31. They detected CO at five of the telescope positions, all of which appeared to be at locations of prominent dust lanes, described by the authors as at the inner sides of the southern arms. The association with the galaxy and, in particular, with the gas layer at the detected positions, was confirmed by the close agreement between the radial velocities and velocity widths of the CO line and that of the HI at the same position. Note that in most of the early papers on CO in M31, authors refer to the southern and northern "arms", meaning not the thin arms defined by optical or high-resolution HI maps, but the thicker HI ring, where it intersects the major axis.

This pioneering work was followed by a second survey by the same scientists (Combes *et al.* 1977b), in which they extended their major axis exploration by adding further points to the southern axis and observing 18 positions along (or, at least near) the northern major axis. For the northern positions, which were situated at 200 pc intervals, CO was only detected at five positions, all contiguous and located in the deep dust lanes associated with the HI ring, especially the arm in the ring with the highest HI density (Chapter 5). As in the southern data, the CO maximum appeared to lie at the inside of the HI arm. Also, the CO kinematics agreed well with HI data.

These results constitute the early history of the field. The next stage in the development involved follow-up studies, covering more of the galaxy and using, in some cases, better resolution, to allow both better comparison with HI and other components and better knowledge of the characteristics of the clouds containing the molecules.

16.2 Spatial Distribution

Early Results

The galaxy has only recently been globally mapped at CO wavelengths, so far only with relatively low spatial resolution (Koper *et al.* 1991, described below). The first global clues about the CO distribution in M31 had come during the initial pioneering work of Combes *et al.* (1977b), described above. Fig. 16.1 shows the radial distribution of CO, compared to that of HI, as plotted by Combes *et al.* (1977b). On the southwest axis the two CO maxima coincide closely with the inside shoulders of the HI profiles of arms, while on the northeast axis, the one CO peak is on the inside of one of the HI arms.

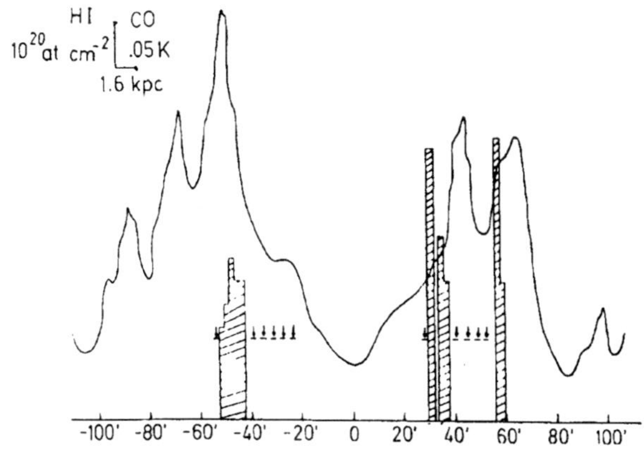

Figure 16.1 An early comparison of the intensity of CO (histogram) and HI along the major axis of M31. The arrows are 3 sigma upper limits for CO (from Combes et al. 1977b).

TABLE 16.1
CO Maps of M31*

Reference	Line	Beam	Map Location	No. of Pts.
Blitz 1985	$(1 \rightarrow 0)$	65″	H II regions	491
Boulanger *et al.* 1984	$(1 \rightarrow 0)$	33″	Stark *et al.* peak	24
Casoli *et al.* 1987	$(1 \rightarrow 0)$	22″	(+22′, +11′)	41
Casoli and Combes 1988	$(1 \rightarrow 0)$	23″	(+32′, +9)	49
Combes *et al.* 1977b	$(1 \rightarrow 0)$	65″	Major axis strip	31
Ichikawa *et al.* 1985	$(1 \rightarrow 0)$	14″	Stark *et al.* peak	14
Kutner *et al.* 1990	$(2 \rightarrow 1)$	30″	Stark *et al.* peak	31
Lada *et al.* 1988	$(1 \rightarrow 0)$	15″	(−32′, +2′)	41
Nakano *et al.* 1987	$(1 \rightarrow 0)$	14″	Stark *et al.* peak	30
Ryden and Stark 1986	$(1 \rightarrow 0)$	1′.7	(+28′, +10′)	77
Sandqvist *et al.* 1989	$(1 \rightarrow 0)$	33″	Minor axis strip	30
Stark 1979	$(1 \rightarrow 0)$	1′.7	Minor axis strip	20
Stark *et al.* 1981	$(1 \rightarrow 0)$	1′.7	(36′, +5′)	143
Vogel *et al.* 1987	$(1 \rightarrow 0)$	7″	Lada *et al.* peak	1

*Adapted from Kutner *et al.* (1990)

Stark (1979), using the Bell Labs 7 m telescope, observed many points along the minor axis of M31, showing that the central bulge is apparently devoid of molecules and that the conspicuous optical spiral arms have a molecular counterpart, though the high inclination of the galaxy led to poor resolution of the structure. Stark *et al.* (1981) reported a more extensive survey, covering a 20 × 15 arc min area of the southern portion of the arm structure. Boulanger *et al.* (1981), also using the Bell Labs telescope, discussed the CO distribution in the southwest portion of the galaxy. Their results indicated that the CO is located in the area where both HI and optical dust clouds are present. In this region the CO seems to be a better tracer of the arms than the HI.

CO in HII Regions

In an attempt to learn about the nature of CO complexes in M31, Blitz (1985) mapped 49 HII regions in the galaxy, using the NRAO 11-m telescope at Kitt Peak. For the CO J = 1-0 transition, the spatial resolution was 1.1 arcmin. Seventeen of the HII regions were detected, with a maximum number detected at a galactocentric radius of about 9 kpc. There was a suggestion that the integrated line strength shows a radial decrease outward in the galaxy, possibly related to an abundance gradient.

Vogel *et al.* (1987) looked at two M31 HII regions, BA 292 and P248, with the Owens Valley interferometer, giving spatial resolution of 7 arcsec for the J = 1-0 emission. Rather than pointing to the positions of the HII regions, they mapped the entire area and detected two giant molecular clouds (GMCs), each close to the positions of optical HII regions. The characteristics (brightness, velocity dispersion, size, and mass) are very similar to those of a typical GMC in our Galaxy, such as the Orion GMC. Figure 16.2 shows the CO map for BA 292.

CO in Spiral Arms

Several of the studies of the 1980s examined the CO found in portions of the spiral arm structure. Ryden and Stark (1986), for example, mapped the CO (J = 1-0) along the spiral structure in a 4 × 20 arcmin strip in the northeast portion of the galaxy (Fig. 16.3). They found that the CO emission is confined to a fairly narrow lane, 1 to 2 kpc wide, morphologically much like the optical dust lane in that area of the galaxy. An indication that the velocity deviates systematically in the outer portion of the arm suggests that the molecular gas may be accelerated as it crosses the arm.

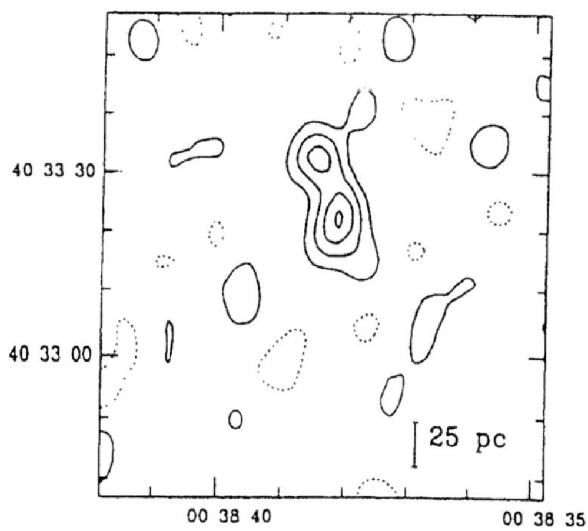

Figure 16.2 The integrated CO emission from the area of the HII region BA 292. Contour intervals are 2.6 Jy km s^{-1} (from Vogel et al. 1987).

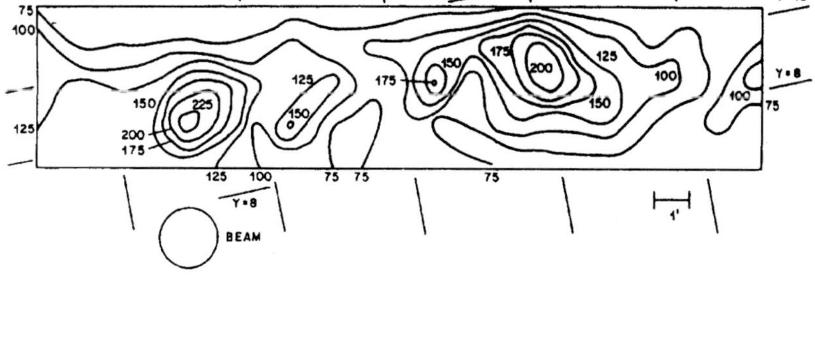

Figure 16.3 Map of the integrated CO emission in a northeast spiral arm (from Ryden and Stark 1986).

A portion of this same area was subsequently observed at higher resolution by Casoli *et al.* (1987), using the IRAM 30m telescope, giving a resolution of 22 arcsec for the J = 1-0 frequency. CO was detected at almost half of the 41 positions observed (Fig. 16.4), with all lines showing the same LSR velocity. This fact indicates that the emission comes primarily or entirely from a single GMC complex, with five or six HII regions superimposed (possibly embedded). There is not a good correlation in position with the peaks of emission at CO and Hα wavelengths; the two components tend to be offset from each other, though at the 500 pc level they agree in position. The molecular hydrogen mass is approximately 1.3×10^6 solar masses and the diameter of the complex is 160 pc.

Using the 45 m telescope at the Nobeyama Radio Observatory, Ichikawa *et al.* (1985) had 14 arcsec resolution for their survey of the CO emission in a small portion of the southern arm area. They determined a CO arm width of approximately 500 pc. Positions of the CO peaks were not coincident with the positions of bright HII regions in the area.

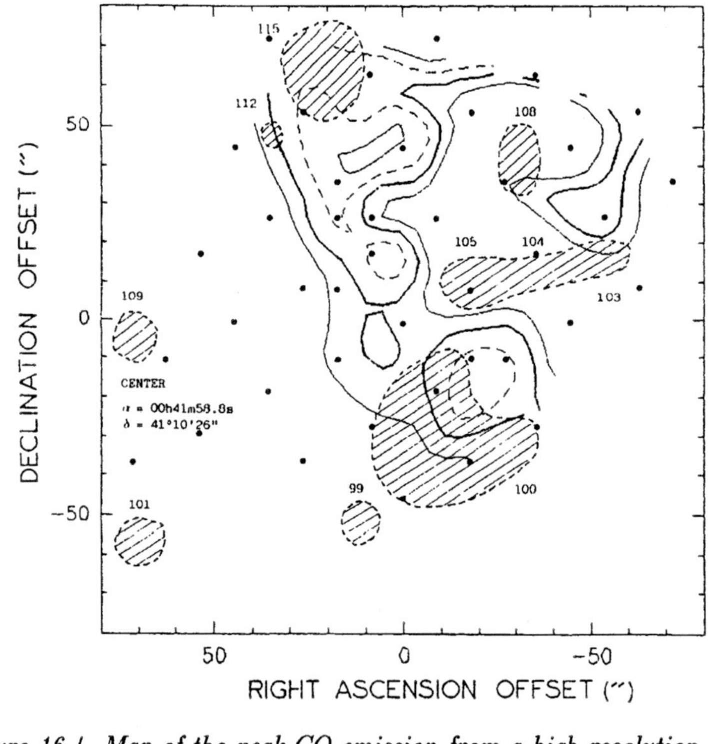

Figure 16.4 Map of the peak CO emission from a high-resolution survey of a part of the southeast arm region. HII regions are indicated by the cross- hatched areas; their BA numbers are given. Dots are the positions of the CO observations (from Casoli et al. 1987).

The Nobeyama telescope was also used in a study of a portion of a spiral arm by Lada *et al.* (1988), who, in an important additional step, matched the CO map with observations with the same resolution at 21 cm, to allow direct comparison with atomic hydrogen. The general area is rich in HI, optical dust clouds, HII regions and OB associations. The dust cloud D118 lies near the center of the region looked at with high resolution, and the HII regions BA 292 and P248 are included (Fig. 16.5). They found GMC complexes with total masses of the order of 10^6 solar masses, made up of one or more giant and intermediate-sized clouds. About 30% of the mass of the complexes is in the form of atomic hydrogen. The spatial distributions of HI and molecular hydrogen (inferred from the CO) are similar, and the peaks of molecular density are adjacent to the HII regions.

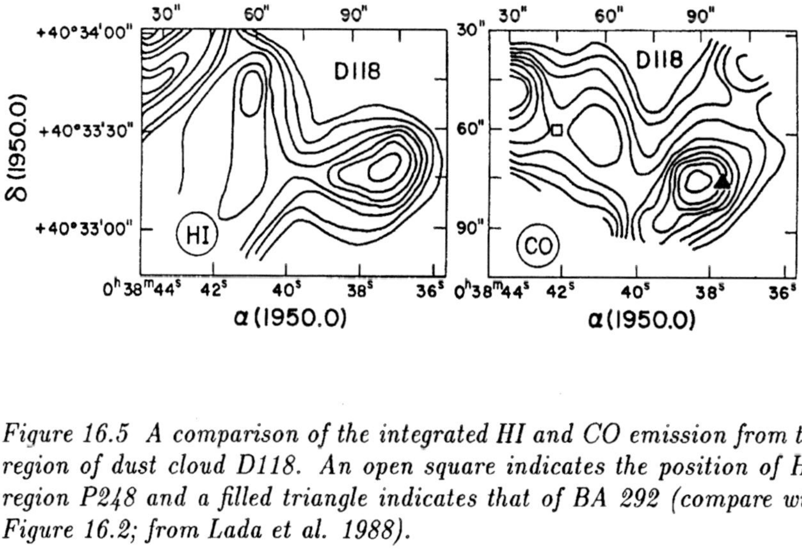

Figure 16.5 A comparison of the integrated HI and CO emission from the region of dust cloud D118. An open square indicates the position of HII region P248 and a filled triangle indicates that of BA 292 (compare with Figure 16.2; from Lada et al. 1988).

A limited number of studies have been made using the CO (J = 2-1) transition. For example, Lada *et al.* (1988) reported such observations of the D118 region and concluded that comparison with the J = 1-0 observations indicated space and column densities that are similar to expectations for GMCs like those found in the solar neighborhood. Another survey with the J = 2-1 transition was published by Kutner *et al.* (1990). Their survey covered a portion of the rich southwest arm structure previously studied by Stark (1985) and others. Resolution was 30 arcsec. Fig. 16.6 shows three maps, for the CO peak intensity, the integrated intensity, and the line width. The strong, narrow features are interpreted as GMCs with masses of a few 10^5 solar masses. Between the arms, there are CO features that are weak and that have broad velocity widths, which Kutner *et al.* (1990) interpret as emission from many small clouds with masses of about 10^4 solar masses. The question of the relative importance of GMCs and small clouds has been an issue of debate, which will be discussed in Section 16.3.

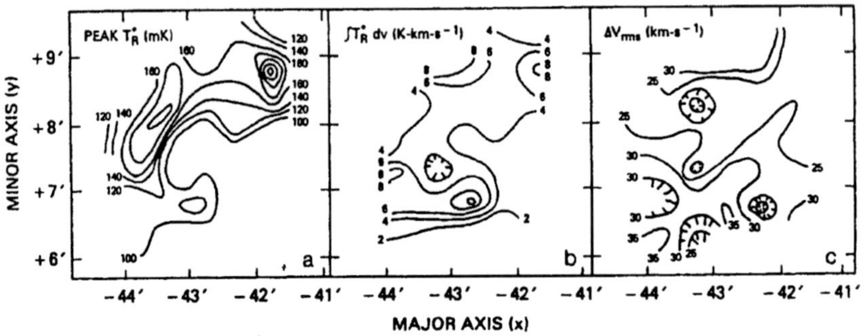

Figure 16.6 CO in a portion of the southwest arm structure, showing contours for the peak intensity, the integrated intensity, and the line width (from Kutner et al. 1990).

A Global CO Map

Using a small enough telescope (the 1.2 m telescope of the Center for Astrophysics) allowed the first global mapping of the CO in M31 by (Koper *et al.* 1991). The angular resolution achieved was 8.7 arcmin and an area of 153 × 45 arcmin was surveyed. Fig. 16.7 shows the results, where they are compared to similar maps of HI and 100-micron infrared emission, convolved to the same resolution. Most of the CO detected is in the area of the 10 kpc ring, where the HI and the brightest star-forming regions lie. However, the CO is displaced inward about 1 kpc compared to the HI. There is very little CO either inside a radial distance of 8 kpc or outside a distance of 15 kpc. Fig. 16.8 shows the radial distribution inferred for molecular hydrogen, where it is compared with the HI. The H_2 appears to be more narrowly distributed in the ring, displaced inwards and more clearly separated into arms (*e.g.*, the 4 kpc optical arm shows up with more density contrast). In addition to providing the first complete map of the galaxy, this study allowed a calculation of the total molecular

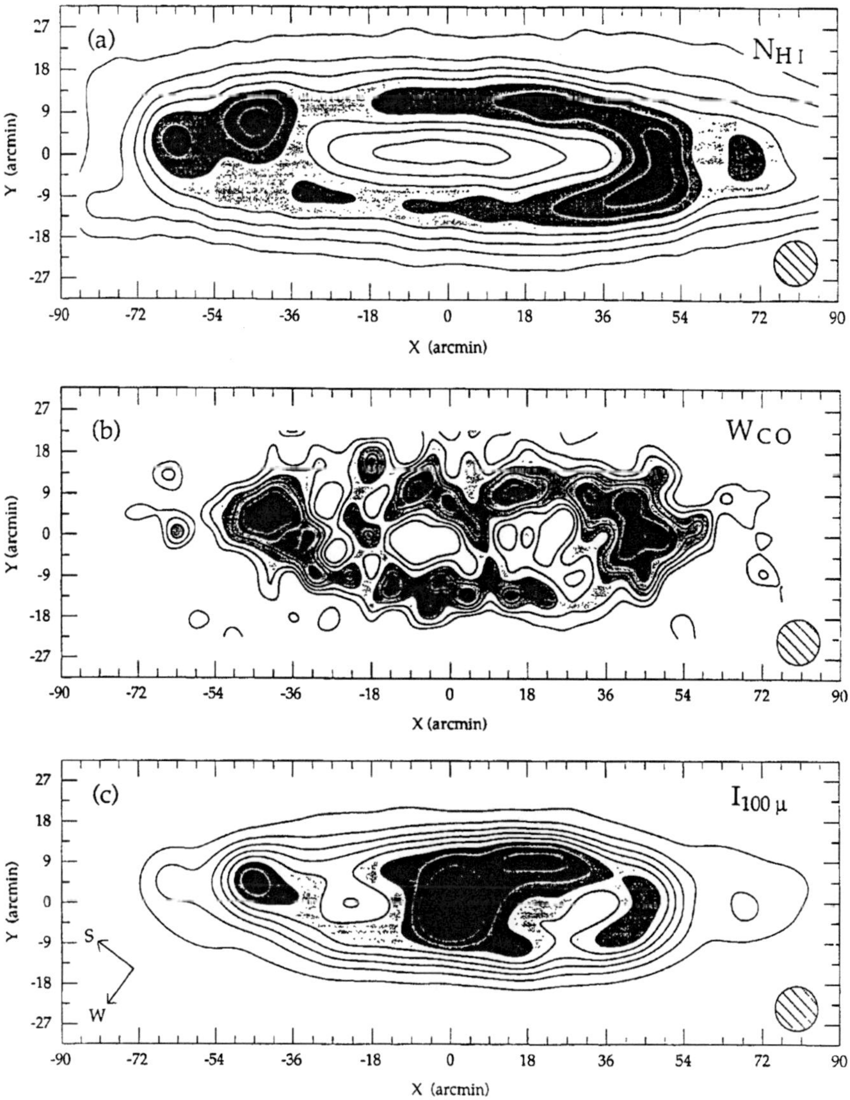

Figure 16.7 Global maps of the CO emission in M31 (center), compared to the HI density and the 100 micron flux, all shown with the same effective resolution. Contour intervals are 2.4×10^{20} cm^{-2} *for HI, 0.18 K km* s^{-1} *for CO and 1.8 MJy* sr^{-1} *for the infrared (from Koper et al. 1991).*

hydrogen mass for M31, which, using standard assumptions, is 2.5×10^8 solar masses. The significance of this result is that it is considerably less than the mass of H_2 calculated for the Milky Way Galaxy, probably a symptom of the earlier Hubble type and the lower rate of star formation of M31. Alternatively, the standard assumptions regarding the conversion factor may not apply in M31's case.

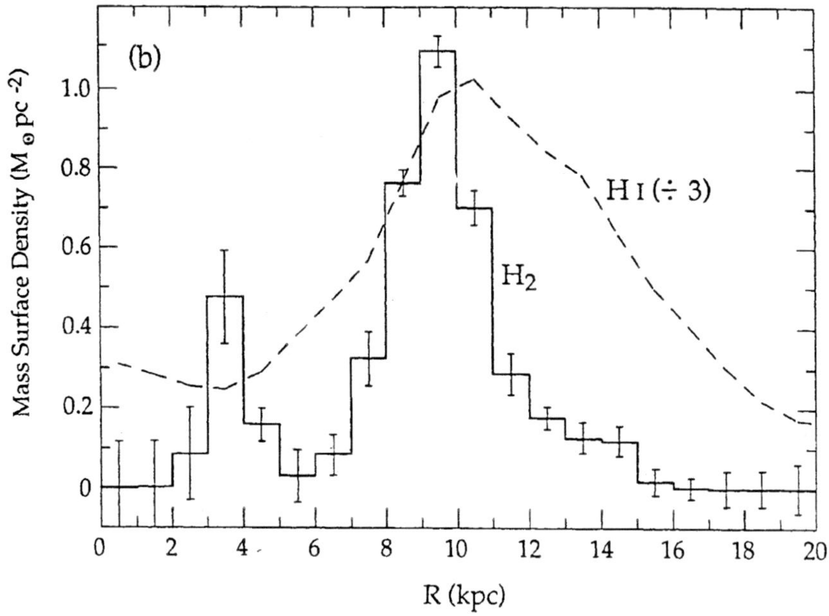

Figure 16.8 Radial distributions of atomic and molecular hydrogen mass surface densities, based on the global CO survey and 21 cm data (from Koper et al. 1991).

16.3 Giant and Not-So-Giant Molecular Clouds

A question that arose early in the study of molecular emission from M31 was the nature of the molecular clouds. Some evidence seemed to show that some, if not all, of the emission came from large numbers of small clouds, rather than from GMCs. For example, Boulanger *et al.* (1981) found CO line widths that seemed too large to be from individual GMCs, and they therefore proposed that the signal came from many small clouds, the different rotational velocities of which caused the observed velocity spread. In their analysis using 33 arcsec resolution data, Boulanger *et al.* (1984) showed that the higher resolution data gave stronger and narrower lines than the 1.7 arcmin data. This is evidence that favors the dominance of GMCs, with only a minor component coming from small clouds. Blitz (1985) argued that the signals he obtained from the directions of HII regions showed anomalously-wide velocity spreads, possibly because of emission coming from several small cloud sources. Ryden and Stark (1986) reported both broad and narrow lines, indicating the presence of both GMCs and numerous small clouds. Casoli *et al.* (1987), using 22 arcsec resolution data, also saw both broad and narrow components.

Vogel *et al.* (1987) and Lada *et al.* (1988) both found no evidence for small clouds. From their data, all emission could be accounted for by assuming that GMCs dominated the signal.

How are these contrary conclusions reconciled? Are the answers different for different regions of M31, or is there a problem with the interpretation of the spectra? Kutner *et al.* (1990) attempted to answer these questions by using the CO ($J = 2\text{-}1$) line, giving them higher resolution than the $J = 1\text{-}0$ line. They had 30 arcsec resolution with the NRAO 12 m telescope. They mapped an area in the southwest arm structure that overlaps the Boulanger *et al.* (1984) and Nakano *et al.* (1987) studies, but achieved higher sensitivity, allowing them to look more reliably at the question of the relative importance of giant vs. small molecular clouds. They were able to conclude that both kinds of clouds are present. The dense arm areas are characterized by having most of the emission coming from GMCs (narrow, strong lines), with masses of nearly 10^6 solar masses, while the interarm regions have the emission dominated by many small clouds (broad, weak lines), with masses of a few $\times 10^4$ solar masses. In the southwest arm structure, about 70% of the emission comes from GMCs, while in the interarm regions, all comes from small clouds.

16.4 The ^{12}CO to ^{13}CO Ratio

The ^{12}CO to ^{13}CO line ratio has been observed in a few areas of M31. Encrenaz *et al.* (1979) observed the lines in the southwest spiral arm and Casoli *et al.* (1987) in a portion of the northern arm structure. Both results indicated that the ratio of integrated emission is approximately 10, which can be compared to a value about half that large found for a comparable region in our Galaxy. This does not necessarily mean, however, that there is a difference between the galaxies. The observed ratio for M31 does not measure the abundance ratios of the isotopes, as there is a signal dilution problem; the observations are received from the entire GMC, but the inner and outer regions emit lines with different ratios of intensities because the saturation of the ^{12}CO line decreases outwards. A definitive measurement of isotopic abundances is still in the future.

16.5 Other Molecules

Although most observational programs have concentrated on CO in M31, other molecules have been detected. For example, formaldehyde (H_2CO) was seen in absorption at a point along the major axis where, surprisingly, the galaxy's continuum emission was weak (Cohen *et al.* 1979). The formaldehyde absorption was suggested to be against the microwave background radiation. Of the four points examined with the Mk II Jodrell Bank telescope, only one point, 42 arcmin on the major axis from the nucleus, was detected. Its strength was consistent with a surface density of formaldehyde that is similar to the highest values found in the Milky Way Galaxy.

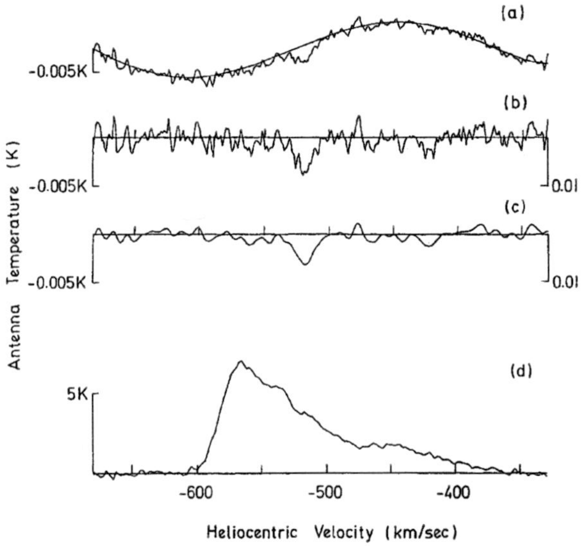

Figure 16.9 The absorption spectrum of formaldehyde in M31. (a) the unprocessed spectrum with a polynomial baseline fitted to it, (b) the spectrum after subtraction, (c) the spectrum smoothed to a velocity resolution of 6 km s^{-1}, and (d) the HI spectrum for the same position (from Cohen et al. 1979).

16.6 Summary

Molecules in M31 account for only 2.5×10^8 solar masses, a considerably smaller fraction of its total mass than in the case of our Galaxy. The CO in the spiral arms is emitted from areas towards the inside of the arm region and primarily from giant molecular clouds of total mass $\sim 10^6$ solar masses. CO from interarm regions appears to be emitted mainly from small clouds with masses of $\sim 10^4$ solar masses.

CHAPTER 17

X-RAY SOURCES

The study of the X-ray emission from M31 is a young field, having its beginnings in the mid-1970's. By 1991, however, a fairly good understanding of the various sources is at hand. The importance of these sources is the result of the happy fact that M31 is a normal, high-luminosity galaxy; it therefore provides a good opportunity to study the X-ray emission in detail, allowing a better understanding of X-rays from more distant galaxies.

17.1 Discovery Observations

X-rays from M31 were first detected by Bowyer *et al.* (1974) and Giacconi *et al.* (1974). Bowyer *et al.* used two X-ray detectors mounted on a sounding rocket, which was programmed to point at M31 and at nearby points to establish background levels. The beam size was nearly the same as M31's optical image size (Figure 17.1) and pointing was very close to being centered on the galaxy's nucleus. The integrated flux in the 0.5 - 5.0 keV band was found to be $(2.2 \pm 0.6) \times 10^{39}$ ergs s^{-1}. The X-ray spectrum, although poorly-defined by the weak signal, was found to be consistent with that found for bright X-ray sources in the Milky Way Galaxy.

Giacconi *et al.* (1974) reported an X-ray source from the Uhuru satellite all- sky scans that had a position close to M31 (Figure 17.1). From their flux measurement at 2 - 6 keV, combined with their data for the 0.5 – 5 keV band, Bowyer *et al.* (1974) find a spectrum of the form F(E) $= (0.023 \pm 0.006)\ E^{-2.4} \exp\left(-4 \times 10^{21}\sigma(\mathrm{ISM})\right)$ photons $cm^{-2}s^{-1}keV^{-1}$, where the absorption cross section of the interstellar medium to soft X-rays is denoted as σ(ISM). The mean slope for bright Galactic sources is very similar, and it was concluded by Bowyer *et al.* (1974) that the sources in M31 are probably mostly binary X-ray stars similar to those in the Milky Way.

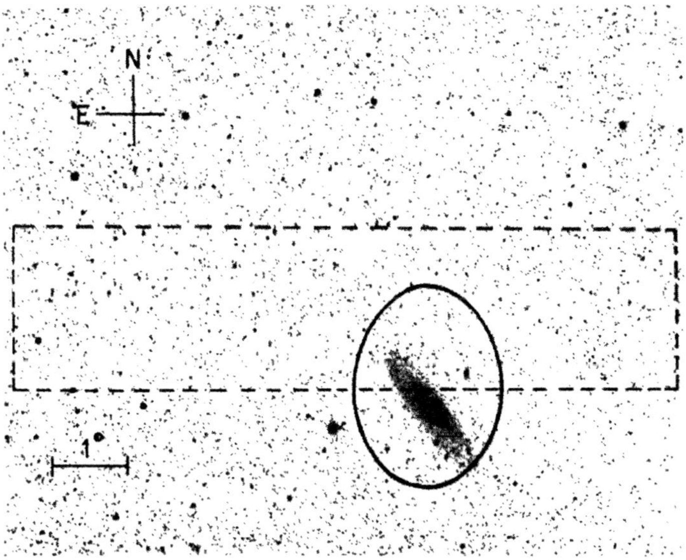

Figure 17.1 Early X-ray observations of M31. The box is the Uhuru detection (Giacconi et al. 1974) and the ellipse is the aperture of the XR1 detector used by Bowyer et al. (1974).

Additional early observations include that of Cooke *et al.* (1978), who listed it as a probable identification of one of the Ariel V detections. Its intensity in the 2 - 10 keV band is given as 4×10^{-11} erg cm^{-2}s^{-1}. This converts to 2.8×10^{39} ergs s^{-1}, in good agreement with the integrated luminosity found by Bowyer *et al.* (1974).

17.2 Einstein Observations

The first detections of individual X-ray sources in M31 came with the imaging capability of the Einstein satellite (van Speybroeck *et al.* 1979). The initial analysis showed the presence of 69 sources (more were extracted from the data later, as discussed below). These were found to include 40 objects that, on the basis of their spatial distribution, belonged to the star-forming regions of the disk. Many were located along the heavily-populated ring at 10 kpc (Figure 17.2). Initially, seven certain and 4 possible identifications were made with globular clusters and one with a supernova remnant (both categories are now more plentifully-identified, as discussed below). At the nuclear position, van Speybroeck *et al.* (1979) found a bright source with an X- ray luminosity nearly 10^3 times as bright as the X-ray source at the nucleus of our Galaxy. They interpreted it as a thermal source, probably from hot gas falling onto the supermassive object at the center of M31.

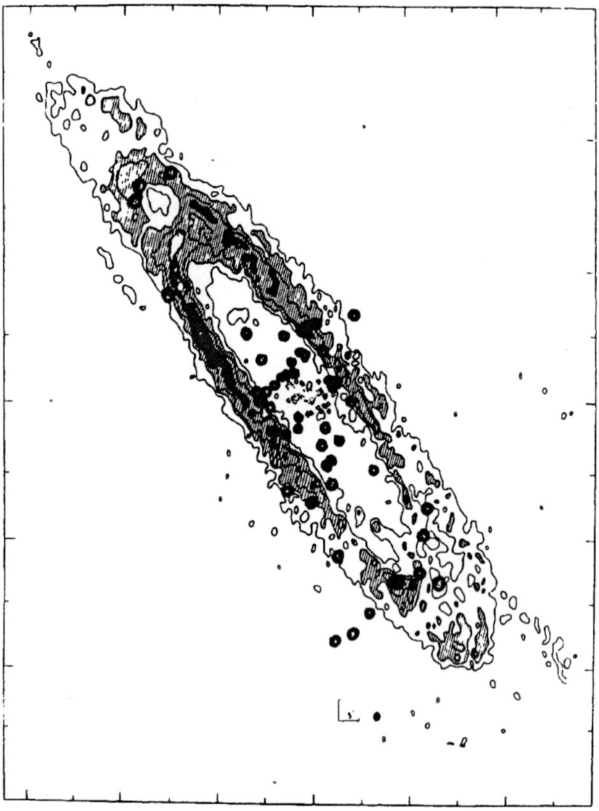

Figure 17.2 The X-ray sources detected by the Einstein satellite (circles), plotted on the HI map of M31 of Emerson (1976) (from van Speybroeck and Bechtold 1981).

Table 17.1 provides a listing of the X-ray sources from the Einstein data, as compiled and reduced by Trinchieri *et al.* (1991). The numbers there will be used throughout this chapter to identify individual sources.

TABLE 17.1
X-ray Sources in M31 Detected with Einstein
(from Trinchieri *et al.* 1991 and Crampton *et al.* 1984)

Src No.	R.A. (1950) (h m s)	Dec (1950) (° ′ ″)	L_x (0.2-4.0 keV) (c.g.s.)	Ident.
1	0 36 51.9	40 19 35	$\geq 3.45 \times 10^{36}$	
2	0 37 18.7	40 15 53.3	$\geq 9.43 \times 10^{36}$	blue star
3	0 37 30.31	40 33 39.2	1.45×10^{38}	
4	0 37 36.88	40 27 33.2	9.40×10^{37}	G52
5	0 37 42.3	40 13 23	$\geq 5.69 \times 10^{36}$	blue stars
6	0 37 42.4	41 12 36	–	G59?
7	0 38 02.8	40 35 06	$\geq 3.79 \times 10^{36}$	
8	0 38 40.7	40 42 34	$\geq 1.02 \times 10^{37}$	foreground star
9	0 38 58.99	41 17 55.1	5.01×10^{37}	G108
10	0 39 08.2	40 57 38	–	
11	0 39 08.5	40 04 57	–	galaxy
12	0 39 23.57	41 01 49.4	1.47×10^{37}	
13	0 39 28.8	40 22 44	9.42×10^{36}	
14	0 39 29.11	41 02 10.3	–	foreground star
15	0 39 31.3	40 03 17	$\geq 2.81 \times 10^{38}$	G145?
16	0 39 31.65	40 44 49.3	2.23×10^{37}	G144
17	0 39 31.65	41 00 54.9	7.24×10^{36}	
18	0 39 32.39	40 39 27.3	9.58×10^{36}	
19	0 39 34.26	40 55 58.5	2.45×10^{37}	
20	0 39 34.51	40 57 37.1	2.83×10^{37}	G148
21	0 39 37.42	40 59 36.2	7.86×10^{36}	
22	0 39 38.37	40 57 08.8	9.13×10^{36}	
23	0 39 38.82	40 59 10.6	3.95×10^{37}	
24	0 39 39.03	40 57 43.2	5.00×10^{36}	
25	0 39 41.04	40 57 14.9	5.62×10^{36}	
26	0 39 41.90	41 02 50.3	4.54×10^{36}	G158
27	0 39 44.27	40 55 58.5	1.94×10^{37}	
28	0 39 44.51	40 48 10.9	1.76×10^{37}	
29	0 39 44.66	41 00 52.5	4.40×10^{36}	
30	0 39 45.6	41 09 35	9.84×10^{36}	
31	0 39 46.75	41 00 24.3	5.11×10^{37}	
32	0 39 46.89	40 59 57.2	2.08×10^{37}	
33	0 39 47.09	41 03 17.7	4.74×10^{36}	G169
34	0 39 47.98	40 56 49.6	1.75×10^{37}	near nucleus
35	0 39 48.89	41 00 05.4	–	near nucleus
36	0 39 49.78	40 59 59.4	5.15×10^{36}	
37	0 39 50.3	40 48 08	1.50×10^{37}	
38	0 39 50.3	41 16 22	8.99×10^{36}	G178?
39	0 39 50.61	41 00 20.5	2.24×10^{37}	
40	0 39 54.14	41 13 45.3	9.15×10^{36}	blue star?

Src No.	R.A. (1950) (h m s)	Dec (1950) (° ′ ″)	L_x (0.2-4.0 keV) (c.g.s.)	Ident.
41	0 39 54.38	40 59 39.5	2.37×10^{38}	
42	0 39 55.29	40 58 03.7	$< 8.1 \times 10^{36}$	
43	0 39 55.32	40 59 50.1	3.40×10^{37}	
44	0 39 55.68	40 59 23.3	1.58×10^{37}	
45	0 39 57.11	40 59 01.1	6.17×10^{36}	vd B2
46	0 39 57.15	40 55 34.2	5.00×10^{36}	
47	0 39 57.79	41 00 13.3	1.46×10^{37}	
48	0 39 58.04	40 58 20.1	5.59×10^{36}	
49	0 39 58.07	40 59 29.0	1.11×10^{37}	
50	0 39 58.22	41 00 35.0	7.05×10^{36}	
51	0 39 58.31	40 35 29.6	5.27×10^{37}	in M32
52	0 39 58.72	40 59 19.4	9.29×10^{36}	
53	0 39 58.97	40 58 45.8	5.64×10^{36}	
54	0 39 59.08	41 00 19.7	1.04×10^{37}	
55	0 39 59.52	41 00 07.3	1.91×10^{37}	
56	0 40 00.19	40 59 43.6	5.21×10^{37}	
57	0 40 00.53	40 59 52.7	2.35×10^{37}	
58	0 40 00.68	40 55 13.6	9.17×10^{36}	blue star
59	0 40 02.56	40 59 52.6	1.04×10^{37}	
60	0 40 02.90	41 00 04.6	4.07×10^{37}	
61	0 40 03.76	40 59 08.5	1.04×10^{37}	
62	0 40 04.01	41 08 59.0	1.98×10^{37}	blue star
63	0 40 04.35	40 58 57.3	4.62×10^{37}	
64	0 40 04.99	41 01 51.5	1.07×10^{37}	
65	0 40 06.20	40 58 54.1	4.67×10^{36}	
66	0 40 06.70	40 59 18.0	5.24×10^{36}	
67	0 40 07.48	41 14 44.3	6.87×10^{37}	G192
68	0 40 08.14	41 02 29.5	3.24×10^{37}	
69	0 40 08.26	40 59 15.8	1.03×10^{38}	
70	0 40 10.1	41 09 31	2.17×10^{37}	
71	0 40 10.65	40 59 39.7	4.21×10^{37}	
72	0 40 11.60	41 00 45.6	4.41×10^{36}	
73	0 40 13.66	40 54 42.2	1.80×10^{37}	blue star?
74	0 40 14.09	40 59 05.4	1.20×10^{37}	
75	0 40 14.16	40 52 02.3	2.31×10^{37}	
76	0 40 15.30	41 02 55.4	9.40×10^{36}	G198
77	0 40 15.56	40 59 41.3	1.21×10^{37}	Bo 144
78	0 40 17.52	40 58 11.8	9.93×10^{36}	near nucleus
79	0 40 18.69	40 58 59.1	1.88×10^{37}	Bo 146
80	0 40 19.55	41 01 42.1	1.10×10^{37}	G200
81	0 40 23.3	40 31 55	–	foreground star

Src No.	R.A. (1950) (h m s)	Dec (1950) (° ′ ″)	L_x (0.2-4.0 keV) (c.g.s.)	Ident.
82	0 40 25.38	41 02 34.3	1.17×10^{37}	
83	0 40 26.21	40 58 27.5	4.32×10^{37}	Bo 153
84	0 40 27.16	41 01 43.4	1.14×10^{37}	
85	0 40 30.15	40 50 58.3	2.45×10^{37}	G213
86	0 40 34.3	41 42 50	$\geq 5.49 \times 10^{36}$	
87	0 40 34.50	41 03 54.4	1.24×10^{37}	
88	0 40 38.8	41 01 46	7.53×10^{36}	blue stars?
89	0 40 43.56	41 02 08.1	1.43×10^{37}	blue stars?
90	0 40 43	40 51 22	9.44×10^{36}	blue star
91	0 40 49.98	40 57 04.2	–	A star
92	0 40 52.96	40 58 23.8	2.72×10^{37}	G235
93	0 40 59.97	41 07 48.1	1.64×10^{37}	B star
94	0 41 09.11	41 00 33.2	2.33×10^{37}	blue cluster
95	0 42 13	40 30 07	9.41×10^{36}	
96	0 41 18.74	41 14 15.4	2.01×10^{37}	G264
97	0 41 44.78	41 05 13.8	4.94×10^{37}	OB assoc.
98	0 42 12	40 43 34	1.19×10^{37}	
99	0 42 28.3	41 20 40	$\geq 5.91 \times 10^{36}$	
100	0 42 37.5	41 29 13	4.10×10^{36}	
101	0 42 54.94	41 51 41.6	2.87×10^{37}	blue star
102	0 43 00.49	41 23 21.9	2.48×10^{38}	G307
103	0 43 24.2	41 52 17	5.46×10^{36}	
104	0 43 37.9	41 48 26	$\geq 4.29 \times 10^{37}$	
105	0 43 41.38	41 45 30.0	3.17×10^{37}	G322
106	0 43 42.6	41 31 30	4.10×10^{36}	
107	0 44 15.0	42 04 25	$\geq 2.04 \times 10^{37}$	
108	0 45 14.2	41 24 37	$\geq 2.56 \times 10^{37}$	

17.3 X-Ray Globular Clusters

A discussion of the Einstein sources was given by van Speybroeck and Bechtold (1981). By this time, seventeen sources were found to be coincident with globular clusters. The mean X-ray luminosity of the globulars that they identified is 5.2×10^{37} ergs s^{-1}, while the brightest X-ray globular in the local Galaxy has a luminosity of only 3.5×10^{37} ergs s^{-1}; the M31 globulars reach much higher luminosities than our local sample, either because of a physical difference in the populations or because of sampling effects. The latter is probably at least partly a cause, as we now have about 300 globular cluster identifications in M31 (Chapter 9), while only about 150 globulars are estimated for the Milky Way Galaxy. Einstein would only have detected one of our globulars, but with a population of clusters that is two times larger, at least some of the excess X-ray clusters would be expected, were the X-ray luminosity functions similar. Long and van Speybroeck (1983) suggest that the higher luminosities of the M31 sources may also be related to the greater metals abundance range observed for them.

Battistini *et al.* (1982), using the Einstein positions, examined their optical images of M31 globular clusters in order to look for structural properties of the X-ray and non-X-ray globulars, to see if there is any difference. They examined 21 globulars and concluded that the M31 clusters with X-ray sources are more luminous than those in the Galaxy. They suggest that this may be a population effect, as discussed above, and that it may also be caused by possibly different dynamical interactions of the clusters with the nucleus. For both galaxies, the X-ray globulars tend to be nearer the galaxy's center and of higher metallicity than the average globular cluster. This last conclusion, however, may not hold up when larger samples of clusters are included. For example, Crampton *et al.* (1984) found identical metal abundances for their sample of clusters when they compared X-ray clusters with non-X-ray clusters, using Sharov and Lyuty's (1983) colors, transformed into abundance parameters.

Crampton *et al.* (1984) used the CFHT to obtain slitless spectra of the Einstein sources and found 23 objects that could be identified with globular clusters; nineteen are considered very strong identifications. Twelve of the sources are among the known or suspected variable X-ray sources, a fact consistent with the prevalence of variability among the X-ray sources in Galactic globulars.

A detailed look at one of the M31 X-ray clusters, X-ray source No. 85, cluster G213, also known as Bo 158 and V64 (Figure 17.3), was made by Cacciari *et al.* (1982), who used the X-ray data, ultraviolet spectra from the IUE and optical data (Figure 17.4). They concluded that this cluster is exceptionally luminous ($M_V = -10.0$) and massive (10^6 solar masses). Its optical, UV and X-ray spectra are similar to those of X-ray globular clusters in the Milky Way. Its spectral type (F8) is early, suggesting the possible presence of a well-populated blue horizontal branch, but the color and metallicity (B-V = 0.80 and [Fe/H] = –0.75) suggest a red horizontal branch cluster. This apparent discrepancy remains unresolved.

17.4 Stellar Sources

In the Crampton *et al.* (1984) search for optical counterparts for the Einstein sources, 13 were found to lie near blue and very blue stars. The luminosities of the stars correspond to those of the brightest supergiant X-ray binaries, like Cen X-3, in our Galaxy, and thus it is (statistically, at least) probable that the M31 objects are also such stellar sources. Some of the identifications are for variable X-ray sources, further supporting the idea that they may be similar to Galactic massive X-ray binaries.

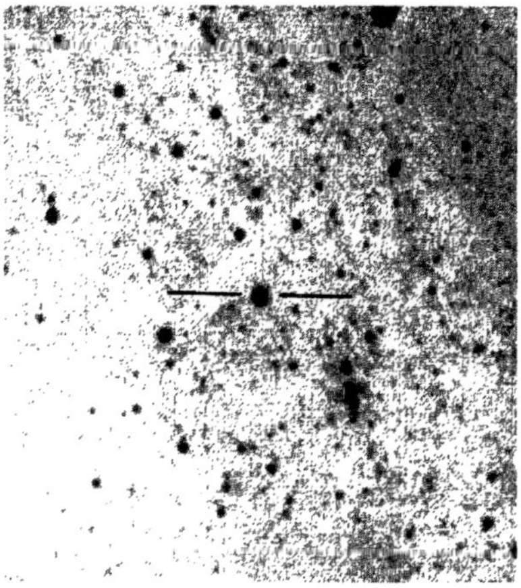

Figure 17.3 One of the brightest X-ray globular clusters in M31, G280. It has an apparent magnitude of $V = 14.18$ and lies superimposed upon the densely populated spiral structure. The bright stellar complex below it and to the right is cluster C410. (Kitt Peak National Observatory 4m telescope photo).

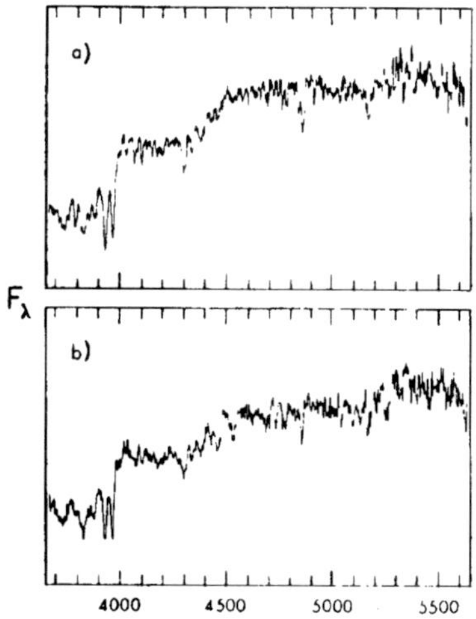

Figure 17.4 Optical spectra of two M31 globulars: (a) Bo158, an X-ray cluster and (b) G72, not identified as an X-ray source (from Cacciari et al. 1982).

Collura, Reale and Peres (1990) looked in detail at the Einstein data for two X-ray sources for indications of variability. They found that one of them, source 25I, which is number 101 in Table 17.1, has properties that strongly indicate that it is a massive X-ray binary. It is identified with a faint OB star with $m_{pg} = 20.5$ (Figure 17.5). Its X-ray luminosity is 10^{37} ergs s^{-1}, appropriate for such an object. Collura *et al.* (1990) attempted to find a periodicity in its variability and they detected a possible systematic variation with a period of 49,500 sec (13.75 hours). However, the period is very poorly constrained (Figure 17.6) and this result is tentative at best. If truly the period of a massive binary, it would be unusually short.

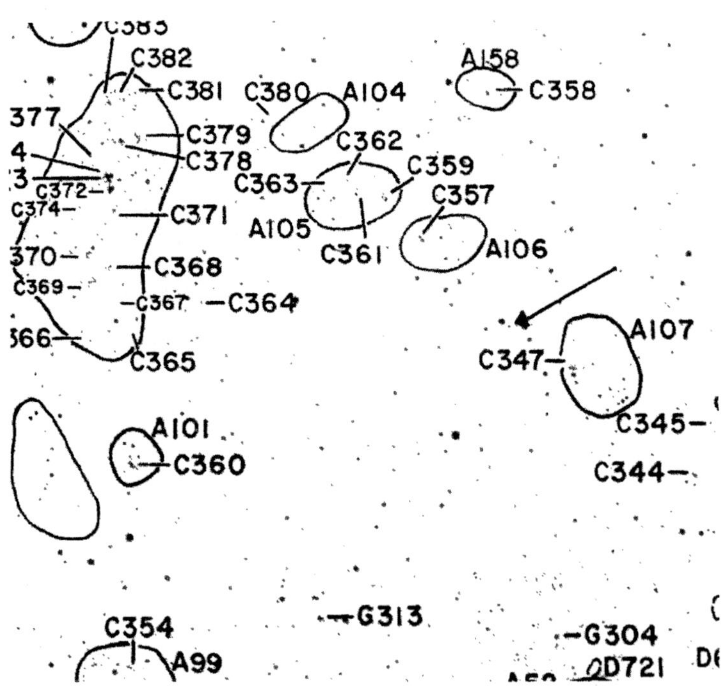

Figure 17.5 The northeast portion of the M31 disk with the candidate for X-ray source 101 indicated by the arrow. It lies in an interarm area (from the ATLAS OF THE ANDROMEDA GALAXY).

Trinchieri *et al.* (1991), using a complete array of Einstein data, concluded that, of the 108 individual sources, sixteen are found to vary, although considerably more observational data will need to be gathered to determine the nature of their variation.

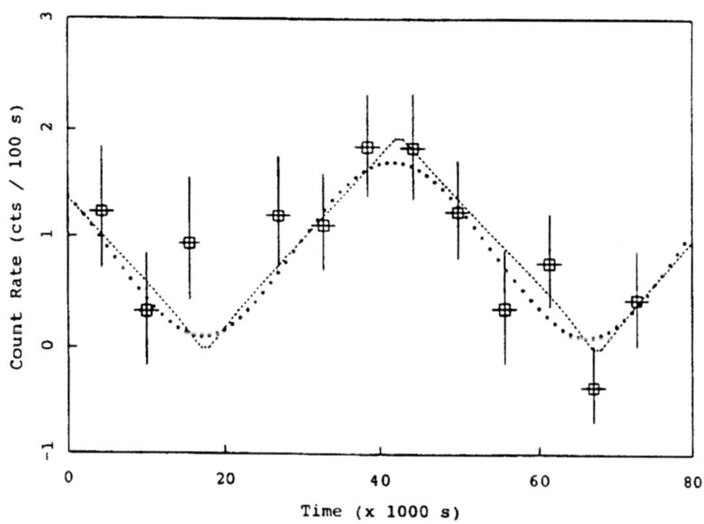

Figure 17.6 The tentative X-ray light curve of source 101, shown in Figure 17.5. The dashed and dotted lines represent attempts to model the curve (from Collura et al. 1990).

17.5 A Gaseous Component?

There is a significant amount of unresolved X-radiation coming from the bulge of M31 (Figure 17.7) and this is now thought to be mostly emitted by many low-mass X-ray binaries, although there is also a possibility that some of the X-rays come from a hot gas component, as for high-luminosity X-ray galaxies (see Section 17.6). The X-ray spectrum of the bulge has been examined from the Einstein soft X-ray data (Fabbiano *et al.* 1987) and the Ginga harder X-ray data (Makishima *et al.* 1989, Figure 17.8). Both analyses agree that the spectrum from 0.5 to 20 keV indicates that the source is most likely a composite of the spectra of several low-mass X-ray binaries. Makishima *et al.* (1989) found that of the total energy emitted in the 2-20 keV band, 5×10^{39} ergs s^{-1}, less than 20% could be coming from diffuse X-ray components. Optical and radio data show that there is some but not much cool gas in the bulge. From the X-ray data, Trinchieri *et al.* (1991) estimate that there can be no more than

2×10^6 solar masses of hot gas in the bulge. Combined with the HI measurements, this upper limit indicates that there must be a galactic wind present; otherwise more gas should be there; the present mass loss

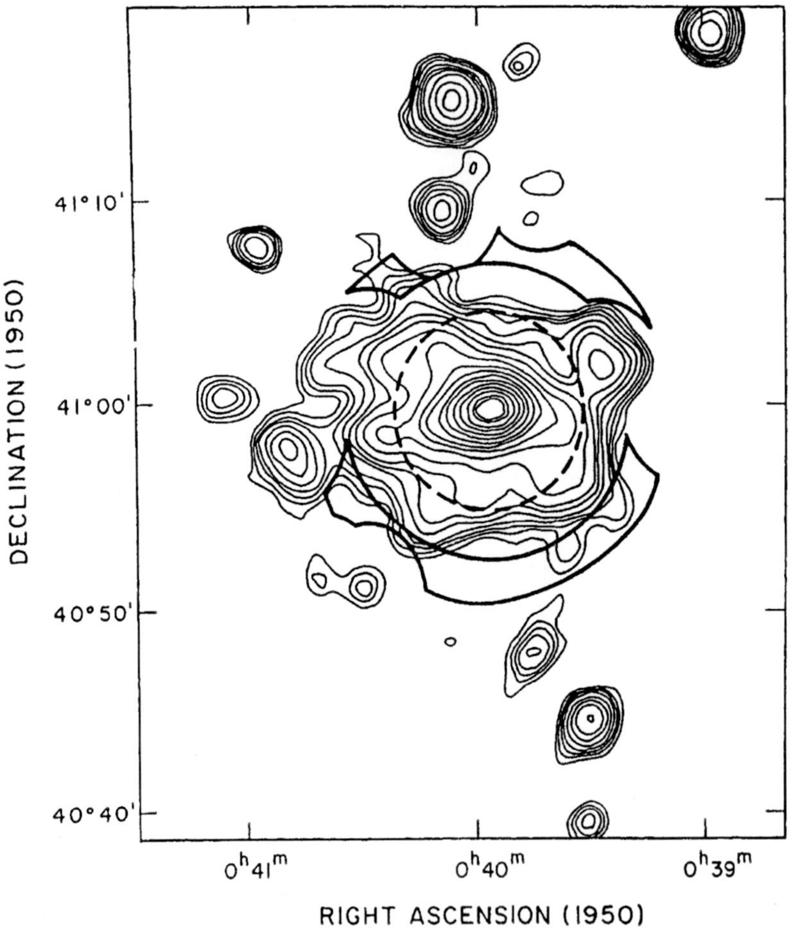

Figure 17.7 Contour of the X-ray emission from the bulge of M31. The first contour is 2σ above the background level and the dashed circle is the area used for spectral analysis described in the text. The solid lines delineate the adopted area used for background values (from Fabbiano et al. 1987).

rate for the bulge is 0.05 solar masses per year (Jacoby 1980), and therefore the total mass of gas released into the bulge over the lifetime of M31 (if at this rate) should be 8×10^9 solar masses. Apparently, most of this has been blown away.

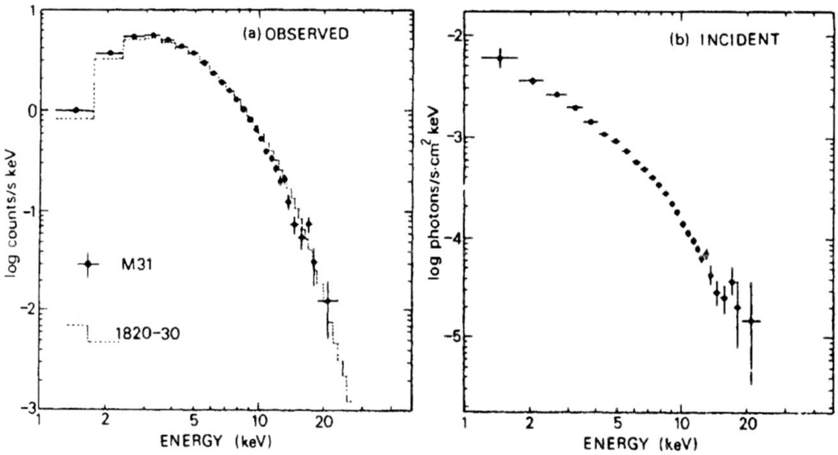

Figure 17.8 X-ray spectra of M31 from the Ginga satellite. (a) raw pulse-height spectrum compared to that of the Galactic low-mass X-ray binary 4U 1820-30; (b) inferred deconvolved incident spectrum (from Makishima et al. 1989).

17.6 Supernova Remnants

Blair *et al.* (1981) found that two of the sources were coincident with known supernova remnants: one source (number 82 in Crampton *et al.* 1984, but not listed in Table 17.1) corresponds to Baade-Arp (BA) No. 521 and source 89 corresponds to BA No. 23. No source could be found at the position of the supernova of 1885.

17.7 Summary

There are about 100 known X-ray sources in M31. Many sources are located in the 10 kpc ring area, where star formation is going on. Some sources are coincident in position with blue stars that may be supergiant X-ray binaries. Nearly twenty sources appear to be in globular clusters including several that are brighter in X-rays than any globular cluster source in our Galaxy. Sixteen sources observed by the Einstein satellite appear to be variable; these also may turn out, on further study, to be massive X-ray binaries. The central, unresolved X-ray emission is probably emitted by many low-mass binaries, but some may also come from hot central gas. There are two known coincidences with positions of supernova remnants.

CHAPTER 18

STARS AND STELLAR POPULATIONS

Many of the previous chapters have dealt with stars of various kinds and with the mix of stars and interstellar material that makes up the stellar population of M31. This chapter looks at the remaining aspects of these subjects: ordinary stars in Andromeda and what they tell us about the population of stars in the galaxy.

A particularly rewarding review of the entire subject of stellar populations in M31 has recently been published by van den Bergh (1991).

18.1 Stellar Populations

In some sense, the modern idea of stellar populations had its origins in M31. It was there that Baade (1944a, b) first obtained the evidence that two distinct populations exist, leading to the concepts of Populations I and II. His ability, during 1944, to resolve both the elliptical galaxy companions and the central bulge of M31, showed him the distinctive nature of Population II and made clear its differences from the already-well-resolved Population I of the spiral arms and disk (Baade 1958).

In the two-population framework, a study by Baum and Schwarzschild (1959) was an important step in the interpretation of quantitative integrated data, which preceded the later attempts at more detailed population synthesis (Section 18.2). Their ingenious scheme took advantage of the fact that the upper part of the luminosity function of Population I differs greatly from that of Population II (as defined by globular clusters). They showed that one could compare star counts to a certain (bright) absolute magnitude with the integrated magnitude of the same region, and one would obtain significantly different results for a Population I region compared to Population II. They tested the scheme on several regions in outer parts of M31 and in the E companion galaxy

NGC 205, with the conclusion that the M31 population agreed with a solar neighborhood Population I relation, while the NGC 205 data indicated good agreement with Population II (Figure 18.1).

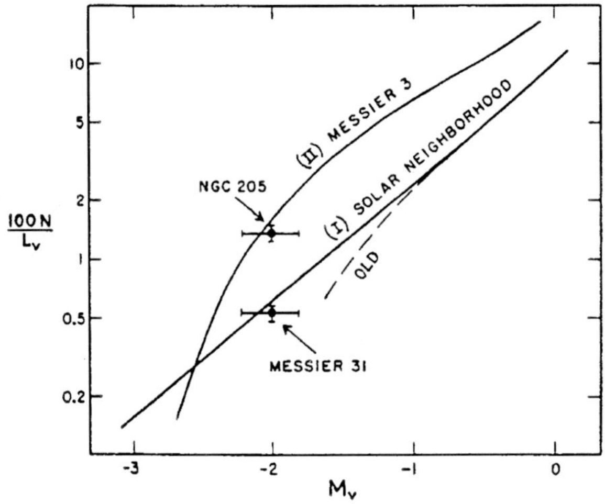

Figure 18.1 Population differences between M31 and NGC 205, detected by Baum and Schwarzschild (1959).

The concept of two populations was an extremely important step in the development of our understanding of both stars and galaxies. It helped to make sense out of the growing field of stellar evolution and it helped to lay the groundwork for an understanding of the Hubble sequence of galaxies and for theories of galaxy evolution. However, as in the case of many great, unifying concepts, time eventually showed that the truth is more complicated and that Baade's simple scheme is inadequate to explain the details that we can now learn about stars and galaxies (*e.g.*, see Hodge 1989).

Recent work on the stellar populations in M31 has served both to expand our understanding of the history of the galaxy and to refine the ways in which the simple Population I/Population II dichotomy fails. Figure 18.2 displays the simple two-point Population Diagram, used originally to describe the stellar population of M31, while Figure 18.3 shows a "Population Box", in which three parameters are allowed to vary continuously (time, heavy-element abundance, and star-formation rate). A schematic population box for M31, based on ideas prevalent in 1989, but not on details of stellar numbers, is shown in Figure 18.4. A more realistic population box for M31 would have a highly-textured surface, representing the many different and differently-timed sites of star formation throughout the galaxy.

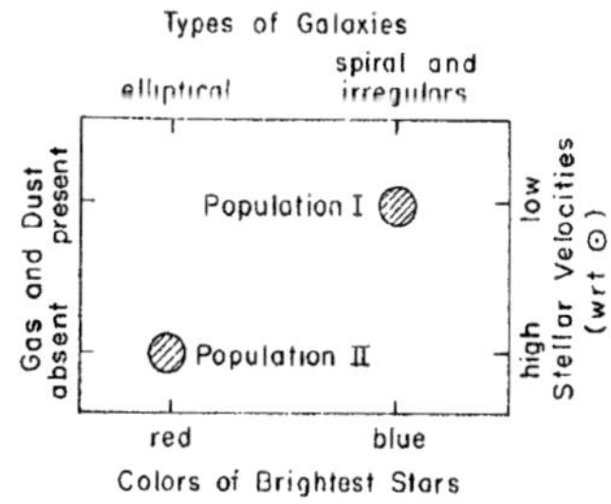

Figure 18.2 The two-point population diagram, as originally formulated by Baade (from Hodge 1989).

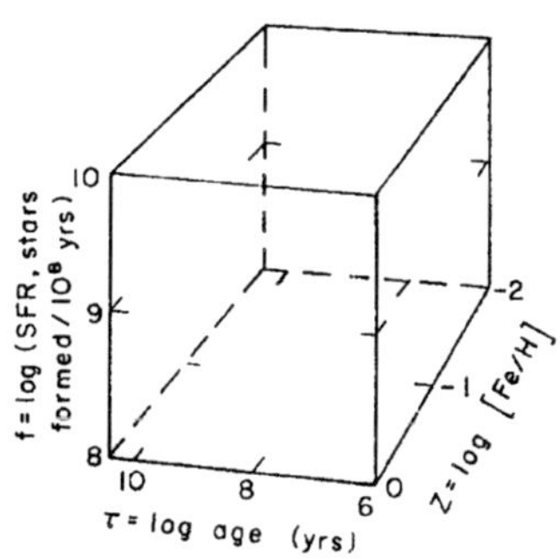

Figure 18.3 The "population box" (Hodge 1989).

18.2 Population Synthesis

Quantitative attempts to determine the details of the stellar population of M31, beyond mere population types, began with intermediate- and narrow-band photometry and attendant attempts to fit the spectral energy distribution, including absorption line intensities, with various mixes of stars. For example, an early population synthesis for M31 was based on five-color photometry of the central region (McClure and van den Bergh 1968). A model for M31 was derived by assuming the presence of a reasonable assortment of star types, and by fitting the energy distributions of these types to the observed galaxy energy distribution. Table 18.1 summarizes their results for the relative numbers of different kinds of stars in M31.

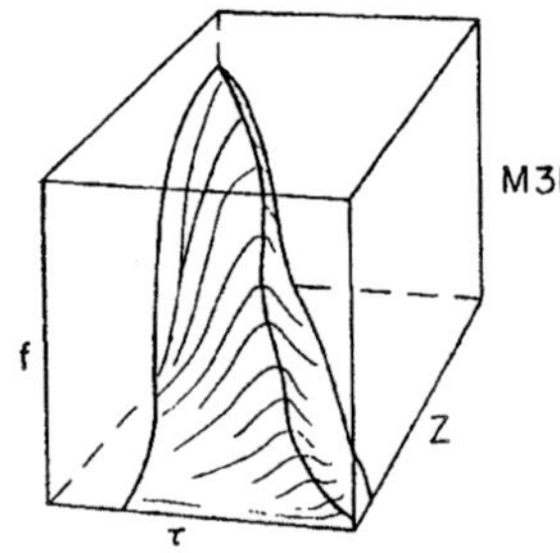

Figure 18.4 A schematic representation of the stars in M31 in the population box (Hodge 1989).

Table 18.1
Population Models for M31 (from McClure and van den Bergh 1968)

Star Type	percent contribution to V light	number of stars (percent)
F5 V	4	0.0464
G0 V	7	0.204
G8 V	9	0.601
K2 V	10	1.53
K6 V	15	10.0
M1 V	6	17.5
M5 V	2	70.1
G8 IV	2	0.0161
K0 III	40	0.0509
K5 III	5	0.00211

Spinrad (1962 and many subsequent references) had already pioneered the process of deriving population data from galaxy spectra, an effort which culminated to some extent in the famous paper of Spinrad and Taylor (1971) on the stellar content of the nuclear region of M31. They used spectrum scanner observations obtained with the Lick Observatory 3-m telescope to measure the energy distribution in the inner 10 arcsec of M31 from wavelengths of 3300 Angstroms to 10700 Angstroms, supplemented by infrared photometry to 3.4 microns. Spectral features that were included in the analysis were atomic lines of Na I, Ca I, Ca II, Mg, and H and molecular bands of TiO, CN, CH, CaH, and MgH. Assuming that the luminosity function is smooth and continuous, the line strengths and the continuum energy distribution were found to give reasonably unique fits. One somewhat surprising (and controversial) result of the analysis was that the M31 central region was found to have a strongly dwarf-enriched lower main sequence, with the dwarfs contributing about 30% of the light, rather than a few per cent, as in the solar neighborhood. Old stars were indicated, with a main sequence turn-off at a magnitude somewhat above the sun's luminosity; only a few stars younger than 4 billion years were detected. The line indices indicated that the metal abundance of the stars was greater than solar and no metal-poor Population II stars contributed appreciatively to the integrated light. Figure 18.5 shows Spinrad and Taylor's comparisons of data with two models of the M31 nucleus.

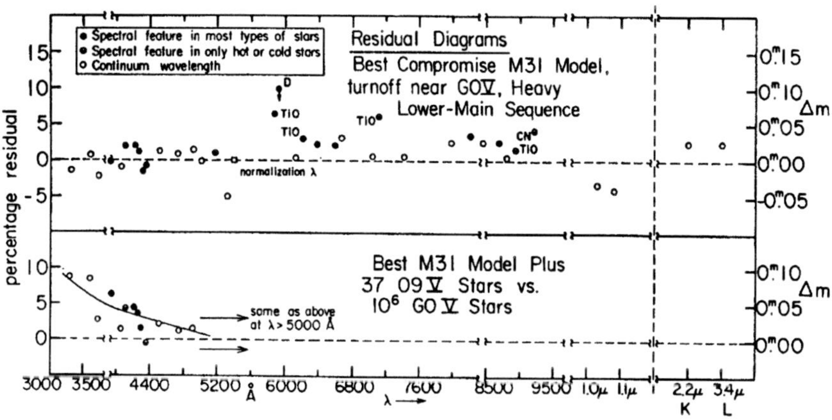

Figure 18.5 A comparison of M31 spectrophotometric data with two models of M31. Large residuals are seen for the sodium D lines, red Ti0 bands and the near-infrared CN band (from Spinrad and Taylor, 1971).

While the Spinrad and Taylor (1971) result seemed to raise some interesting questions about the stellar population of the M31 nucleus, it was clear from McClure and van den Bergh's (1968) work that the nucleus was not necessarily representative even of the stars in the bulge. The latter paper showed that the strong cyanogen lines found in the nuclear region were not found in the amorphous bulge just outside of the nucleus. Following up this result, Spinrad *et al.* (1971) examined the spatial variation in a number of lines outwards to a distance of about 1 kpc from the nucleus and concluded that the CN lines decreased abruptly, the Na lines dropped slowly, and the Mg triplet remained nearly constant with distance. They constructed a model of the disk population of M31, concluding that the stars have abundances and numbers very like the stars in the solar neighborhood.

O'Connell (1976), in his careful spectral synthesis work on elliptical galaxies, included M31's central region (with $r < 50$ pc) for comparison. He found that the nucleus of M31 is very similar to those of giant ellipticals, with most of the light coming from old stars (age $>\sim 8$ billion years), with high heavy element abundances. Only marginal evidence was found for an enhanced population of low-mass main sequence stars. Table 18.2 summarizes the characteristics of the M31 solution.

Table 18.2
Populations in M31, Compared to E Galaxies (from O'Connell 1976)

	E galaxies	M31
M/L_V (solar units)	11.4	9.7
Percent at λ5050 from K0-3 III SMR	25.3	35.5
Other important groups	K0-1 V	G8 V
	K3-4 V	F5-8 V
Percent at λ5050 from		
"hot" stars (earlier than F5)	3.7	0.2

A more recent attempt to establish whether or not there is a dwarf star enrichment in the M31 nuclear area was made by Faber and French (1980), who obtained new observations of the Na I λ8190 line. They concluded that the line is stronger in the semi-stellar nucleus than in the surrounding bulge, supporting Spinrad and Taylor's (1971) claim for dwarf-enrichment. Other indicators, such as FeH (Whitford 1977) and H_2O and CO (Baldwin *et al.* 1973) have pointed to a giant-dominated nucleus, however.

Various concerns have been expressed about the population synthesis results that suggest that we should treat the conclusions with some caution. Faber (1972), Taylor and Kellman (1978), Peck (1980), and others have examined the problems of incorporating the many variables, including line strength anomalies and observational errors, with the conclusion that this kind of analysis is more difficult than might be thought.

18.3 High-luminosity Stars

The most conspicuous Population I stars in M31 are, of course, the stars of high luminosity: OB stars, WR stars, and red supergiants. Especially with the advent of high-efficiency detectors and sites of good seeing, these kinds of stars in M31 are readily observable spectroscopically, and a growing amount of quantitative information is being obtained about them. Particular questions being asked, for which they are especially useful, include the question of heavy-element abundance (which probably decides the mix of types of WR stars and which determines the stellar winds observed), the question of the maximum luminosities reached (which, for both OB stars and red supergiants, calibrates one of the extragalactic distance parameters), the issue of the shape of the upper initial mass function, and the star-formation rate.

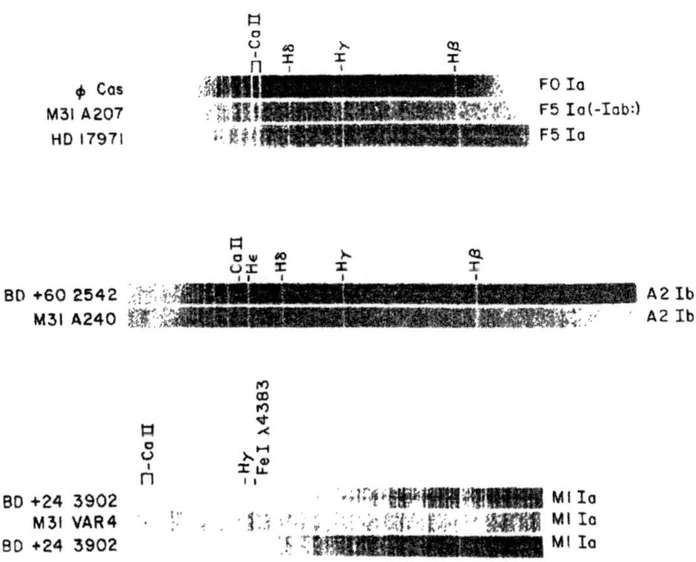

Figure 18.6 Spectra of three M31 supergiants, compared to Galactic stars (from Humphreys 1979).

OB Stars

Although bright blue stars were recognised in M31 long ago, the pioneer in their detailed study, especially by spectroscopic means, was Humphreys (1979). Her initial study examined the spectral properties of stars in

Baade's Field IV, where the color-magnitude diagram plotted by Swope (Baade and Swope, 1963) allowed a preselection on the basis of color. Of the 14 stars for which spectral types were obtained, nine turned out to be foreground stars. The five M31 members included three early-type stars, an F supergiant and an M-type variable. No O-type stars were found, the early-type stars being of types B0 I, B2 Ia, and A2 Ib. Figure 18.6 shows spectra of three M31 supergiants from her sample. Earlier-type stars tend to be fainter, main sequence objects, while these were all supergiants, with absolute magnitudes determined to be in the range −6.0 to −7.4. Humphreys also found, from colors, a number of suspected supergiant members in the area, for which absolute magnitudes were a magnitude or so fainter. Figure 18.22 (in Section 18.6 below) shows her color-magnitude diagram for the stellar associations in the area and shows the positions of the two B supergiants and two of the fainter suspected supergiants. Clearly, the supergiants lie above the association's main sequence, in a position expected for stars just evolved off of the high-mass end.

Additional information became available when Massey *et al.* (1985) obtained ultraviolet spectra of M31 stars. The star examined first, known as CHFT 3, was a double or multiple image with a composite spectrum. Balmer lines, He I and He II lines, and OII lines are present in the optical spectra. The UV spectrum shows C IV and Si IV lines. Although the separation of the components was not certain because of uncertainties about the reddening, the cooler star was thought to be of type B2, while the hotter component was an early or mid-O type. The reddening law for this star fits the LMC law rather better than the Galactic law. Figure 18.7 shows the IUE spectrum of M31 CFHT 3, compared to that of a similar star in M33.

A later study of two M31 blue supergiants by Hutchings *et al.* (1987) explored the unexpected lack of emission lines in the UV spectrum of CFHT 3. Using 16- hour long exposures, Hutchings *et al.* (1987) obtained reasonably well-exposed spectra of two stars in M31's OB associations, OB 78-277 and OB 69-WR2. The former had been found to have an optical spectral type of O6-O7 by Massey *et al.* (1986). However, its line spectrum did not seem to be similar to those of Galactic supergiants of this spectral type. In particular, the line of N III at λ4634-41 was seen in absorption, rather than emission, and the ultraviolet lines of C IV and Si IV did not show the normal P Cygni profiles that would be expected for such a hot Galactic star. Further work on this star by Humphreys *et al.* (1990), however, indicated that its spectral type must be B1 I, somewhat later than found by Hutchings *et al.* (1987), resolving at least some of the star's peculiarities. Figure 18.8 shows the optical spectrum of OB78-277 obtained with the MMT by Hutchings *et al.* (1987).

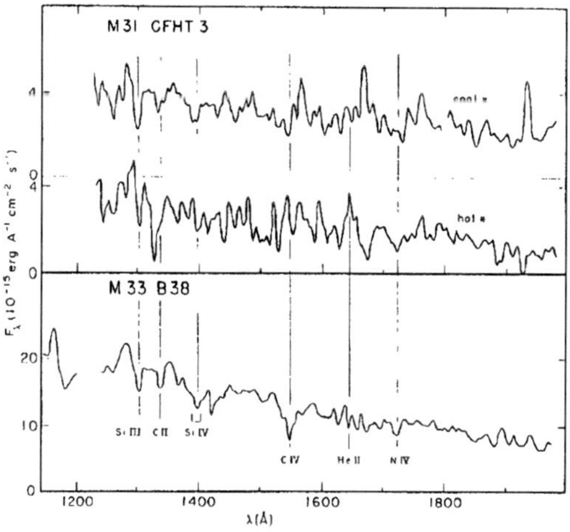

Figure 18.7 Ultraviolet spectra of the composite star, M31 CFHT 3, in which the two components are shown separately. Both are compared to the supergiant star B38 in M33 (from Massey et al. 1985).

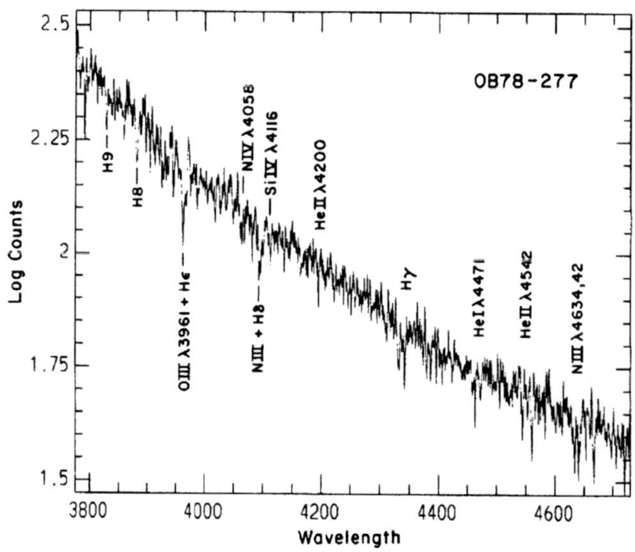

Figure 18.8 The optical spectrum of the O6 star, OB 78-277, in M31, showing Balmer, He I and He II lines (from Hutchings et al. 1987).

A more comprehensive spectroscopic study of the luminous blue stars in M31 was completed by Humphreys *et al.* (1990). A total of 53 candidates was examined spectroscopically and 26 stars were identified as early-type members of M31, with spectral types ranging from O8 to F8. Figure 18.9 shows the temperature- luminosity diagram of these stars, which suggests that there may be a lack of massive stars in M31, especially for masses greater than ~60 solar masses. In any case, the numbers of visually-brightest supergiants is smaller than would be expected from the numbers of fainter main sequence stars, possibly indicating a smaller upper main sequence cut-off mass than in the Galaxy or the LMC.

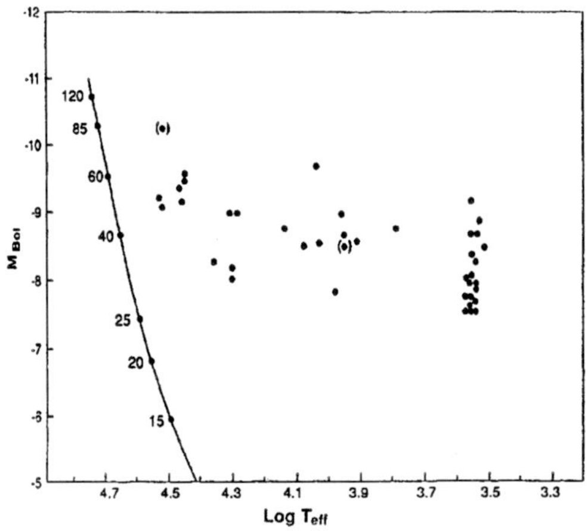

Figure 18.9 The temperature-luminosity diagram for the supergiants studied by Humphreys et al. (1990). The positions of appropriate masses for main sequence stars are plotted on the zero-age main sequence.

The ability to locate, and study in detail, the OB stars of M31 was greatly increased by the completion of a global survey of the stars of M31, performed by an automated plate scanner (Berkhuijsen *et al.* 1988). The catalog includes magnitudes, colors, and positions of 11438 stars, ranging in V from 11.5 to 20. Scans of two neighboring fields to the same limits turned up 8778 stars, suggesting that approximately ~2700 stars in the catalog are members of M31, the rest being foreground stars. In their analysis of the catalog, Berkhuijsen and Humphreys (1989) estimated that there are approximately 300 O stars in M31 among the nearly 700 blue stars detected. Figure 18.10 shows the spatial distribution on the sky of the bluest stars in the sample. The survey detected about 30% more blue stars in the northeastern half of the galaxy than in the southwestern, suggesting a difference (at least temporary) in the star formation rate across the disk (see Figure 18.11 for a schematic illustration of how the differential rotation of M31 will redistribute the different areas of active star formation in the future). A statistical analysis of the data gives a luminosity function for stars with U – V ≤ -0.9 mag that has a slope that is very similar to the slopes found for other galaxies of the Local Group (Freedman 1985). The Berkhuijsen *et al.* (1988) survey provides a rich source of candidate stars for further detailed spectroscopic work.

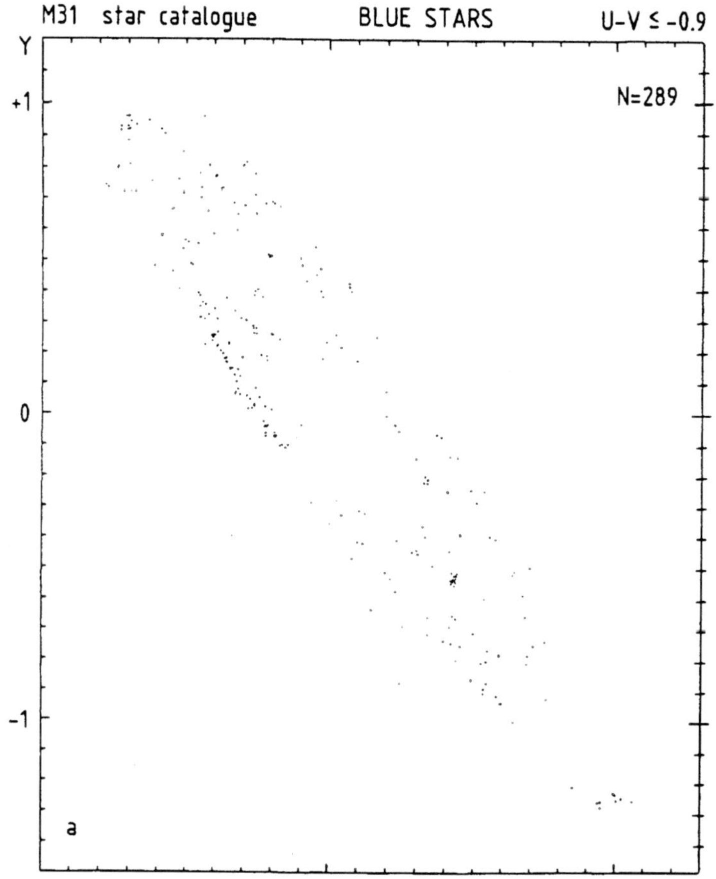

Figure 18.10 The positions on the sky of the bluest stars found in the Berkhuijsen and Humphreys (1989) survey, showing their concentration to the spiral arm structures.

A related survey, based on blink microscope searches of four plates of M31 taken with the 2-m telescope of the Rozhen Observatory, which has a larger plate scale than the Schmidt plates used by Berkhuijsen *et al.* (1988), was published by Nedialkov *et al.* (1989). The 113 blue stars with U–V < 0 are all fainter than V $= 16$; only two are brighter than V $= 17$. The authors point out that this is a very different result from that for M33, where there are 66 such stars. The difference is another indication that the star formation rate and/or the initial mass function of M31 differs from that of later type galaxies of the Local Group, a result also touched on by several other indicators (*e.g.*, see Chapters 10, 11, 13, and 14). Approximately 50% of the blue stars are in common with the catalog of Berkhuijsen *et al.* (1988); the Rozhen survey had a brighter faint limit, but was more complete in detecting and separating blue stars in the more crowded regions.

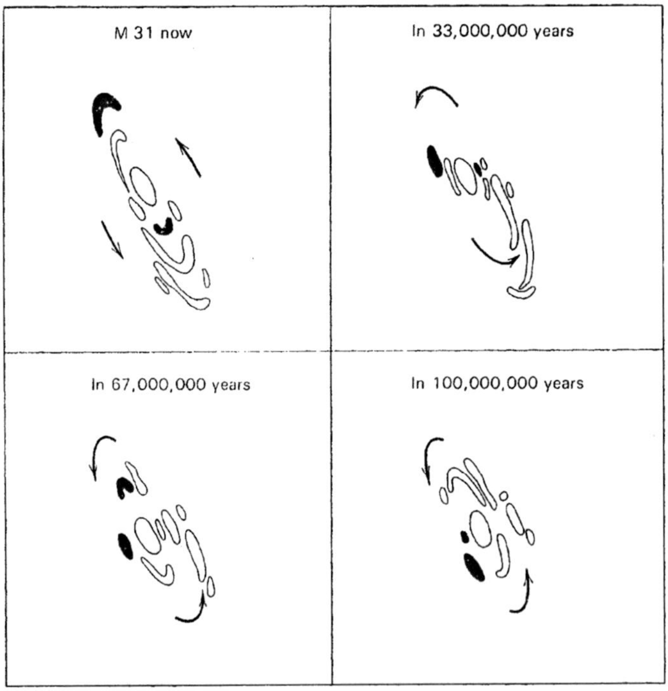

Figure 18.11 An illustration of the effect of differential rotation on the radial distribution of concentrations of OB stars in M31, showing that the present concentrations to the NE and SE will be noticeably rearranged in 10^8 *years (from Hodge 1979). Two star clouds are blackened to make them easy to follow around.*

Wolf-Rayet Stars

The first systematic searches for Wolf-Rayet stars in M31 were carried out by Moffat and Shara (1983, 1987). Figure 18.12 shows the areas searched and plots the positions of the spectroscopically-confirmed W-R stars found. They concluded that the W-R stars of M31 preferentially lie in the ring of most active star formation at ~10 kpc from the nucleus and that a gradient exists in the ratio of early to late-type WC stars as a function of galactocentric radius, which they tentatively attributed to a composition gradient. Furthermore, they estimated that the number density of W-R stars is an order of magnitude smaller in M31 than in M33.

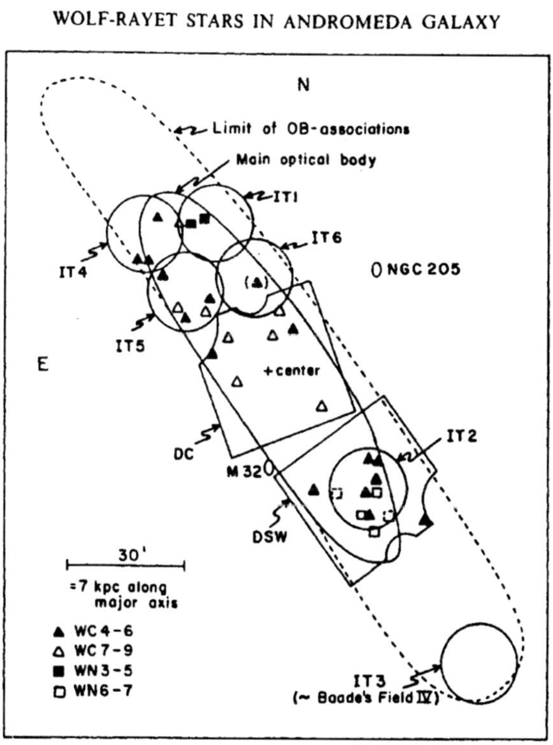

Figure 18.12 The positions of WR stars in M31 as plotted by Moffat and Shara (1987). Boxes and circles indicate the areas searched and defined symbols show positions of the WR stars discovered. Two likely candidates are identified with dashed squares.

Somewhat different results were reported by Massey and Armandroff (1990), who spectroscopically examined a large number of candidates chosen from the wide- band photometry of Massey *et al.* (1986). They reported that 34 W-R candidates were selected from the eight fields in M31 studied and that 19 were spectroscopically confirmed as W-R stars. Only 7 had been discovered in the Moffat and Shara (1983, 1987) searches, suggesting that the latter had an incompleteness problem, especially for WN stars. The WC/WN ratio for the M31 stars is 0.9 from Massey and Armandroff's (1990) statistics, much lower than was previously thought.

The absolute number density of W-R stars was found by Massey and Armandroff (1990) to be smaller in M31 than in M33, a result similar to that reported by Moffat and Shara (1983, 1987), but the more recent searches indicated that the deficiency is a factor of three, rather than a factor of ten. The surface density in the areas searched (which were dominated by OB associations) was calculated to be 1.5 W-R stars per kpc^2 in M31, which is less than that found for M33 (5 W-R stars per kpc^2) but which compares favorably with that for the solar neighborhood (2 W-R's per kpc^2) and for the LMC (also 2 W-R's per kpc^2). These numbers, of course, depend on the nature of the areas searched; a later calculation (Armandroff and Massey, 1991), for example, gave somewhat different numbers for these densities in active fields of the two galaxies: 1.2 W-R's per kpc^2 for M31, compared to 7.8 for M33.

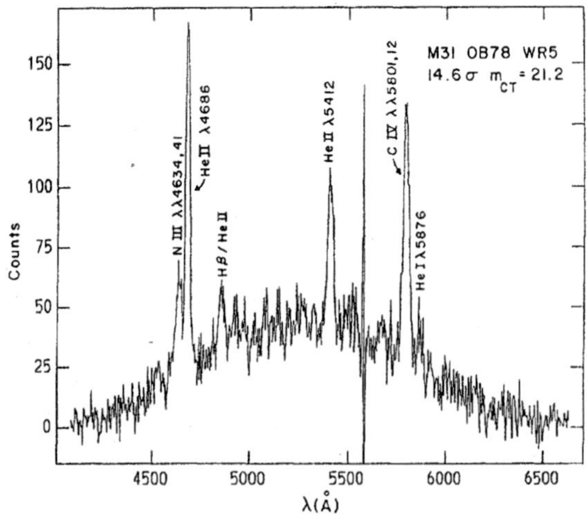

Figure 18.13 The optical spectrum of a WR star in the M31 OB association, OB78, published by Armandroff and Massey (1991).

Armandroff and Massey (1991) compared their results for M31 with their surveys of W-R stars in other Local Group galaxies. The WN stars of M31 were found to be spectroscopically similar to those of the Galaxy, M33 and the LMC, whereas the WC stars of M31, like those of M33, are anomalous, having wider and weaker lines than most Galactic examples. Furthermore, there is a lack of WC8 and WC9 stars in both M31 and M33, which make these galaxies more similar to the outer than the inner regions of our Galaxy. No such separation with radial distance was found for M31.

Additional work on the M31 W-R stars has been reported by Schild *et al.* (1990) and Cananzi and Azzopardi (1990). The former reported on the spectral characteristics of 6 W-R stars, while the latter examined W-R stars (as well as O and B stars) found in selected OB associations (OB 33, 42, 59 and 69). Their derived WC/WN ratio is 1.3, in reasonable agreement with that found by Massey and Armandroff (1990).

Red Supergiants

Many of the early studies of red supergiants in M31 were prompted by the hope of using their magnitudes to calibrate their bright luminosity limit as a distance indicator (Sandage and Tammann 1974; Humphreys 1979, 1985). A result of Humphreys' early searches, concentrated in Baade's Fields III and IV, is that the confirmed red members of M31 did not appear to exceed absolute magnitudes of $M_V = -7.5$. She found that this limit may be relatively independent of the luminosity and type of the host galaxy, as the red stars in M33 and the Magellanic Clouds, for example, also reached approximately this value. Further, more complete searches were carried out in the Berkhuijsen *et al.* (1988) Schmidt plate survey and by Efremov and Ivanov (1987) and Nedialkov *et al.* (1989). In the latter global search, no red candidate member was found that was brighter than $M_V = -8.0$, out of a total of 43 stars with $B-V > 1.8$ and $V < 18$.

A detailed look at individual red supergiants in M31 was made for stars in a portion of Baade's Field III by Richer and Crabtree (1985). Using the CFHT and special filters, they obtained CCD images that allowed them to distinguish carbon stars (with prominent CN bands) from M stars (with prominent TiO bands). Five carbon stars and 41 cool M giants were discovered. Compared to the ratio of C/M stars found for other Local Group galaxies, the ratio of 5/41 found is very low, matched only by the ratio in the solar neighborhood. The authors suggested that this fact may be the result of a similar value of [Fe/H] for the sampled portion of M31 (which is 11 kpc from the center) to that for our part of the Milky Way Galaxy. The other galaxies for which this ratio has been measured similarly (for example, the Magellanic Clouds) have lower metals abundances. Otherwise, the carbon stars found have normal properties, as far as they could be determined; their absolute magnitudes average I $= -4.60 \pm 0.54$. The AGB luminosity function agrees with that found for the LMC (Figure 18.14).

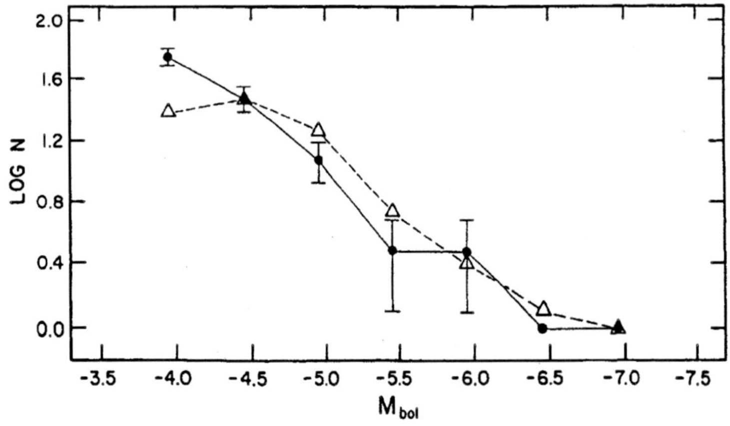

Figure 18.14 The luminosity function for the asymptotic giant branch for M31 (solid line) compared to that for the LMC (dashed line) by Richer and Crabtree (1985).

Humphreys *et al.* (1988) obtained JHK photometry of 57 red supergiant candidates, taken from a global search for the reddest bright stars in the M31 field. Spectra in the red were obtained for 45 of the candidates. Using both spectroscopic and photometric luminosity discriminators, they identified 23 probable red supergiants. Figure 18.15 shows spectra of three of the stars and compares them to Galactic standards. Humphreys *et al.* (1988) found that the red supergiants are strongly concentrated to the spiral arms, almost all lie within the boundaries of OB associations, and all are found in the north or northwest sectors of M31. Comparisons with Berkhuijsen *et al.* (1988) and with Efremov and Ivanov (1987) shows excellent agreement.

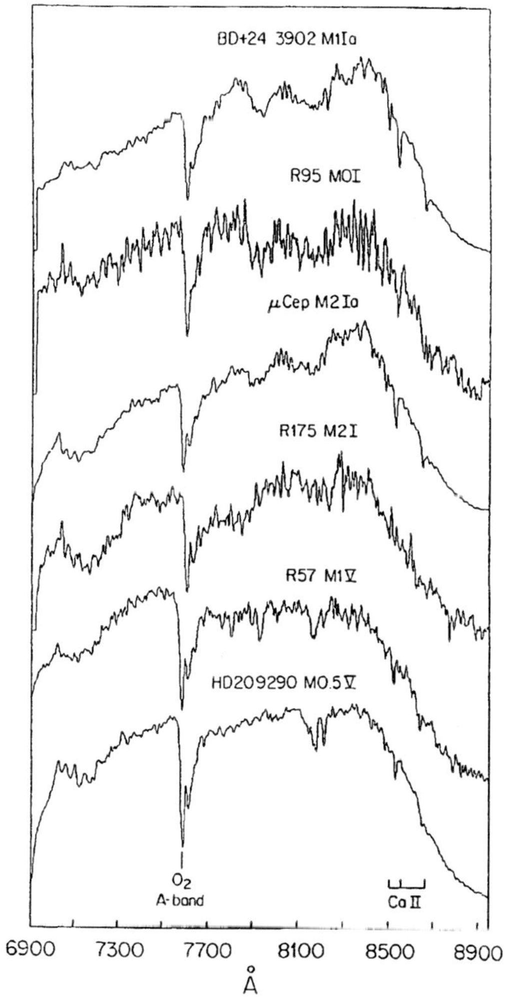

Figure 18.15 Near infrared spectra of three M31 red supergiants, R95, R175, and R57, compared to spectra of Galactic standards (from Humphreys et al. 1988).

18.4 Red Giants and Population II Stars

Some of the red stars found in the searches discussed above, plus many more, of course, are more properly described as ordinary red giants, rather than supergiants. The red giants of M31 were first clearly recognized by Baade (1963), who interpreted the images that he had resolved in the

1940's on red plates as being the top of the Population II component of the galaxy. In the intervening years, the clear distinction between Baade's two populations has blurred, however, and we now believe that many of the red giants seen in the disk of M31 are a collection of evolved stars of a variety of ages, masses, and compositions, including disk stars and superimposed halo giants.

Baade's conclusions were based on a statistical trend for many red stars to "suddenly" resolve when his plates reached a certain limiting magnitude. However, to pick out individual red giant members proved to be a little more difficult, as many of the stars in the field of M31 are foreground red dwarfs. Cohen (1978) employed a method for discriminating giants in M31 from dwarfs that used the strengths of the Na I doublet at λ8183-8195, the Ca II triplet, the FeH band at λ9910, and TiO bands (this technique was later used by Humphreys *et al.* (1988), as described above).

Deep studies of the halo red giants of M31 were carried out by Mould and Kristian (1986). They found color-magnitude diagrams that indicated that the population of the halo, even at their most distant position along the minor axis, is predominantly like a metal-rich globular cluster rather than a metal-poor one, such as would be found at a corresponding position in our Galaxy. The giants suggested a spread in metals, including some stars that might be metal-poor (*e.g.*, with [Fe/H] = −2), but with the majority having properties indicating values of [Fe/H] of approximately −0.6. On the other hand, Crotts (1986) found mostly metal-poor giants in the population that he examined at closer locations to M31's center (a distance from the center of 4 kpc). This apparent discrepancy was not cleared up until more data were obtained for color- magnitude diagrams to fainter limits and from spectroscopy of individual giants.

The nature of the halo giants was further investigated by Pritchet and van den Bergh (1988), who obtained deep CCD frames of a halo field about 9 kpc from the center along the minor axis. They concluded that the mean abundances of the giants, as indicated by the giant branch position and morphology, is intermediate between the values quoted by Mould and Kristian (1986) and by Crotts (1986). The mean value of [Fe/H] is approximately −1.0, but there is evidence for a fairly large spread (Figure 18.16), with a dispersion of 0.3 in [Fe/H]. The blue horizontal branch is weak, even though there are many RR Lyrae variables in this part of M31's halo (Chapter 13). Thus, the Population II stars of the halo of M31 differ from those of the Galaxy in having a somewhat greater mean heavy element abundance. This conclusion is similar to what was found for the globular clusters (Chapter 9) and it indicates that the early history of M31's halo must differ from that of the Galaxy; enrichment of the interstellar gas by the initial stellar formation event must have been more complete and more wide-spread for M31.

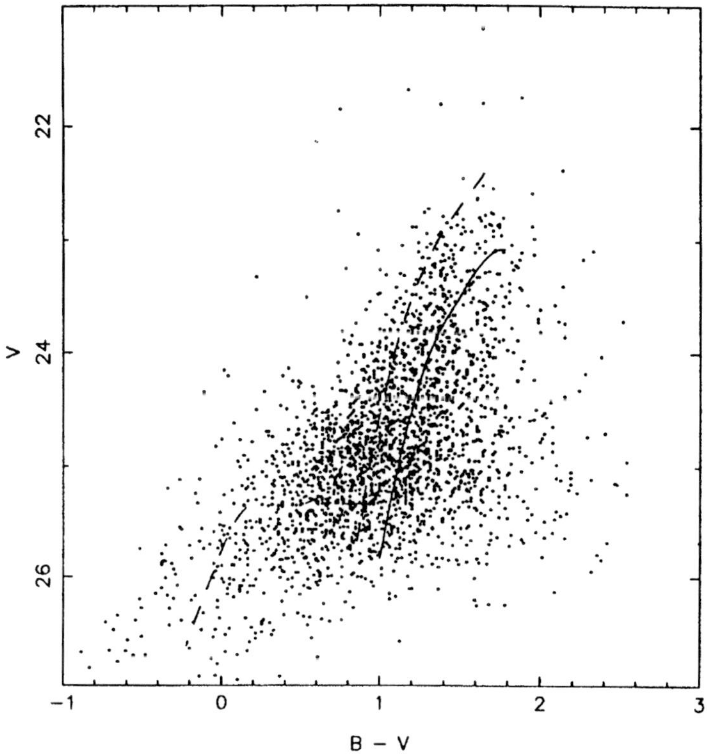

Figure 18.16 A color-magnitude diagram of stars in the halo of M31 compared to the ridge-line giant branches of M5 (dashed line) and 47 Tuc (solid line) (from Pritchet and van den Bergh 1988). Most of the M31 halo giants lie at an intermediate level.

The red giants of the bulge of M31 at approximately 1 kpc from the nucleus were observed by Mould (1986). Spectra of several of the bulge red giants were obtained by Rich *et al.* (1989), as shown in Figure 18.17. The giants are very similar in properties to our Galaxy's bulge population, as found, for instance, in Baade's window.

Infrared photometry for over 500 red giants in the bulge were reported by Rich and Mould (1991). Initially, the result of the photometry seemed to indicate that there was a fundamental difference between the M31 bulge stars and those of the Milky Way; while the Galactic bulge has no giants brighter than a bolometric absolute magnitude of –4.2, the bulge of M31 showed the presence of stars as bright as –5.5.

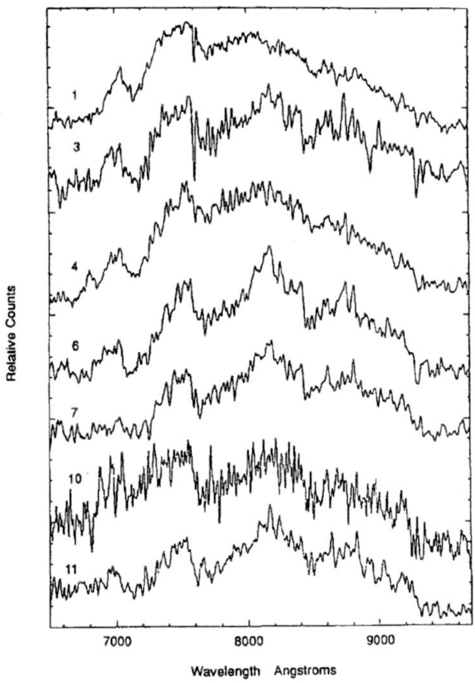

Figure 18.17 Spectra of seven red giants in the M31 bulge, obtained by Rich et al. (1989). Strong bands of TiO are visible at λ7054, 7598 and 8430.

A later analysis of a different, but nearby field in the bulge by Davies *et al.* (1991) showed excellent agreement with the Rich and Mould results, but reached a different conclusion. Using Kent's (1989) deconvolution of the bulge and disk of M31 to estimate the relative contribution of bulge and disk to their samples, they concluded that the unusually-bright giants detected could all be Population I giants (of greater mass and luminosity) from the disk, superimposed on the bulge field. If this is correct, then the M31 bulge giants agree in properties with those of the Milky Way Galaxy's bulge stars.

A close look at the red giants of the disk was taken by Richer and Crabtree (1985, discussed above in Section 18.3) and Richer *et al.* (1990), who examined the late-type stellar content in the disk at three distances from the nucleus: 4 kpc, 11 kpc, and 20 kpc. The widths of the giant branches in all three fields are much greater than could be explained by observational effects, and the authors conclude that the disk giants represent the funneling of stars of a range of masses into the same temperature-luminosity locus, much as we find for the Milky Way disk giants. The reddest giants include stars that are significantly redder than the reddest giants in even the most metal-rich globulars. Probably these stars are simply the reddest examples of normal metal-rich Population I giants.

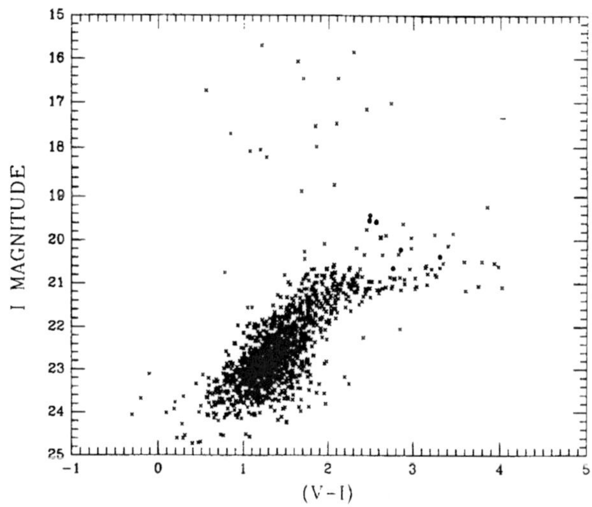

Figure 18.18 A color-magnitude diagram for 1465 stars in an outer field of M31 (Richer et al. 1990). Carbon stars are indicated by filled circles.

18.5 Stellar Populations in the Ultraviolet

One of the surprises of the early space-based observations of galaxies was the discovery that the ultraviolet energy distribution of M31's central area shows an ultraviolet upturn (Code 1969; Figure 18.19). This fact was confirmed by several successive observations (Code and Welch 1979; Johnson 1979; Wu *et al.* 1980; Davis *et al.* 1982). The question immediately arose as to whether this upturn is the result of the presence of a population of hot, young stars that contaminates the Population II of the central area, or whether it merely indicates the presence of horizontal branch and related hot, evolved stars.

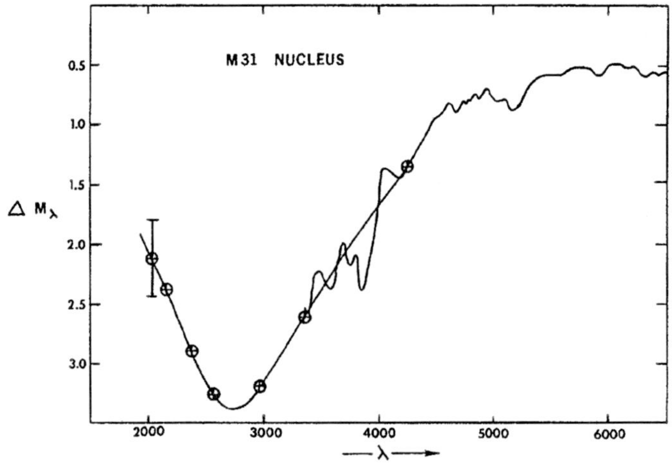

Figure 18.19 The spectral energy distribution of M31's nucleus. Circles are data in the UV from OAO-2 (from Code 1969).

The first clear answer to this puzzle came from the combination of twelve exposures of the nuclear area of M31 made with the IUE (International Ultraviolet Explorer) telescope (Welch 1982). The mean energy distribution derived by Welch was found to agree reasonably well with previous results (Figure 18.20). By using the combination of many spectra, he was also able to detect several spectral absorption lines, ranging from the Si II line at λ1260 to the Ca II line at λ3180, though these lines are very weak and in some cases rather uncertain. Strong emission lines, especially C IV at λ1550 and Si IV at λ1395/1403, were absent. Together with the spatial distribution of the ultraviolet light, these data argue that the ultraviolet upturn is primarily the result of the presence of hot evolved stars rather than young, massive stars.

Ultraviolet images of M31 have been harder to obtain. An early example is the image obtained from a rocket-borne telescope by Carruthers *et al.* (1978). Taken in the wavelength range λ1230-2000, their image showed the presence of the nucleus, the stellar association NGC 206, and a faintly-discernible spiral arm pattern. Deharveng *et al.* (1980) obtained a balloon-borne telescope image of M31 at λ2000. It shows the nucleus as a well-resolved, centrally-peaked feature. The 10 kpc "ring", where the young stars are concentrated, is also conspicuous. NGC 206 is a peak in the distribution of ultraviolet light from the southwest portion of the ring. Deharveng *et al.* (1980) find that the surface brightness of the spiral arms and NGC 206 in the ultraviolet are close to what would be predicted from their known massive stellar populations.

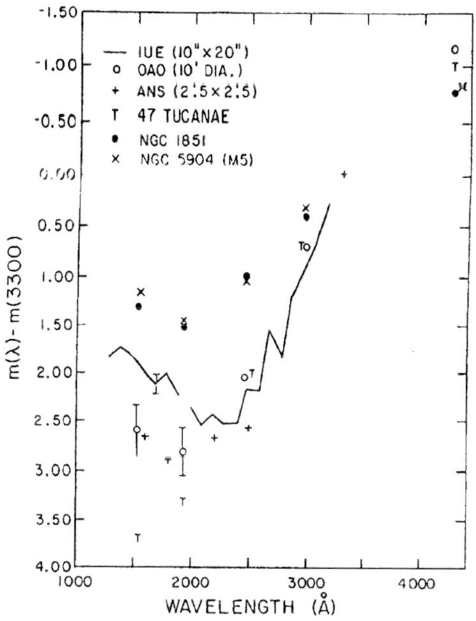

Figure 18.20 Ultraviolet energy distributions for the M31 bulge, as determined from three space-borne telescopes, IUE, OAO, and ANS. These are compared to OAO spectra of three globular clusters (from Welch 1982).

18.6 Color-Magnitude Diagrams

The first color-magnitude diagram of M31 was obtained by Hubble (Chapter 2), who plotted colors and magnitudes for the stellar association NGC 206. Although Hubble's use of these data was clearly important and informative, the practice of studying the individual stars of M31 in this way was not continued. The next color-magnitude diagrams (henceforth CMD's) to be plotted were not done until Baade and Swope (1963) published their results for Baade's Field IV, his most distant field, 96 arcmin from the nucleus. The CMD showed a weak main sequence, as well as some red giants and some foreground stars. Small stellar associations in the field showed an excess in their population of bright main sequence stars, as expected.

Four subsequent studies of Field IV followed Baade and Swope's pioneering paper. Van den Bergh (1964) extended the coverage of the field by carrying out five-color photometry of the stars, better defining the reddening and segregating the foreground stars more reliably. Using the technique of photometry of composite plates, Racine (1967) combined Palomar and Tautenburg Schmidt plates to derive a somewhat more accurate CMD for Field IV (Figure 18.21). When the foreground stars are removed by reference to the three color diagram, the result is a very clean main sequence, reaching to absolute magnitudes of $M_V = -6$, with only four red giants present. Later, Humphreys (1979) carried out a thorough investigation of Field IV, including spectroscopy and four-color photometry. Her results for bright members of M31 are discussed in Section 18.3, above. Figure 18.22 shows her complete CMD for the stellar associations in Field IV; it compares well with Racine's diagram (Figure 18.21), showing a similarly well-populated main sequence and having several more supergiant members, Humphreys having selected them spectroscopically.

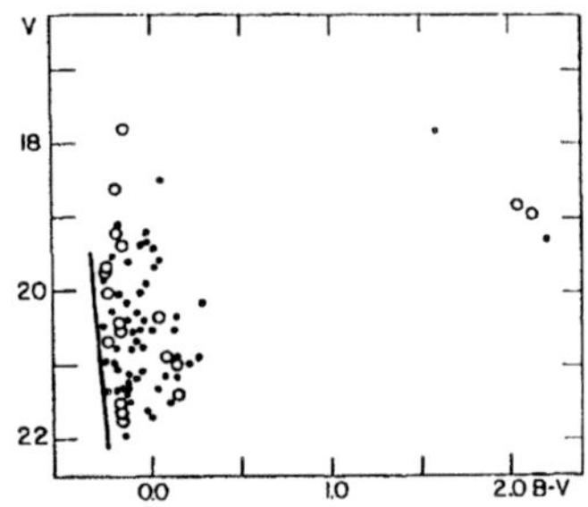

Figure 18.21 An early CMD of M31 field stars, based on composite Schmidt plates of Baade's Field IV by Racine (1967). Some of the foreground stars have been removed by using the three-color diagram.

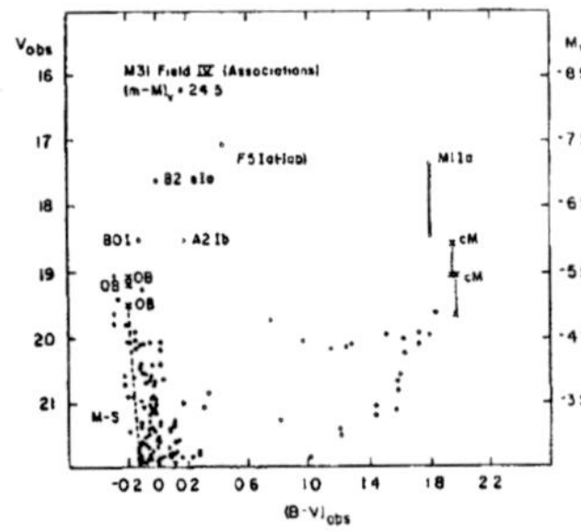

Figure 18.22 A CMD for Field IV obtained by Humphreys (1979) for the stellar associations. Her spectroscopic types for several of the stars are also indicated.

An additional look at the CMD for stars in Baade's Field IV was published by Hodge *et al.* (1988), based on CCD images in four colors obtained with the Kitt Peak 4-m telescope. The CMD reached to V = 24, extending the Humphreys (1979) result to fainter magnitudes. The CMD for field stars in this area is similar to what Humphreys (1979) found for it; there is a main sequence that extends from the faint limit up to $M_V = -6.2$ and there is a red giant region of older, evolved stars (Figure 18.23).

Nearly 40 years after Hubble's CMD for NGC 206, van den Bergh (1966) did a modern study of its properties. Using the Kitt Peak 84-inch telescope (as it was then called), he measured UBV magnitudes for the entire stellar association (which his paper refers to by its other name, OB 78). The CMD derived shows a thickly-populated main sequence reaching at least to $M_V = -7$, with several supergiants at brighter magnitudes (Figure 10.9). His comparison with h and χ Persei indicates that the age of NGC 206 is probably similar to the age of that young Galactic cluster, which is a few million years old. In agreement with this young age, he argued that the difference in velocity between the two subclusterings in the object implies an expansion age of 15 million years. A more recent CMD of NGC 206 was published by Odewahn (1987; see Figure 10.10).

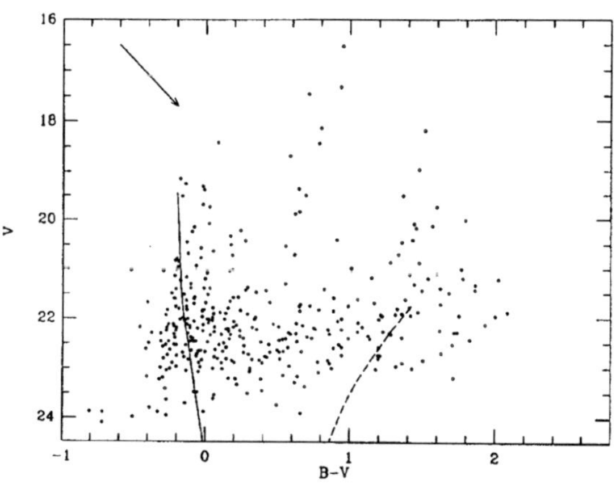

Figure 18.23 Another CMD for Field IV, this one obtained by CCD photometry to fainter limits (Hodge et al. 1988). The solid line is the zero age main sequence, the dashed line is the position of a M15-type red giant branch, and the arrow is the direction of reddening.

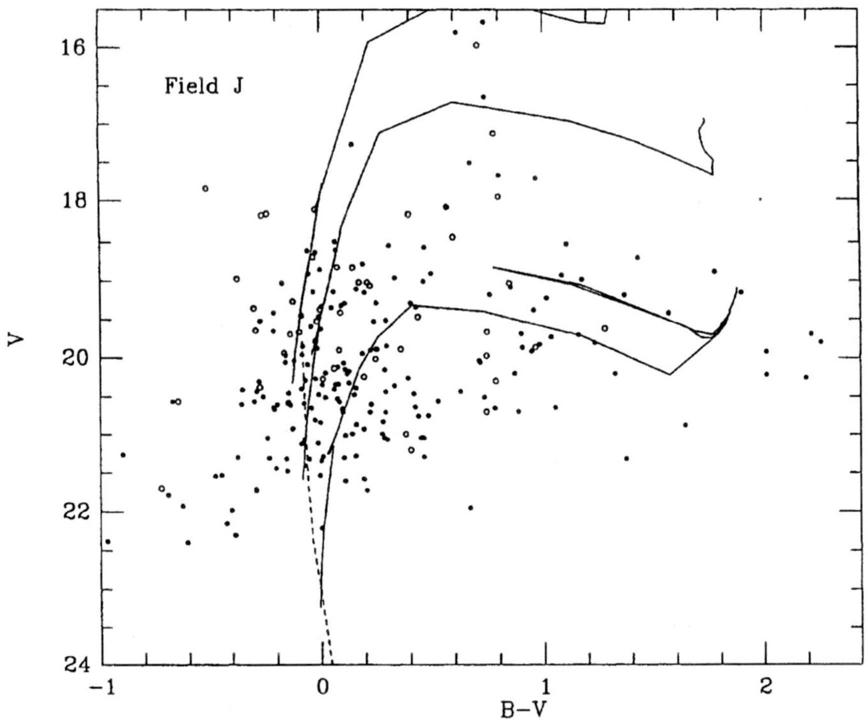

Figure 18.24 A CMD for a disk field in M31 (from Hodge and Lee 1988). The dashed line is the zero age main sequence and the solid lines are evolutionary tracks for stars of 40, 20 and 9 solar masses, respectively. Open circles are for lower accuracy data while filled circles are for the higher accuracy data.

The photometry of Massey *et al.* (1986) of four small fields in the M31 disk, in addition to exploring the properties of several M31 OB associations, also examines the CMDs of field stars. A similar photometry of five fields, one nearly coincident with a Massey *et al.* (1986) field, by Hodge and Lee (1988) provides very similar results. Both report that the disk field CMDs are quite similar at different locations, with the observed differences mostly attributed to modest differences in the recent history of star formation rates at the different locations. In all CMDs there is a prominent main sequence and a population of red giants (Figure 18.24).

The relative number of red giants is strikingly different for the inner disk CMDs from that for the remote disk, as can be seen by comparing Figure 18.23 with Figure 18.24. Although the main sequences in these two fields are very similar, the outer disk field has relatively many more red giants. The authors believe that this may be partly a population difference, with the percentage of older stars higher in the outer parts of the disk, but it also is at least partly the result of the fact that more faint red stars could be sampled in the outer, sparser regions. A contribution from intermediate-color stars, mostly from the foreground, shows up most conspicuously in the R vs. (V - R) CMDs (Figure 18.25).

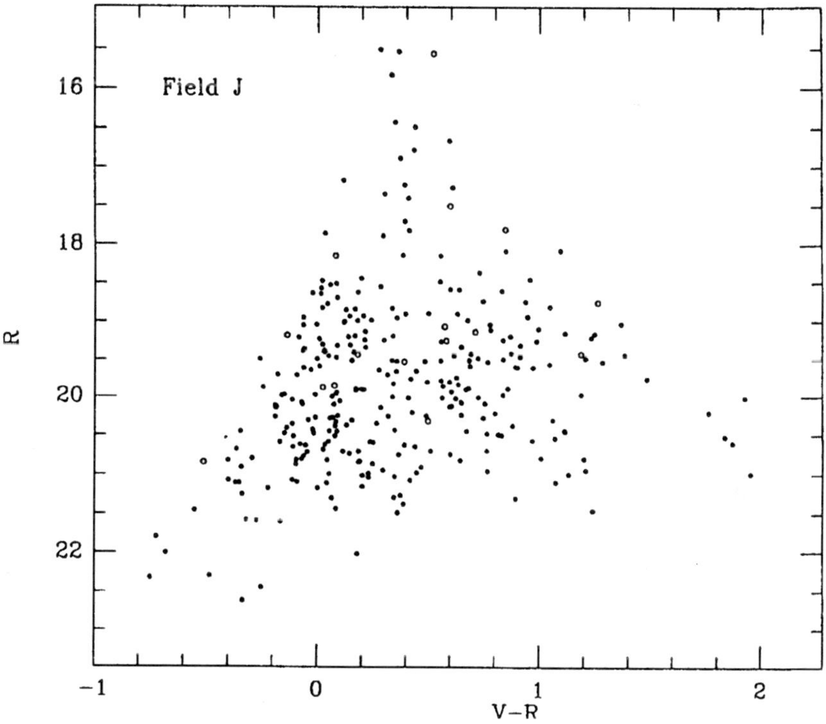

Figure 18.25 An R vs. V–R diagram for the stars in Figure 18.25, showing the conspicuous presence of intermediate-color foreground stars that show up well in such a plot.

Nikolov and Tasheva (1989) reported an extensive photographic photometry of colors and magnitudes of stars inside and outside of OB associations. Figure 18.26 shows their composite CMD for all stars outside of OB association boundaries. There are many intermediate-color stars, probably nearly all foreground Galactic stars, and the tip of a thick main sequence at $M_V = -4$. A few red giants are found above the plate limits, ranging from $M_V = -4.5$ to $M_V = -6.5$. These are presumably evolved, massive giants, rather than old, low mass stars such as found in the halo and bulge, which would be too faint to be detected in this program.

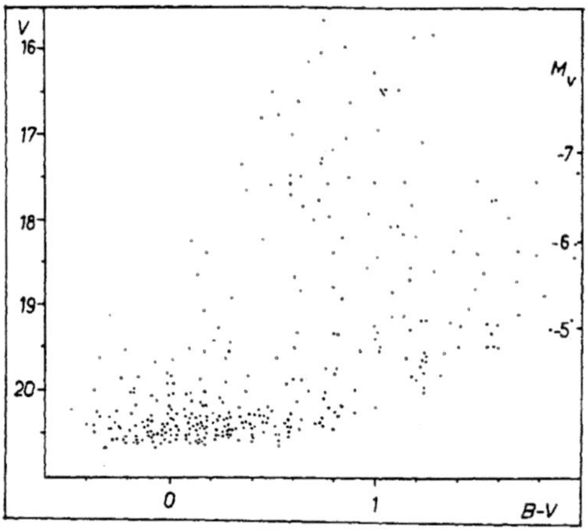

Figure 18.26 A CMD for M31 disk stars outside of OB associations, as reported by Nikolov and Tasheva (1989).

18.7 Luminosity Functions

Obtaining reliable luminosity functions for M31 disk stars has been an onerous task. The main problem is crowding at faint magnitudes and the difficulty of correcting for the effects of missed stars in an inhomogeneous star field. For Baade's Field IV, where stars are relatively well-separated and uniformly-distributed, the problem has been less difficult than for more central disk areas. For example, Hodge *et al.* (1988) derived a luminosity function, corrected for crowding, foreground stars and reddening. Stars in the sample ranged from $M_V = -4.5$ to -0.5; for the fainter stars, corrections for completeness were large. Comparing their data for the main sequence with that of Freedman (1985), found for many thousands of stars in several nearby galaxies, the authors concluded that the LF of Baade's Field IV is very similar (Figure 18.27).

Luminosity functions for other areas have been obtained, but usually to brighter limits. Differences are small; they indicate probable variations in the detailed star-formation histories of different portions of the disk, agreeing in these general terms with the conclusions based on the morphologies of the CMDs. A comparison of six field luminosity functions is given in Hodge and Lee (1989), in which the relative predominance of very young stars is shown to be high in three fields, intermediate in two, and low in one.

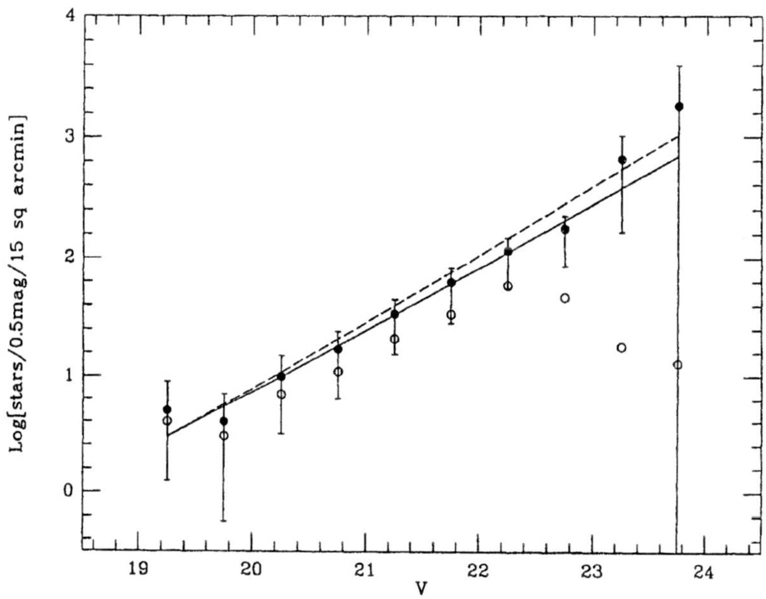

Figure 18.27 The luminosity function for main sequence stars in Baade's Field IV. Open circles are for the measured stars, filled circles (and the solid line) are for the data after correction for completeness, and the dashed line is the general LF found for galaxies by Freedman (1985).

18.8 The Star Formation Rate

The rate of star formation in M31 has long been thought to be lower than in our Galaxy, largely because of the lack of high surface brightness arms and the scarcity of giant HII regions. In this sense, M31 has been a prototype for its Hubble class, largely responsible for the impression that Sb galaxies have comparatively low star formation rates.

The first good estimate of the star production rate was made by van den Bergh (1964), basing the calculation on the total number and star populations of the OB associations. He derived a value of 0.1 solar masses per year.

A more recent look at M31 star formation was taken by Walterbos (1988), who calculated the star formation rate from four different lines of evidence. He used Kennicutt's (1983) relation between the Hα luminosity and the star formation rate for galaxies, as well as Kennicutt's measurement of the Hα equivalent width of 4 Angstroms, which, when corrected for a [NII] contribution through the filter used, implies a total luminosity of 2.6×10^{40} erg s^{-1}. This results in a calculated global star formation rate of 0.23 solar masses per year.

From infrared data, especially the IRAS measurements of M31, Walterbos calculated a star formation rate of 0.10 to 0.20 solar masses per year, assuming a reasonable scenario for the production and disruption of the clouds in star-forming regions.

A third method was based on the thermal radio emission from M31, which is approximately 600 mJy at 10.7 GHz. This translates into a star formation rate of 0.34 solar masses per year.

Walterbos also estimated the star formation rate by examining the number and nature of the HI holes found by Brinks and Bajaja (1986). The argument in this case is a little more complicated than for the above estimates. From the expansion rate within the holes and the mean size, the mean age of the holes is estimated. The mass missing from the hole can be measured and the total energy required to form the hole can thus be calculated. From the total number found in M31 and the average age, the production rate is found to be one hole per 10^5 years. Approximately 100 supernovae are required to form an average hole. Combining this with an adopted IMF leads to the conclusion that the star production rate in M31 is 0.1 solar masses per year.

From these diverse lines of evidence it can be concluded that the star formation rate in M31 is fairly low, on the order of 0.1 to 0.4 solar masses per year, which is significantly less than the rate for the Milky Way Galaxy, recently estimated to be between 3 and 10 solar masses per year (Dopita 1987).

18.9 Summary

The general populations of stars in M31 have been examined by a wide variety of techniques, including spectral synthesis, spectra of individual stars, CMD's and luminosity functions. Spectral synthesis has been primarily applied to the nuclear and bulge areas, but even there, where light levels are high, the results are somewhat controversial.

Spectra of individual stars reveal many early and late-type supergiants that resemble Galactic luminous stars. A possible difference in the number of very luminous supergiants may indicate a lower cut-off mass for the IMF in M31 than in the Galaxy.

The WN stars in M31 are similar to those in the Galaxy, whereas the WC stars are different, having wider and weaker lines.

The red giants of the halo indicate a predominately intermediate abundance and/or age and a spread in luminosity indicating a spread also in abundance and/or age. The bulge and nuclear population has an ultraviolet excess that is probably the result of the presence of hot evolved stars.

CMDs and luminosity functions for field stars in M31's disk are similar to what would be found in the Galaxy for disk stars near the sun. However, the star formation rate in the disk is low, probably about one-tenth that in the Milky Way Galaxy.

CHAPTER 19

A COMPARISON OF M31 WITH THE MILKY WAY*

Although not a close match to our Galaxy, M31 is, nevertheless, a useful galaxy for comparing with the Milky Way, as it is the only large spiral for which we can obtain detailed information of certain kinds. Recent work at many wavelengths has made it one of the most thoroughly observed galaxies beyond the Milky Way. The following is a discussion of some of the more challenging areas of similarity and of contrast between them. Insight into certain questions is certainly gained by their comparison, but it must also be confessed that, frustratingly, some problems that confound students of the Milky Way because of our location in the plane, similarly confound those who study M31 because of its nearly edge-on inclination angle.

19.1 Integrated Properties

Table 19.1 compares some of the basic properties of Andromeda and the Milky Way, as best we can determine them so far. For M31, the Hubble type is clearly Sb; Hubble himself used it as a typical example and there has been relatively little controversy on the subject, other than to argue about a possible weak bar-like structure in the central bulge (Lindblad, 1956; Sharov and Lyutyi, 1980). For our Galaxy, however, the Hubble type is not so clearly known. Of eleven different methods recently used to gauge this parameter, two conclude that it is Sb, two that it is Sbc, and seven that it is Sc (Baade, 1951; Arp, 1964, 1965; Becker, 1964; van den Bergh, 1968; Georgelin and Georgelin, 1976; de Vaucouleurs and Pence, 1978; and Hodge, 1983).

The relative sizes of M31 and the Galaxy can be compared in a number of ways. Table 19.1 summarizes five of the various measures

---

* Adapted from Hodge (1985).

of the diameters. No matter how the diameter is defined, that of M31 is consistently larger than that of the Galaxy by about 40%. For M31 the most distant open cluster, for example, is 139 arcmin from the center (Hodge, 1979), which corresponds to 29 kpc, while for the Milky Way, Christian and Janes (1979) find Be 20 to be the most distant open cluster, with a distance of 20 ± 3 kpc. The objects are rather comparable. From Christian and Janes' (1979) CM diagram, the absolute magnitude of Be 20 is $M_V = -7.4 \pm 0.2$. At M31 this would be V = 17.2. The integrated brightness of M31's most distant open cluster, C1, is estimated from KPNO 4-m plates to be V = 17.9.

Table 19.1
Comparisons of M31 and the Galaxy

Parameter	Galaxy	M31
Adopted distance (kpc)	8.5	725
Hubble type	Sc	Sb
M_B (face-on)	$-20.5 \pm .5$	$-21.1 \pm .4$
$(B-V)^o_T$	$0.53 \pm .04$	$0.74 \pm .06$
V(R = 8.5 kpc) (km/sec)	220 ± 20	265 ± 10
σ_v (O) (sph. comp.) (km/sec)	130 ± 10	160 ± 30
D_o @ V=25.0 mag/sec (kpc)	24 ± 5	38 ± 2
Effective diam. (D_e) (kpc)	11	15
D (from open cluster system) (kpc)	40	58
D (from HII region system) (kpc)	45	47
$D_{26.6}$ (kpc)	34	47
No. globular clusters	150	300
Mass ($M_\odot$)	2×10^{11}	4×10^{11}
M_H ($M_\odot$)	5×10^9	4×10^9

The outermost luminous blue stars in the anticenter direction are estimated to lie at a distance of 20 kpc (Chromey, 1978). For M31, Richter (1971) has cataloged aggregates of OB stars at ~26 kpc.

The neutral hydrogen diameters of the two galaxies are also in about this ratio. Emerson's (1976) outermost H I contours for M31 are at 30 kpc, for example, while Baker (1976) finds an H I cut-off for our Galaxy at about 25 kpc from the Galactic center.

When these data are combined with the various photometric measures of size (Table 19.1), we must conclude that the Milky Way is M31's inferior, being roughly two-thirds as large.

19.2 Structure

There are both similarities and striking differences between M31 and the Milky Way in structure. This discussion will concentrate in particular on the spiral structure, but mention should be made of other components, as well. Both galaxies have compact nuclei, with that of M31 being somewhat the less interesting (that is, the less active). The galaxies both have central bulges made up largely of old, metal-rich stars, with that of the Galaxy being somewhat the smaller and bluer of the two (Sharov and Lyutyi, 1980; Arp, 1965; de Vaucouleurs and Pence, 1978). The latter reference attempts to isolate the spheroidal component, assuming it to be structurally governed by an $r^{1/4}$ projected density law, and finds the Galaxy's bulge to have a blue absolute magnitude about 1.2 mag fainter than that of M31. The central bulge in each case contains numerous examples of planetary nebulae, novae, X-ray sources and dust clouds.

The haloes of the two galaxies can also be compared. Both are surrounded by a nearly spherical and extremely extended system of globular clusters. Andromeda has about twice as many cataloged globular clusters as the Milky Way. Its most distant globular lies at a projected distance of 130 kpc (Sargent *et al.* 1977), while our Galaxy's system appears to extend out to 116 kpc (Aaronson *et al.* 1984). Both also are surrounded by a small population of dwarf elliptical galaxies (van den Bergh, 1972).

It is the disk component, especially the spiral arm structure, however, that is especially interesting to compare. Most attempts to disentangle the spiral structure of M31 have used familiar, traditional "tracers", the OB stars, young clusters, H II regions, and H I, in much the same way the local spiral structure is sought in our Galaxy. The difference, of course, lies in the type of difficulty encountered: obscuration and distance uncertainties for the Milky Way and a steep angle of inclination for M31. All agree that both galaxies have spiral arms, but the literature is full of contradictory interpretations of the detailed spiral structure. For M31, arms are outlined by the counts of brightest stars (van den Bergh, 1958; Reddish, 1962), by population morphology (Baade 1963), by OB associations (van den Bergh, 1964), from radio continuum data (Pooley, 1969; van der Kruit, 1972; Berkhuijsen and Wielebinski, 1974; Beck, 1982), from H I surveys (Roberts, 1966; Byrd, 1978; Brinks, 1984; Sofue and Kato, 1981; and many others), from CO data (Stark, 1983), using H II region surveys (Arp, 1964; Simien *et al.* 1978), from ultraviolet images (Deharveng *et al.* 1982), from Cepheids (Efremov *et al.* 1981), from surface photometry (Hodge and Kennicutt, 1982), from dust lanes (Hodge, 1980), from IRAS data (Walterbos and Schwering, 1987), and from young open clusters

(Hodge, 1979). Although the raw data available for these studies are fairly consistent, their interpretation has been very divergent. While most fits have been of a two-armed trailing spiral pattern, Simien *et al.* (1978) instead argued for a one-armed leading spiral, with some theoretical justification provided by the work of Kalnajs (1974). Such a contradiction seems to indicate that the situation has been, at least recently, even worse than in the case of our Galaxy.

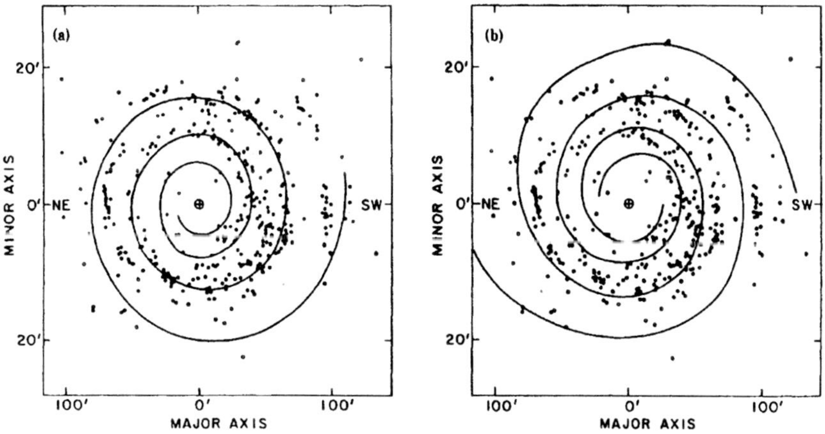

Figure 19.1 Comparison of the M31 open clusters' distribution (corrected to face-on) with (a) the one-armed leading model given by Simien et al. (1978) and with (b) Arp's (1965) two-armed trailing model.

The leading one-armed spiral pattern, based primarily on H II regions, fits rather poorly the young open clusters' positions. Although the fit within ~50 arcmin of the center looks reasonable, in the outer parts, where the open clusters better define a pattern than the sparser H II regions, the fit is seriously deficient. The clearly defined arm segments near the major axis at 70-90 arcmin clearly have a slope opposite to that of the model. Arp's two-armed spiral also does not fit these segments, but at least they are roughly parallel, with pitch angles of the same magnitude and the same sign. It is clear that the arms of M31 are not to be fit to a simple, perfect logarithmic spiral pattern. Byrd's (1978) gravitationally distorted spiral pattern looks much more promising. Perhaps the moral of this comparison is that students of the Milky Way's spiral structure should not be too discouraged if their data do not fit a perfect mathematical model, as M31 does not attain such perfection, nor do many other galaxies.

As a graphic demonstration of some of the similarities and differences in the spiral arms of the two galaxies, Fig. 19.2 shows a section of the NE portion of M31 that corresponds in location to the solar neighborhood of the Milky Way. Plotted is a rectified face-on map of the area in question, in which are identified positions of young open clusters, dust clouds, OB associations, and H II regions, from the sources listed above. Two spiral arm crossings are obvious. The inner has a pitch angle of perhaps 5°. They are separated by approximately 4 kpc at the position of the optical major axis, which vertically crosses the center of the diagram. All of the optical tracers agree remarkably well.

Figure 19.3 compares this map with a similar one for the Milky Way plotted to the same scale and for the same distance from the nucleus (based on the summary diagram in Bok and Bok (1981)). Although the diagrams show a general similarity, there are three conspicuous differences: (1) the arms of M31 are much better defined, with interarm areas almost empty of tracers, (2) the M31 arms are at least twice as widely spaced, and (3) the pitch angle for the M31 arms is much smaller. These facts, of course, have been known for many years; Fig. 19.3 merely supplies an especially graphic demonstration of these important differences.

Many other interesting comparisons could be made, especially about the radial distributions of different components of the galaxies, rotational parameters, kinematics, warping of the plane, velocity dispersions, and so on. Much progress is being made in these fields. Questions do remain, however. It is particularly challenging to realize that for both galaxies we still cannot answer firmly the following rather basic questions:

1. How many arms are there?
2. What are the shapes of the arms?
3. What dominates the spiral pattern, generally and in detail?
4. What is the history of star formation in the plane and in the halo?
5. What is the total mass?
6. How is the mass distributed?

19.3 Summary

A comparison of a few of the basic properties of M31 and the Milky Way indicates that in almost every respect M31 is larger than the Galaxy. It is more luminous, redder, more massive, and of earlier Hubble type. A detailed comparison of the spiral structure, based on optical tracers, for comparable areas in the outer parts of each galaxy show differences in the arm spacings, in density enhancement, and in pitch angle.

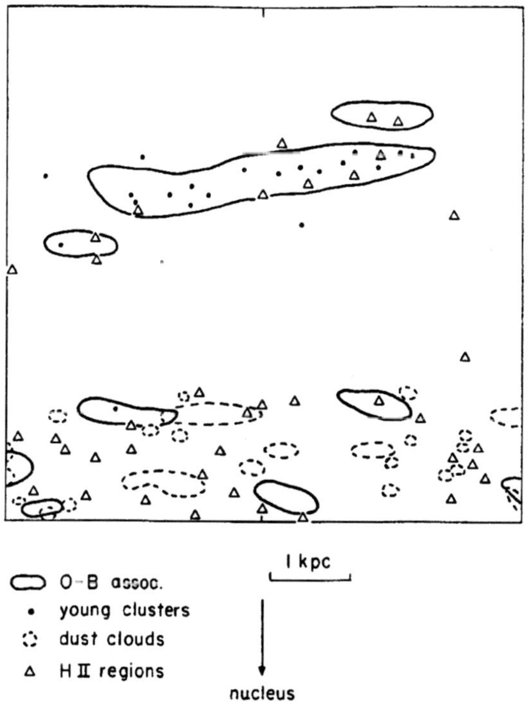

Figure 19.2 Map of part of the portion of M31 along the NE axis, rectified to face on, and extending from 7 to 14 kpc. Spiral arm tracers are identified. Baade's arms N_3 *and* N_4 *are conspicuous.*

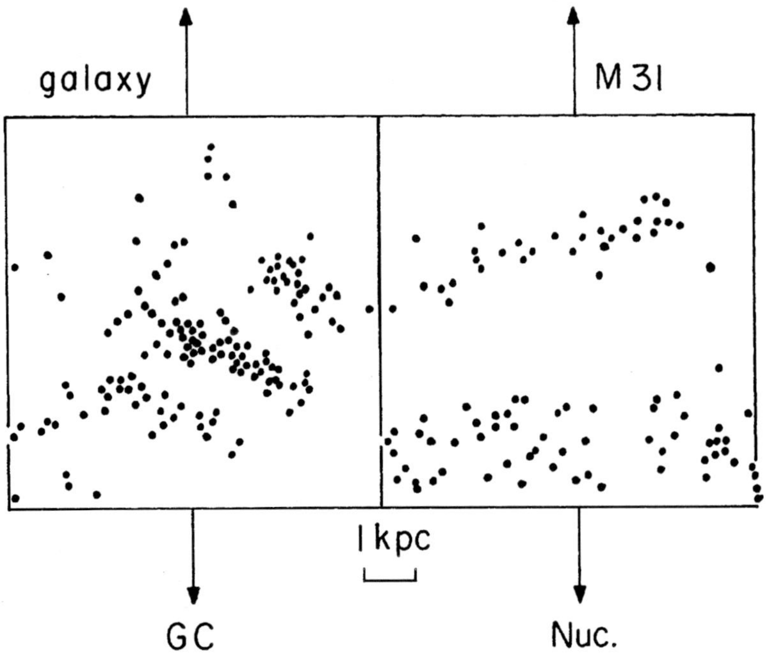

Figure 19.3 A comparison of the area in Figure 19.2 and a comparable area in the Milky Way galaxy, with the same kinds of tracers plotted.

CHAPTER 20

ANNOTATED BIBLIOGRAPHY, 1885–1950

References to papers dealing with M31 are listed here in order of the year of appearance. Brief annotations are given for most of them. The first part of this compilation is based in part on work done many years ago by students at the University of California, Berkeley, as a class project. Especially hard-working were Conrad Sturch, Ralph Robbins, K. S. Krishna Swamy, Carol Webb, and Ann Merchant (Boesgaard). Not all of the references could be checked directly in available library collections. Thus, especially for some of the older references, we had to rely on the data in the *Astronomische Jahresbericht* or other secondary sources.

Following this chapter is a list of references to all papers published since 1950 and used in the text of this monograph. These are arranged alphabetically and are in standard reference format.

1885

Most references in this year deal with the supernova, eventually named S Andromedae, discovered in August 1885, and called a nova at that time.

Bakhuysen, H. *Astr. Nachr.*, **112**, 323. Gives brightness and position of S And.

Bigourdan, I. *Astr. Nachr.*, **112**, 286 and 404. Gives brightness and position of S And.

Cacciatore, G. *Astr. Nachr.*, **112**, 387. Describes a 12^m or 13^m object near S And about 15″ to 18″ away.

Charlier, G. V. L. *Astr. Nachr.*, **112**, 389. Photometer measurement and position of S And.

Clark, J. E. *Nature*, **32**, 1885. On September 6 the nebula was brighter than normal.

Common, A. *Nature*, **32**, 522. A finding chart for comparison stars in the neighborhood.

Copeland, R. *Astr. Nachr.*, **112**, 286. S And was observed as early as August 19 by I. Ward at 7.5^m and had a continuous spectrum. Gives position relative to nucleus.

Denning, W. F. *Nature*, **32**, 465. General description of S And as seen through a 10-inch telescope.

Engelhardt, W. *Astr. Nachr.*, **112**, 285. Day-by-day description of S And.

Folio, *Astr. Nachr.*, **112**, 248. Observations of S And.

Gothard, E. V. *Astr. Nachr.*, **112**, 390. Spectrum on September 6 and 12 of both S And and nebula. Both are continuous with no lines. Also photographs of the region.

Hartwig, E. *Astr. Nachr.*, **112**, 245, 285, 355, 358, 360. Discovery of S And at Dorpat on August 31. Measured relative position with heliometer.

Huggins, W. and Rosse, Lord *Nature*, **32**, 465. One-or-two-sentence reports of a dozen observations of the nebula since 1848. Describes spectrum of S And as continuous from C to F with bright lines probable.

Kammermann, A. *Astr. Nachr.*, **112**, 299, 321, 387. Believes S And to be a new star, but not connected with nebula. Measured position relative to center.

Konkoly, *Astr. Nachr.*, **112**, 286. No star observed on August 9 or 13. Podmaniesky observed a faint one on August 22. Observed spectrum without collimator lens on 254-mm telescope.

Lamp, E. *Astr. Nachr.*, **112**, 245. Brightness and position of S And compared to star observed in nebula in 1836.

Millosovich, E. *Astr. Nachr.*, **112**, 321. Brightness and position of S And.

Oppenheim, H. *Astr. Nachr.*, **112**, 245. Discovered S And independently while searching for comets.

Ricco, A. *Nature*, **32**, 523. S And is 8^m with a continuous spectrum and suspected bright bands.

Schroder, C. *Astr. Nachr.*, **112**, 246. Magnitude of S And; continuous spectrum with red brightest.

Schultz, H. *Astr. Nachr.*, **112**, 302. Position and description of 10^m star in nebula in 1867 and 1875.

Seabroke, G. *Nature*, **32**, 523. Spectrum of S And described.

Spitaler, R. *Astr. Nachr.*, **112**, 284. Position of S And given.

Tempel, W. *Astr. Nachr.*, **112**, 301. Nova not observed on August 16. Position and brightness on September 9 given.

Tromholt, W. *Nature*, **32**, 579. Speculation about the nova being a very long period variable since a bright star in this nebula is told about in folklore.

Valentiner, W. *Astr. Nachr.*, **112**, 403. Position with meridian circle.

Vogel, W. *Astr. Nachr.*, **112**, 283, 302, 387. Summarizes previous observations of S And. It has a continuous spectrum, strongest in red and yellow, with a dark band between green and yellow and another in the blue between F and G. By September 10 star was down to 9^m. Continuous spectrum.

Ward, I. *Astr. Nachr.*, **112**, 404. Claims first discovery of S And on August 19, 9.5^m at that time.

Wolf, M. *Astr. Nachr.*, **112**, 284. Established time of appearance of S And as between August 16 and 25.

1886

Barnard, E. E. *Astr. Nachr.*, **113**, 31. A small faint nebula near the northeast end of M31 is described.

Bartfay, J. A. von *Astr. Nachr.*, **115**, 253. Descriptive observations at Budapest.

Charlier, C. V. L. *Astr. Nachr.*, **113**, 165. All nova magnitudes are off by 0.4^m.

Dreyer, J. *Astr. Nachr.*, **113**, 270. Hartwig had claimed central part of nebula had changed. Dreyer argues that previous observers always disagreed about the central portion and no change had really occurred.

Engelhardt, W. *Astr. Nachr.*, **115**, 252. No sight of nova this year; only its 11^m companion.

Engleman, R. *Astr. Nachr.*, **113**, 269. Position of nova relative to 11^m star, and brightness (September 22: 8.92, to November 18: 10.8).

Gothard, E. *Astr. Nachr.*, **115**, 252. Photographed in August 1885, but plates too dark to see nova. Spectrogram also taken, but not detailed. In October got better results with plate of different sensitivity. Spectrum of S And resembles that of a Wolf-Rayet star.

Gully, L. *Astr. Nachr.*, **113**, 45. Reprint of article claiming discovery on August 17, 1885.

Hall, A. *Nature*, **33**, 566. No parallax with respect to 11^m star from September 25 to February 7, when nova could no longer be seen at Washington.

Hall, M. *Obs.*, **9**, 69. Description of drawing of nebula and comparison to other drawings. Brief account of nova.

Hartwig, E. *Astr. Nachr.*, **113**, 21, 387. Changes in the center of the nebula. Defends observations against Dreyer's attack.

Hasselberg, B. *Astr. Nachr.*, **113**, 19. S And is like novae of 1848, 1866, and 1876. Argues that it is not connected with nebula.

Kövesligethy, R. von *Astr. Nachr.*, **115**, 231, 303, 305, 307, 308. Variations in sharpness, color, and spectrum of nucleus of nebula where nova appeared. Compares brightness of nova with 11^m comparison star. The variations of the nebula prove that the nova was in the nebula.

Konkoly, N. von *Astr. Nachr.*, **115**, 253, 267. Says that Bartfay is wrong; he was probably looking at 11^m star and not the nova.

Lamp, E. *Astr. Nachr.*, **115**, 265. Agrees with drawing of Trovelot of nebula in 1874.

Lamp, J. *Astr. Nachr.*, **118**, 21. Position of nova relative to 11^m star and nucleus. Gives position of nucleus with respect to 11^m star.

Lynn, W. T. *Obs.*, **9**, 69. Discusses discovery of the nebula. First noted on charts around 1500.

Millosevich, E. *Astr. Nachr.*, **113**, 15. Position of nova relative to nucleus.

Müller, G. *Astr. Nachr.*, **113**, 23. Gives brightness (September 2, 7.95 to October 13, 10.04). Notes a systematic error in Charlier's work.

Müller, G. *Astr. Nachr.*, **115**, 265. Poor observing, but no star at position of Nova (S And), although a concentration is observable.

Schönfeld, E. *Astr. Nachr.*, **115**, 265. If there is a star near the nucleus now, it must be below 11^m.

Seeliger, H. *Astr. Nachr.*, **113**, 353; *Nature*, **33**, 397. Mathematical theory of cooling of a hot ball of gas. Predicted results agree well with observations of S And. Original heating may have been due to stellar collision.

Sherman, O. T. *Astr. Nachr.*, **113**, 45. Spectrum of S And. Bright lines at λ5315 and 5575 and M, which is probably nebulous. Lines close to coronal lines.

Tarrant, K. *Obs.*, **9**, 397. Cannot see any change in the nucleus.

Spitaler, R. *Astr. Nachr.*, **114**, 325. Meridian circle observations at Vienna.

Valentiner, W. *Astr. Nachr.*, **115**, 265. Can't find nova this year.

1887

Franz, J. *Astr. Nachr.*, **118**, 123. Measured parallax of nova relative to three neighboring stars and found negative parallax, showing that nebula is behind these stars.

Roberts, I. *M.N.R.A.S.*, **49**, 65, 121. Photographs of M31. Confirms nebular theory that it is a new solar system in the process of condensation.

1889

Commons, A. *Obs.*, **12**, 105. Discusses the excellent photograph of M31 by Roberts, which was first to show the spiral structure.

1890

Close, M. *Obs.*, **13**, 54. Inconsistencies in drawings and photographs of M31.

Gore, J. *J.B.A.A.*, **1**, 438. Gives angular dimensions, says it has no parallax, and proceeds to assign one, concluding that the nebula is too small to be another galaxy.

1891

Roberts, I. *M.N.R.A.S.*, **51**, 116. Photographs over a period of five years show variability of stellar nucleus.

1893

Swift, L. *Pop. Astr.*, **1**, 111. Description, brief history of observations.

1898

Barnard, E. *Ap. J.*, **8**, 226, 262. Nucleus not changing. Observed changes probably due to seeing. Photographs do not show changes. Refers to: Rayet, M., Comptes Rendus, September 26, 1898; Rayet verifies Seraphimoff's observation.

Brenner, L. *Astr. Nachr.*, **147**, 287. Lists several observatories where the reported new star was not seen.

Coddington, E. *Pub.A.S.P.*, **10**, 45. Photograph with the Crocker telescope at Lick. Gives a good, concise history of observations.

Comas Sola, J. *Astr. Nachr.*, **148**, 11. No changes of nucleus in photographs.

Espin, E. *Br. Astr. Assn. J.*, **9**, 85. New star (8.4^m) in nebula.

Hartwig, E. *Astr. Nachr.*, **148**, 11. Gives a review of the reported variations of nebula and presents a chart showing Nova 1885 and comparison stars.

Müller, J. *Astr. Nachr.*, **147**, 287. Lists several observatories where the reported new star was not seen.

Pickering, E. *Astr. Nachr.*, **147**, 363, and *Harvard Circular*, **34**. Comparison of photographs from 1893 to 1898 shows no variation of nucleus.

Seraphimoff, *Astr. Nachr.*, **147**, 319, and *Nature*, **58**, 515, 605. Sharp central star, not condensation, of 11^m; proves variability of nucleus.

Scheiner, J. *Astr. Nachr.*, **148**, 325. Photographic spectrum indicative of a cluster of solar-type objects.

1899

Barnard, E. *Obs.*, **22**, 376. The nebula is farther than the fixed stars.

Hale, G. *Ap. J.*, **9**, 184. Negative parallax for nebula. Speculates as to whether the comparison stars are in the nebula.

Hussey, W. J. *A.J.*, **19**, 152, and *Obs.*, **22**, 137. Note concerning the central condensation of M31; brightness and spectrum.

Wilson, H. C. *Pop. Astr.*, **7**, 507. Description and reproduction of recent photographs with a brief history of earlier photography. Tells of difficulties in obtaining parallax.

1900

Roberts, Mrs. Isaac, "Photographs of Stars, Star Clusters, and Nebulae," Vol. II, Knowledge Office, London. Plates taken with a 20" reflector, with descriptions. Plates 10–18 are of spiral nebulae, including four exposures of M31. Author finds that a 10-hour exposure shows no more stars than a 90-minute exposure — her conclusion is that the part of the universe that we can see from the Earth is finite. Tentative suggestion that stellar systems may evolve from nebulous matter, since certain groups of stars seem to fall on lines or curves, indicating more than a casual association.

1902

Roberts, Mrs. Isaac, *J.B.B.A.*, **12**, 109. Describes another plate of M31. Nucleus resembles a small bright star surrounded by nebulosity. She cannot tell for sure what stars in the area are connected with it.

1904

Asmussen, O. *B.S.A.F.*, **18**, 49. Reproduction of a photograph taken by Nielsen in Copenhagen.

Ritchey, G. W. *University of Chicago Decennial Publications*, **8**, 389. An excellent photo of M31 and a brief description of spiral structure.

Wesley, W. K. *M.N.R.A.S.*, **64**, 237. Author concludes that spurious details in Ritchey's photographs had not been introduced by plate processing.

1905

Smith, A. *E. M.*, **82**, 367, 402. Popular article describing the nebula.

1907

Götz, P. *Heidlb. Astrophys. Publ.*, **3**, Nr. 1–39. Positions and magnitudes for 1,259 stars involved in the Andromeda Nebula, together with the positions of 54 recognizable points, followed by a detailed description of the nebula, a discussion of the relation of the star-density to the form and brightness of the gaseous mass, and the results of a statistical investigation of the distribution of stars. All stars are fainter than the 9th magnitude, 64 fainter than 16th.

Bohlin, K. *Astronomiska Iakttagelser och Undersokningar a Stockholms Observatorium*, **Vol. III**, 4, p. 66. From 15 photos of M31, three separate determinations of the parallax were made, with a mean of 0.17″.

1908

Bererich, A. *Nat. Rund.*, **23**, 1–3, *Weltall*, **8**, 147. Description and discussion of work of Götz and Bohlin.

Denning, W. F. *Pop. Astr.*, **16**, 197. Using Bohlin's parallax he derives a distance for M31 of 1.13×10^{14} miles and a linear extent of 3.6×10^{12} miles.

Gore, J. E. *Know. N. S.*, **5**, 71–74. Discusses distance, diameter, thickness, density, volume. Decides that the parallax of Bohlin is too large, for it makes M31 have a mass of 8×10^9 suns. Rejects the external galaxy notion. He explains the nova of 1885 according to a theory of collision and cooling.

Schroeter, J. *Fr. Naturen*, **32**, 18. Article in Norwegian reviewing work of Roberts, Bohlin, and Götz.

Wolf, M. *M.N.R.A.S.*, **68**, 626. Discusses the lengths of axes and the position angles of 52 oval nebulae.

1909

Fath, E. A. *Lick Bull.*, **149**, 71. Using the Crossley reflector fitted up as a nebular spectrograph, spectra of M31 and other spirals are determined. Spectrum of M31 found to be "of solar type," with 14 identifiable absorption lines and an intensity maximum at λ4640. Author unable to understand why all the stars in M31's nucleus should be of one spectral type. Also derives from Bohlin's parallax the result that the stars in the nucleus are the size of asteroids.

Gore, J. E. *Know. N. S.*, **6**, 147. Argues against the similarity of Milky Way and M31.

Kapteyn, J. *Ap. J.*, **30**, 284. In a footnote at the end of this article, Kapteyn uses a recent observation by H. D. Babcock on M31 to strengthen his conclusion that "there must be an appreciable amount of absorption in space." Babcock's observation compared two photos of M31, one through a red glass plate, to conclude that M31 was 1 magnitude redder than a star of the same spectrum.

Scheiner, J. *Ap. J.*, **30**, 69. Defends his 1899 spectrum of M31.

1911

Sutherland, A. *Ap. J.*, **33**, 251. Author expresses Bode's Law mathematically as the sum of two logarithmic spirals. This suggests to him that M31, the solar system, and Saturn's rings are all similar examples of a fundamental law of nature governing the condensation of matter into systems.

1912

Wolf, M. *Sitzungsberichte der Heidelb. Akad. der Wissenschafter Abt. A 1912* 3 Abhandlung, Nr. 15. Wolf's studies of spiral spectra, in particular M31, agree with those of Fath.

1913

Reynolds, J. H. *M.N.R.A.S.*, **74**, 132–136. Measurement of plate density as a function of distance from the center. Obtains a central bulge and wings to which the curve $(x+1)^2\ y =$ const. is a good fit. Believes that the nucleus is one star, much involved with the surrounding nebulosity. He feels that if the nucleus contained more than one star, we should be able to resolve them photographically. The inverse-square nature of the light curve lends support to the hypothesis that this is simply a reflection nebula — measures in polarized light are needed.

Slipher, V. M. *Lowell Bull. #58*, **2**, 56. Discusses how to build a spectrograph for a faint source. Gets a mean velocity for M31 of −300 km/sec. This high velocity suggests to him that a study of proper motions of spirals should be done. Proposes as a solution to the origin of the 1885 nova the hypothesis that the nebula "encountered a dark star" in its rush towards us at such high velocity.

1914

Slipher, V. M. *Pop. Astr.*, **23**, 21. Summary of work on nebulae, including rotation of M31. Found rotation greater near the nucleus, inclination of the lines indicating a speed of 100 km/sec at 20″ from the nucleus. Author states that the spectrum of M31 shows no composite features such as those shown by star clusters.

1915

Curtis, H. D. *Pub.A.S.P.*, **27**, 214. Redid Keeler's original photos for proper motion studies. The results for nebulae (including M31) is an average of 0.033″, which, using Slipher's velocities, implies distances on the order of 10,000 light years. No evidence has been found for rotation through positional measurements.

Pease, F. G. *Pub.A.S.P.*, **27**, 133. Radial velocity obtained with 60" is −329 km/sec.

1917

Barnard, E. E. *A.J.*, **30**, 175. Over a period of seven years the proper motions of M31 are undetectable. Positional values are available since 1836 but are too inaccurate to be used.

Curtis, H. D. *Pub.A.S.P.*, **29**, 108, 145. Absorption in spiral nebulae.

Lundmark, K. and Lindblad, B. *Astr. Nachr.*, **205**, 161; *Ap. J.*, **46**, 206. Effective wavelengths are obtained and converted to spectral types and color indices. M31 is G4.

Ritchey, G. W. *Pub.A.S.P.*, **29**, 210. On novae in spiral nebulae.

Ritchey, G. W. *Pub.A.S.P.*, **29**, 257. Discovery of a faint nova.

Shapley, S. *Pub.A.S.P.*, **29**, 213. Discovery of a faint nova.

Wolf, M. *Vierteljahrschrift*, **51**, 115. Radial velocity of −450 km/sec for M31.

1918

Curtis, H. D. *Lick Pub.*, **13**. Author emphasizes that dark lanes are often found in spirals, formed by obscuring matter and not just an absence of stars.

Duncan, J. C. *Pub.A.S.P.*, **30**, 255. Discovery of a nova.

Pease, F. G. *Proc. Nat. Acad. Sci.*, **4**, 21. (Mt. Wilson Comm. 51.) Obtains a radial velocity of −316 km/sec. Rotation measures required an exposure of 79 hours on the 60". His results along the major axis are fairly well represented by a straight line, implying that any theory of orbits obeying an inverse square law must be abandoned.

Ritchey, G. W. *Pub.A.S.P.*, **30**, 162. Three additional novae discovered.

Sanford, R. *Pub.A.S.P.*, **30**, 341. Two more novae discovered.

Van Maanen, A. *Pub.A.S.P.*, **30**, 307. A very carefully done parallax of M31 gives $0.004 \pm 0.005''$.

1920

Seares, F. H. *Ap. J.*, **52**, 162. Calculates the surface brightness of the Galactic system as viewed from a distant point in the direction of the galactic pole for various distances from the center and finds the brightness of the central part to be of visual magnitude 23 per square second of arc, whereas Andromeda is more than 100 times brighter. Concludes that our Galaxy is not a typical spiral.

1921

Reynolds, J. H. *Obs.*, **44**, 368. M31 shows a *recession* of 278 km/sec. Photographs show very little ultraviolet in M31. The angular dimensions of M31 are roughly $130' \times 40'$.

Slipher, V. M. *Pop. Astr.*, **29**, 272. Evidence for the rotation of NGC 224 was obtained from the inclination method, *i.e.*, keeping the slit of the spectrograph over the major axis of the nebula.

1922

Doig, P. *J.B.A.A.*, **32**, 138. Gives a short account of the novae in M31. Taking the absolute magnitude of the nova of 1885 as −14.0, gets a distance of 540,000 light years and a diameter of about 18,000 light years. Comes to the conclusion that the possibility is large enough to admit the hypothesis that it is an external universe.

Hopmann, F. *Astr. Nachr.*, **214**. A photometric study of nebulae, including M31.

La Place-Janssen and Haarh, G. E. H. *Astr. Nachr.*, **215**, 285. Discusses the parallax of the Andromeda Nebula.

Öpik, E. *Ap. J.*, **55**, 406. Assumes that the ellipsoidal shape of the inner parts of the nebula is due to rotation, and then applies Kepler's third law and gets an estimate of the distance to be 450,000 parsecs.

Wirtz, C. *Astr. Nachr.*, **215**, 349. Gives −316 km/sec for M31's systemic velocity.

1923

Lundmark, K. *Pub.A.S.P.*, **35**, 95. Assuming that the mean absolute magnitude of the 22 known novae in M31 is equal to that for the novae in Sagittarius, he gets the distance for M31 to be 63 times the distance for the Sagittarius region. This gives a distance of about 4×10^6 light years.

1924

Reynolds, J. H. *M.N.R.A.S.*, **85**, 142. Finds that spirals vary greatly in the matter contained, both in their nuclei and arms and concludes that though M31 and M33 may be compatible in dimensions with our Galaxy, most spirals are relatively quite insignificant.

1925

Hubble, E. P. *Obs.*, **48**, 139; *Pop. Astr.*, **33**, 252. From the observed Cepheid variables in M31 a Shapley period-luminosity curve has been constructed on the basis of visual magnitudes. From these a distance of 285,000 parsecs (= 930,000 light years) is obtained. Assumptions are (1) variables are actually connected with spirals, (2) no serious amount of absorption due to amorphous nebulosity is in the spiral, and (3) the nature of Cepheid variation is uniform throughout the observable portion of the universe.

Jeans, J. H. *M.N.R.A.S.*, **85**, 531. Considers the hypothesis that nebular condensations are formed by gravitational instability in a gas. Up to 90% of matter in the arms might be in solid or liquid state. Assuming the lenticular shape of M31 to be due to rotation, gets a period of 5.7×10^{14} sec, a mean density of 9×10^{-22} gm/cm^3, a diameter of the nucleus of 10^{21} cm and a mass of 5×10^{42} gm. Suggests that M31 exemplifies a state intermediate between the typical spiral and the Galactic system.

Landmark, K. *M.N.R.A.S.*, **85**, 865. Discusses various methods by which the distances of the spiral nebulae can be estimated. Charlier showed the distance of NGC 224 to be 28 times the diameter of the Galactic system. On the assumption that Galactic and Andromeda nebulae have equal absolute magnitudes, Lundmark finds the distance of NGC 224 to be 32 times the diameter of the Galactic system in good agreement.

1926

Hubble, E. P. *Ap. J.*, **64**, 321. Classifies M31 as Sb.

Lee, O. J. *Pop. Astr.*, **34**, 492. Gives an account of proper motion studies by various people and gets $0''.0184$ for the annual proper motion from his own data.

Reynolds, J. H. *M.N.R.A.S.*, **87**, 112. Assuming that the nebula is roughly circular, a comparison of the major and minor axes of the apparent ellipse shows that the inclination is about 70°. It is of massive type with arms of considerable breadth, one arm more irregular than the other. The smaller globular nebula NGC 221 has the same radial velocity as M31 and so they may be connected. Star counts show 10 times as many stars at the extremities of the ellipse as near the center.

1927

Lundmark, K. and Ark, F. *Mat. Astron. och Fysik.*, **20b**, No. 3. Assuming the dispersion in the absolute magnitudes of the separate stars is small, it is possible to compute the distances without making any assumption as to their size or total brightness. For 30 objects, including M31, gives total magnitude, apparent diameters, magnitudes of brightest stars, relative distances and parallaxes.

Luyten, W. J. *Harvard Bull.*, **851**. Two variable stars found in the nebula varied from 16.5^m to 15.3^m and 13.5^m to 14.5^m.

1928

Duncan, J. C. *Pub.A.S.P.*, **40**, 347. Four novae were discovered on July 16, 1928. Their positions from the nucleus and magnitudes are given.

Markov, A. *Astr. Nachr.*, **234**, 329. From the surface brightness of 19 nebulae, including M31, he comes to the conclusion that the most probable explanation of the spirals is that they are galaxies similar to ours.

1929

Hubble, E. E. *Ap. J.*, **69**, 103. Results of a comprehensive study. Fifty variables and 63 novae were found. The mass density of M31 appears to be about one sun per 20 cubic parsecs and luminosity density about 0.9 magnitudes per cubic parsec. An approximate comparison of sizes, masses, luminosities and densities suggest that the Galactic system is much larger than M31.

Lundmark, K. and Ark, F. *Mat. Astron. och Fysik*, **21a**, 9 and **21a**, 10. From spectroscopic observations of rotation in spiral nebulae, including M31, the masses and absolute magnitudes are deduced.

Perrine, C. D. *Astr. Nachr.*, **236**, 329. Gives distance, diameter, and other properties for spiral nebulae, including M31.

1930

Vinterhansen, J. M. *Nordisk Astronmisk. Tiddskrift Kobenhavn*, **11**, 1. A discussion of M31 in light of the new distance determinations due to Lundmark and Hubble.

1931

Mayall, N. U. *Pub.A.S.P.*, **43**, 217. Contains list of 14 new novae.

1932

Hubble, E. *Ap. J.*, **76**, 44. Identification of 140 nebulous objects in or close to the border of M31 which, from numbers distribution and radial velocities are presumed to be globular clusters associated with the spiral. Comparison is made with globular clusters in our Galaxy and the Magellanic Clouds and similar objects in other nebulae.

Humanson, M. C. *Pub.A.S.P.*, **44**, 381. Spectra of two novae in M31 are described.

Reynolds, J. H. *Obs.*, **55**, 301. A discussion of past observations of possible globular clusters.

1934

Baade, W. and Zwicky, F. *Proc. Nat. Acad. Sci.*, **20**, 254. A distinction is made between common and supernovae. Physics of novae are discussed and S And is used as an example of a supernova.

Shapley, H. *Harvard Bull.*, **895**, 19. Uses densitometer to get values for the major axis of 194 arcmin and of the minor axis of 16 arcmin.

Stebbins, J. and Whitford, A. E. *Proc. Nat. Acad. Sci.*, **20**, 93. Found photoelectric diameter larger than photographic; size more than doubled in the direction north and south from the nucleus.

1935

Bernheimer, W. E. *Wien Urania Zirk 2*, **Nr 4**. Observations of Stebbins and Whitford, Shapley and Vocca are discussed, compared and analyzed and a proposal is made for a set of better observations of the diameter.

Vocca, P. *Memorie della Societa Astrnomica Italiania*, **9**, 75. A confirmation of Hubble's work on the dimensions of M31.

Zanstra, H. *Naturwissenschaften*, **23**, 867. Observations of S And fit Baade and Zwicky's hypothesis about novae.

1936

Hubble, E. *Realm of the Nebula*, Yale University Press, New Haven. General summary of data known about the nebula.

Payne-Gaposchkin, C. *Ap. J.*, **83**, 245. Examines the records of the spectrum of S And and arranges them in a table in chronological order. From the table one can see that the nova spectrum was at first practically continuous and later showed bright lines of no very great intensity. There is also a table of color observations.

Whitford, A. *Ap. J.*, **83**, 424. Integrated magnitudes measured by a photoelectric photometer, particularly for the Andromeda Nebula. With m-M of 23.0 he gets M of −17.5.

1937

Hogg, F. S. *J.R.A.S. Can.*, **31**, 351. Discusses Zwicky's search for supernovae and mentions S And.

Lindblad, B. *Vierteljahrsschrift der Astronomischen Gesellschaft*, **72**. Gives results of a photometric study of the distribution of the dark material in the nebula.

Redman, R. O. and Shirley, E. G. *M.N.R.A.S.*, **97**, 416. Photometry provides luminosity distributions along the axes.

1938

Baade, W. *Ap. J.*, **88**, 285. Compiles photometric data for 18 supernovae, *i.e.*, those known at the end of 1937. Former estimates have been replaced by photometric magnitudes after a redetermination of the magnitudes of comparison stars on the international system. Gets −15 for S And.

Babcock, H. W. *Pub.A.S.P.*, **50**, 174. A linear velocity of rotation of 90 km/sec in the plane of the spiral is measured at r of 4′. It is constant at 150 km/sec until 30′. Systemic velocity is −300 km/sec.

Zwicky, F. *Ap. J.*, **88**, 529. Discusses the frequency of supernovae. A footnote describes a hypothetical case of the calculation of too large a frequency of supernovae for M31.

1939

Becchini, G. and Gratten, L. *Memorie della Astronomica Italiania*, **18**, 303. A statistical study that shows that the novae in our Galaxy agree in frequency with those in M31.

Hoffleit, D. *Harvard Bull.*, **210**, 7. Gives a curve for S And.

1940

Babcock, H. D. *Lick Bull. 498*, **19**, 41. Observations of continuous solar-type spectrum of unresolved stars and of diffuse and emission nebulosities give a complete rotation curve. Finds the mass luminosity ratio and compares M31 with our Galaxy.

Zwicky, F. *Rev. Mod. Physics*, **12**, 66. S And is mentioned as an example of the recognition of supernovae as a separate class of novae.

1941

Wyse, J. D. and Mayall, N. U. *Pub.A.S.P.*, **53**, 269. The distribution of mass in M31 and M33 was determined by comparing a disk model with observed rotation curves.

1942

Danver, G. G. *Ann. Obs. Lund.*, **10**, 7. M31 included in discussion of spiral arm patterns.

Lindblad, B. *Stockholms Obs. Ann.*, **14**, 3. Using Öhman's measures of polarization of a small dark cloud near the nucleus of M31, the conclusion is reached that the brighter edge of the nebula is the nearer.

Öhman, Y. *Pub.A.S.P.*, **54**, 72 and *Stockholms Obs. Ann.*, **14**, 4. Polarization of about 3% observed in a small dark cloud near the nucleus of M31. This polarization may be used to support Lindblad's conclusion about the orientation of the nebula.

Wyse, J. D. and Mayall, N. U. *Ap. J.*, **95**, 24. M31 and M33 are assumed to be composed of flat disks with surface densities represented by 5th-degree polynomials. Assuming circular motion, the observed rotation curve gives the mass distribution. The solutions show little tendency toward central condensation. In both cases the average space density derived from the surface density is about two solar masses per cubic parsec in the main bodies. The total mass of M31 is 9.5×10^{10} solar masses.

1943

Eigenson, M. *Russian A. J.*, **4**, 5. The rotation of M31 as observed by Babcock is interpreted in terms of a spherically distributed uniform system in order to deduce conclusions about our own Galaxy.

Williams, J. and Hiltner, W. *Pub. Obs. Univ. Michigan*, **8**, 103. Used an 18" Palomar Schmidt plate to construct isophotes of M31. The length of the major axis was found to be at least 400′. Faint outer regions tended to spiral in the opposite sense from the arms.

1944

Baade, W. *Ap. J.*, **100**, 137. Photographs on red-sensitive plates resolve the central region of M31 and the companions M32 and NGC 205 into stars. The brightest stars there have photographic magnitudes of 21.3 and color indices of $+1^m.3$.

Chalonge, D. *Bull. Soc. Astron. France*, **58**, 139. A short article on recent research on the Andromeda Nebula.

Leontovski, M. *Bull. de l'Inst. Astron. (Leningrad)*, **53**, 1. Discusses the structure of Stebbins' regions in M31.

1945

Seyfert, K. and Nassau, J. J., *Ap. J.*, **101**, 179. Star counts on blue-sensitive plates made with 24" Schmidt show reasonable agreement with isophotal contours. The luminosity distribution in the main body is similar to that for the solar neighborhood in the observed range of absolute magnitudes. The thickness from high luminosity stars was estimated to be of the order of 200 pc.

Seyfert, K. and Nassau, J. J. *Ap. J.*, **102**, 377. Gives photographic magnitudes for 212 of the 249 nebulous objects in M31 found by Hubble and Baade. The mean absolute magnitude of these objects is about −5.0.

1946

Lindblad, B. and Brahde, R. *Ap. J.*, **104**, 211. The orientation of the Andromeda Nebula is inferred from the relative distributions of novae and variables compared to globular clusters.

1947

Parenago, P. P. *Russian A. J.*, **24**, 178. Babcock's rotation curve for M31 is interpreted as being due to Baade's Population II in the center and Population I in the outer parts; each population having its own velocity curve.

1948

Fricke, W. *Die Naturwis.*, **35**, 52. Discusses the structure of the nucleus of M31 and photometry, rotation, and internal motions of spiral nebulae.

Lindblad, B. *Stockholms Obs. Ann.*, **15**, 4. The orientation of the nebula is discussed using the distribution of globular clusters.

Mayall, N. U. *A. J.*, **54**, 44. Inclination of lines in the spectrum of M31 imply rotation periods for the outer parts in the range 60 to 220×10^6 years.

Parenago, P. P. *Russian A. J.*, **25**, 306. The similarity of our Galaxy to M31 is discussed with respect to rotation and shape of spiral arms.

1949

Artyukina, N. M. *Proc. State Astron. Inst. (USSR)*, **16**, 93. Reviews work that has been done on the distance, Cepheids, mass, etc.

de Vaucouleurs, G. *Obs.*, **69**, 150. Suggests that M31 is close enough that the variation of distance across the nebula may be sufficient for a detectable variation of the period-magnitude relation for the Cepheids from one side to the other.

Hartwig, G. *Die Sterne*, **25**, 7. This article deals with Population II stars in elliptical nebulae and the nucleus of M31.

Ohman, Y. *Stockholms Obs. Ann.*, **15**, 8. Measured the color index of M31 to be −0.046.

1950

Dombrowskiy, W. A. *Pub. Astr. Obs. (Leningrad)*, **15**, 166. Microphotometry of nebulae, including M31.

Hanbury Brown, R. and Hazard, C. *Nature*, **166**, 901. Radio radiation detected.

Haro, G. *A. J.*, **55**, 66. Emission and absorption nebulae in M31 and M33 are described.

Holmberg, E. *Lund. Medd. (2)*, **128**, 56 pp. Gives photometry and light and color distributions.

Radlowa, L. N. *Bull. Astrophys. Obs. Abastumani*, **11**, 91.

REFERENCES

Allen, C., 1974, *Astrophysical Quantities*, (Athlone: London).

Alloin, D., Pelat, D., and Bijaoui, A., 1976, *Astron. Astrophys.*, **50**, 127.

Alloin, D., Pelat, D., and Bijaoui, A., 1977, *Astron. Astrophys.*, **54**, 321.

Argyle, E., 1965, *Astrophys. J.*, **141**, 750.

Armandroff, T. and Massey, P., 1991, *Astron. J.*, **102**, 927.

Arp, H., 1956, *Astron. J.*, **61**, 15.

Arp, H., 1962, *Astrophys. J.*, **135**, 971.

Arp, H., 1964, *Astrophys. J.*, **139**, 1045.

Arp, H., 1965, *Astrophys. J.*, **141**, 43.

Arp, H. and Bruekel, F., 1973, *Astrophys. J.*, **179**, 445.

Baade, W., 1951, *Publ. U. Mich. Obs.*, **10**, 16.

Baade, W., 1955, *Astron. J.*, **60**, 151.

Baade, W., 1958, in *Stellar Populations*, edited by D.J.K. O'Connell (Richerche Astronomische, Specola Astronomisca Vaticana), p. 3.

Baade, W., 1963, *Evolution of Stars and Galaxies* (Harvard University Press: Cambridge), p. 101.

Baade, W. and Arp, H., 1964, *Astrophys. J.*, **139**, 1027.

Baade, W. and Swope, H., 1963, *Astron. J.*, **68**, 435.

Baade, W. and Swope,H., 1965, *Astron. J.*, **70**, 212.

Bajaja, E. and Gergeley, T., 1977, *Astron. Astrophys.*, **61**, 229.

Bajaja, E. and Shane, W., 1982, *Astron. Astrophys. Suppl.*, **49**, 745.

Baker, P., 1976, *Astron. Astrophys.*, **48**, 163.

Baldwin, J., Danziger, J., Frogel, J., and Persson, S., 1973, *Astrophys. Letters*, **14**, 1.

Bash, F., 1981, *Astrophys. J.*, **250**, 551.

Battaner, E., Beckman, J., Mediavilla, E., Prieto, M., Sanchez Magro, C., Muñoz Tuñon, C., and Sanchez Saavedra, M., 1986, *Astron. Astrophys.*, **161**, 70.

Battistini, P., Bonoli, F., Braccesi, A., Fusi Pecci, F., Malagnini, M., and Marano, B., 1980, *Astron. Astrophys. Suppl.*, **42**, 357.

Battistini, P., Bonoli, F., Buonanno, R., Cacciari, C., Corsi, C., and Fusi Pecci, F., 1982, *Astron. Astrophys.*, **113**, 39.

Battistini, P., Bonoli, F., Braccesi, A., Federici, L., Fusi Pecci, F., Marano, B., and Börngen, F., 1987, *Astron. Astrophys. Suppl.*, **67**, 447.

Baum, W. and Schwarzschild, M., 1955, *Astron. J.*, **60**, 249.

Beck, R., 1982, *Astron. Astrophys.*, **106**, 121.

Beck, R., 1990, *Geophys. Astrophys. Fluid Dyn.*, **50**, 39.

Beck, R., Berkhuijsen, E., and Wielebinski, R., 1980, *Nature*, **283**, 272.

Beck, R. and Gräve, R., 1982, *Astron. Astrophys.*, **105**, 192.

Beck, R., Loiseau, N., Hummel, E., Berkhuijsen, E., Gräve, R., and Wielebinski, R., 1989, *Astron. Astrophys.*, **222**, 58.

Becker, W., 1964, *Z. Astrophys.*, **58**, 202.

Berkhuijsen, E., 1977, *Astron. Astrophys.*, **57**, 9.

Berkhuijsen, E. and Humphreys, R., 1989, *Astron. Astrophys.*, **214**, 68.

Berkhuijsen, E., Humphreys, R., Ghigo, F., and Zumach, W., 1988, *Astron. Ap. Suppl.*, **76**, 65.

Berkhuijsen, E. and Wielebinski, R., 1974, *Astron. Astrophys.*, **34**, 173.

Berkhuijsen, E., Wielebinski, R. and Beck, R., 1983, *Astron. Astrophys.*, **117**, 141.

Blair, W. P., Kirshner, R. P., and Chevalier, R., 1981, *Astrophys. J.*, **247**, 879.

Blitz, L., 1985, *Astrophys. J.*, **296**, 481.

Bohlin, R., Cornett, R., Hill, J., Hill, R., Stecher, T., 1988, *Astrophys. J.*, **334**, 657.

Bok, B. J. and Bok, P. F., 1981, *The Milky Way* (Harvard University Press, Cambridge, Massachusetts).

Boulanger, F., Bystedt, J., Casoli, F., and Combes, F., 1984, *Astron. Astrophys.*, **140**, L5.

Boulanger, F., Stark, A., and Combes, F., 1981, *Astron. Astrophys.*, **93**, L1.

Bowyer, S., Margon, B., Lampton, M., and Cruddace, R., 1974, *Astrophys. J.*, **190**, 285.

Brandt, J., 1960, *Astrophys. J.*, **131**, 25.

Brandt, J. and Scheer, L., 1965, *Astron. J.*, **70**, 471.

Braun, R., 1990, *Astrophys. J. Suppl.*, **72**, 755 and 761.

Braun, R., 1991, *Astrophys. J.*, **372**, 54.

Brinks, E., 1984, Ph.D. Thesis, University of Leiden.

Brinks, E. and Bajaja, E., 1986, *Astron. Astrophys.*, **169**, 14.

Brinks, E., Braun, R., and Unger, S., 1989, in *Structure and Dynamics of the Interstellar Medium*, edited by G. Tenorio-Tagle, M. Moles, and J. Melnick (Springer: Berlin).

Brinks, E. and Burton, B., 1984, *Astron. Astrophys.*, **141**, 195.

Brinks, E. and Burton, B., 1985, in *The Milky Way Galaxy*, IAU Symp. No. 106, edited by H. van Woerden, R. J. Allen, and W. B. Burton (Reidel: Dordrecht), p. 437.

Brinks, E. and Shane, W., 1984, *Astron. Astrophys. Suppl.*, **31**, 439.

Buonanno, R., Corsi, C., Battistini, P., Bonoli, F., and Fusi Pecci, F., 1982, *Astron. Astrophys. Suppl.*, **47**, 151.

Burke, B., Turner, K., and Tune, M., 1963, *Annual Report, Dept. of Terrestrial Magnetism* (Carnegie Inst.: Washington), p. 289.

Burstein, D. and Heiles, C., 1984, *Astrophys. J. Suppl.*, **54**, 33.

Byrd, G., 1976, *Astrophys. J.*, **208**, 688.

Byrd, G., 1978, *Astrophys. J.*, **226**, 70.

Byrd, G., 1979, *Astrophys. J.*, **231**, 32.

Byrd, G., 1983, *Astrophys. J.*, **264**, 464.

Byrd, G., Smith, B., and Miller, R., 1984, *Astrophys. J.*, **286**, 62.

Bystedt, J., Brinks, E., de Bruyn, A., Israel, F., Schwering, P., Shane, W., and Walterbos, R., 1984, *Astron. Astrophys. Suppl.*, **56**, 245.

Cacciari, C., Cassatella, A., Bianchi, L., Fusi Pecci, F., and Kron, R., 1982, *Astrophys. J.*, **261**, 77.

Capaccioli, M., Della Valle, M., D'Onofrio, M., and Rosino, L., 1989, *Astron. J.*, **97**, 1622.

Carruthers, G., Heckathorn, H., and Opal, C., 1978, *Astrophys. J.*, **225**, 346.

Casoli, F., Combes, F., and Stark, A., 1987, *Astron. Astrophys.*, **173**, 43.

Chevalier, R. and Plait, P., 1988, *Astrophys. J.*, **331**, L109.

Christian, C. and Heasley, J., 1991, *Astron. J.*, **101**, 848.

Christian, C. and Janes, K., 1979, *Astron. J.*, **84**, 214.

Christian, C. and Schommer, R., 1982, *Astrophys. J. Suppl.*, **49**, 405.

Chromey, F., 1978, *Astron. J.*, **83**, 162.

Ciardullo, R., Ford, H., and Jacoby, G., 1983, *Astrophys. J.*, **272**, 92.

Ciardullo, R., Ford, H., Neill, J., Jacoby, G., and Shafter, A., 1987, *Astrophys.*, **J.**, 318, 520.

Ciardullo, R., Jacoby, G., Ford, H., and Neill, J., 1989, *Astrophys. J.*, **339**, 53.

Ciardullo, R., Rubin, V., Jacoby, G., Ford, H., and Ford, W. K., 1988, *Astron. J.*, **95**, 438.

Ciardullo, R., Shafter, A., Ford, H., Neill, J., Shara, M., and Tomaney, A., 1990, *Astrophys. J.*, **356**, 472.

Code, A., 1969, *Pub. Astron. Soc. Pac.*, **81**, 475.

Cohen, J., 1978, *Astrophys. J.*, **221**, 788.

Cohen, J., 1985, *Astrophys. J.*, **292**, 90.

Cohen, R., Few, R., and Booth, R., 1979, *Mon. Not. R. Astr. Soc.*, **187**, 35P.

Collura, A., Reale, F., and Peres, G., 1990, *Astrophys. J.*, **356**, 119.

Combes, F., Encrenaz, P., Lucas, R., and Weliachew, L., 1977a, *Astron. Astrophys.*, **55**, 311.

Combes, F., Encrenaz, P., Lucas, R., and Weliachew, L., 1977b, *Astron. Astrophys.*, **61**, L7.

Condon, J., 1987, *Astrophys. J. Suppl.*, **65**, 485.

Cooke, B., Ricketts, M., Maccaro, T., Pye, J., Elvis, M., Watson, M., Griffiths, R., Pounds, K., McHardy, I., Maccagni, D., Seward, F., Page, C., and Turner, M., 1978, *Mon. Not. R. Astron. Soc.*, **182**, 489.

Cowley, A. and Starrfield, S., 1987, *Publ. Astron. Soc. Pac.*, **99**, 854.

Cram, T., Roberts, M., and Whitehurst, R., 1980, *Astron. Astrophys. Suppl.*, **40**, 215.

Crampton, D., Cowley, A., Hutchings, J., Schade, D., and van Speybroeck, L., 1984, *Astrophys. J.*, **284**, 663.

Crampton, D., Cowley, A., Schade, D., and Chayer, P., 1985, *Astrophys. J.*, **228**, 494.

Crane, P., Brown, R., and van der Hulst, J., 1981, *Bull. Amer. Astron. Soc.*, **12**, 844.

Crotts, A., 1986, *Astron. J.*, **92**, 292.

Davidge, T., 1990, *Astrophys. J.*, **351**, L37.

Davidge, T., Alloin, D., and Jablonka, P., 1990, *Astrophys. J.*, **358**, L1.

Davies, R., Frogel, J., and Terndrup, D., 1991, *Astron. J.*, **102**, 1729.

Davies, R. and Gottesman, S., 1970, *Mon. Not. R. Astron. Soc.*, **149**, 237.

Deharveng, J. M., Jakosen, P., Milliard, B., and Laget, M., 1980, *Astron. Astrophys.*, **88**, 52.

Deharveng, J. and Pellet, A., 1969, *Astron. Astrophys.*, **1**, 208.

Deharveng, J. and Pellet, A., 1975, *Astron. Astrophys.*, **38**, 15 and *Astron. Astrophys. Suppl.*, **19**, 351.

Dennefeld, M. and Kunth, D., 1981, *Astron. J.*, **86**, 989.

Dennison, B., Balonek, T., Terzian, Y., and Balick, B., 1975, *Publ. Astron. Soc. Pac.*, **87**, 83.

de Vaucouleurs, G., 1958, *Astrophys. J.*, **28**, 465.

de Vaucouleurs, G., 1964, in IAU Symp. No. 20, *The Galaxy and the Magellanic Clouds*, edited by F. Kerr and A. Rodgers (Austr. Acad. Sci.: Canberra).

de Vaucouleurs, G., 1970, *Astrophys. J.*, **159**, 435.

de Vaucouleurs, G., 1978, *Astrophys. J.*, **223**, 351; **223**, 730; **224**, 14.

de Vaucouleurs, G. and Buta, R., 1978, *Astron. J.*, **83**, 1383.

de Vaucouleurs, G. and Corwin, H., 1985, *Astrophys. J.*, **295**, 287.

de Vaucouleurs, G. and Pence, W. D., 1978, *Astron. J.*, **83**, 1163.

Dickel, J., D'Odorico, S., Felli, M., and Dopita, M., 1982, *Astrophys. J.*, **252**, 582.

Dickel, J. and D'Odorico, S., 1984, *Mon. Not. R. Astron. Soc.*, **206**, 351.

Dickey, J. and Brinks, E., 1988, *Mon. Not. R. Astron. Soc.*, **233**, 781.

D'Odorico, S., Dopita, M., and Benvenuti, P., 1980, *Astron. Astrophys. Suppl.*, **40**, 67.

Dopita, M., 1987, in *Star Forming Regions*, edited by M. Peimbert and J. Jugaku (Reidel: Dordrecht), p. 501.

Dressler, A., 1984, *Astrophys. J.*, **286**, 97.

Dressler, A. and Richstone, D., 1988, *Astrophys. J.*, **324**, 701.

Durdin, J. and Terzian, Y., 1972, *Astron. J.*, **77**, 637.

Efremov, Y., 1978, *Astron. Zh. Lett.*, **4**, 125.

Efremov, Y., 1979, *Astron. Zh. Lett.*, **5**, 21.

Efremov, Y. and Ivanov, G., 1987, *Astrophys. Space Sci.*, **129**, 39.

Efremov, Y. N., Ivanov, G. R., and Nikolov, N. S., 1981, *Astrophys. Space Sci.*, **75**, 407.

Efremov, Y., Ivanov, G., and Nikolov, N., 1987, *Astrophys. Space Sci.*, **135**, 119.

Einasto, J., 1969, *Astrofiz.*, **5**, 137.

Einasto, J., 1970, in *Papers presented to the IAU XIV General Assumbly*, **26**, 1.

Einasto, J. and Rümmel, U., 1969, in *The Spiral Structure of our Galaxy*, edited by W. Becker and G. Contopoulos (Reidel: Dordrecht), 51.

Elson, R. and Fall, S., 1985, *Publ. Astron. Soc. Pac.*, **97**, 692.

Elson, R. and Walterbos, R., 1988, *Astrophys. J.*, **333**, 594.

Elvius, A. and Hall, J., 1964, *Lowell Obs. Bull.*, **6**, 123.

Emerson, D., 1974, *Mon. Not. R. Astron. Soc.*, **169**, 607.

Emerson, D., 1976, *Mon. Not. R. Astron. Soc.*, **176**, 321.

Encrenaz, P., Stark, A., Combes, F., and Wilson, R., 1979, *Astron. Astrophys.*, **78**, L1.

Fabbiano, G., Trinchieri, G., and van Speybroeck, L., 1987, *Astrophys. J.*, **316**, 127.

Faber, S., 1972, *Astron. Astrophys.*, **20**, 361.

Faber, S. and French, H., 1980, *Astrophys. J.*, **235**, 405.

Fesen, R., Hamilton, A., and Saken, J., 1989, *Astrophys. J.*, **341**, L55.

Ford, H. and Jacoby, G., 1978, *Astrophys. J. Suppl.*, **38**, 351 and *Astrophys. J.*, **219**, 437.

Ford, H. and Jenner, D., 1975, *Astrophys. J.*, **202**, 365.

Freedman, W., 1985, *Astrophys. J.*, **299**, 74.

Freedman, W., 1986, in IAU Symp. 116, *Luminous Stars and Associations*, edited by C. W. Loore, A. J. Willis, and P. Laskarides (Reidel: Dordrecht), p. 61.

Freedman, W. and Madore, B., 1990, *Astrophys. J.*, **365**, 186.

Frogel, J., Persson, S., and Cohen, J., 1980a, in *Globular Clusters*, edited by D. Hanes and B. Madore (Cambridge Univ. Press: Cambridge), p. 159.

Frogel, J., Persson, S., and Cohen, J., 1980b, *Astrophys. J.*, **240**, 785.

Gallagher, J. and Hunter, D., 1981, *Astron. J.*, **86**, 1312.

Gallagher, J., Kenyon, S., and Hege, E., 1981, *Astrophys. J.*, **249**, 83.

Gaposchkin, S., 1962, *Astron. J.*, **67**, 358.

Georgelin, Y. M. and Georgelin, Y. P., 1976, *Astron. Astrophys.*, **49**, 57.

Giacconi, R., Murray, S., Gursky, H., Kellog, E., Schreier, E., Matilsky, T., Koch, D., and Tannanbaum, H., 1974, *Astrophys. J.*, **27**, 37.

Goodman, J. and Lee, H., 1989, *Astrophys. J.*, **337**, 84.

Gottesman, S., 1970, *Mem. Roy. Astron. Soc.*, **74**, 73.

Gottesman, S. and Davies, R., 1970, *Mon. Not. R. Astron. Soc.*, **149**, 263.

Gottesman, S., Davies, R., and Reddish, V., 1966, *Mon. Not. R. Astron. Soc.*, **133**, 359.

Graham, J., 1988, *Astrophys. J.*, **326**, L51.

Gunn, J., 1975, *Comments Ap. Sp. Phys.*, **6**, 7.

Habing, H., Miley, G., Young, E., Baud, B., Boggess, N., Clegg, P., de Jong, T., Harris, S., Raimond, E., Rowan-Robinson, M., and Soifer, B. T., 1984, *Astrophys. J.*, **278**, L1.

Hanbury Brown, R. and Hazard, C., 1951, *Mon. Not. R. Astron. Soc.*, **111**, 357.

Hanes, D., 1977, *Mon. Not. R. Astron. Soc.*, **179**, 331.

Harris, W., 1974, Ph.D. Thesis, Univ. of Toronto.

Harris, H. and Canterna, R., 1977, *Astron. J.*, **82**, 798.

Hartwick, F. and Sargent, W., 1974, *Astrophys. J.*, **190**, 283.

Haud, U., 1981, *Astrophys. Space Sci.*, **76**, 477.

Henderson, A., 1979, *Astron. Astrophys.*, **75**, 311.

Heasley, J., Christian, C., Friel, E., and Janes, K., 1988, *Astron. J.*, **96**, 1312.

Hiltner, W., 1958, *Astrophys. J.*, **128**, 9.

Hiromoto, N., Maihara, T., Oda, N., and Okuda, H., 1983, *Publ. Astron. Soc. Japan*, **35**, 413.

Hjellming, R. and Smarr, L., 1982, *Astrophys. J.*, **257**, L13.

Hodge, P., 1961, *Astrophys. J.*, **133**, 413.

Hodge, P., 1971, *Ann. Rev. Astron. Astrophys.*, **9**, 35.

Hodge, P., 1975, *BAAS*, **1**, 506.

Hodge, P., 1979, *Astron. J.*, **84**, 744.

Hodge, P., 1979, *Concepts of Contemporary Astronomy* (McGraw-Hill, New York).

Hodge, P., 1980, *Astron. J.*, **85**, 376.

Hodge, P., 1981, *Sci. Am.*, **244**, 92.

Hodge, P., 1983, *Pub. Astron. Soc. Pacific*, **95**, 721.

Hodge, P., 1985, in *The Milky Way Galaxy*, edited by H. van Woerden, R. Allen and W. Burton (Reidel: Dordrecht).

Hodge, P., 1986, in IAU Symp. 116, *Luminous Stars and Associations*, edited by C. Loore, A. Willis, and P. Laskarides, (Reidel: Dordrecht), p. 369.

Hodge, P., 1987, *Publ. Astron. Soc. Pac.*, **99**, 915.

Hodge, P., 1989, *Ann. Rev. Astron. Astrophys.*, **27**, 139.

Hodge, P. and Kennicutt, R., 1982, *Astron. J.*, **87**, 264.

Hodge, P. and Kennicutt, R., 1983, *Astron. J.*, **88**, 296.

Hodge, P. and Lee, M., 1988, *Astrophys. J.*, **329**, 651.

Hodge, P. and Lee, M., 1991, *Publ. Astron. Soc. Pac.*, **102**, 26.

Hodge, P. and Lee, M. G., 1992, in press.

Hodge, P. and Lucke, P., 1970, *Astron. J.*, **75**, 933.

Hodge, P., Mateo, M., Lee, M., and Geisler, D., 1987, *Publ. Astron. Soc. Pac.*, **99**, 173.

Hodge, P., Lee, M., and Mateo, M., 1988, *Astrophys. J.*, **324**, 172.

Hoessel, J. and Melnick, J., 1980, *Astron. Astrophys.*, **84**, 317.

Hubble, E. and Sandage, A., 1953, *Astrophys. J.*, **118**, 353.

Huchra, J., Brodie, J., and Kent, S., 1991, *Astrophys. J.*, **370**, 495.

Huchra, J., Stauffer, J., and van Speybroeck, L., 1982, *Astrophys. J.*, **259**, L57.

Humphreys, R., 1975, *Astrophys. J.*, **200**, 426.

Humphreys, R., 1978, *Astrophys. J.*, **219**, 445.

Humphreys, R., 1979, *Astrophys. J.*, **234**, 854.

Humphreys, R., Blaha, C., D'Odorico, S., Gull, T., and Benvenuti, P., 1984, *Astrophys. J.*, **278**, 124.

Humphreys, R., Massey, P., and Freedman, W., 1990, *Astron. J.*, **99**, 84.

Humphreys, R., Pennington, R., Jones, T., and Ghigo, F., 1988, *Astron. J.*, **96**, 1884.

Humphreys, R. and Warner, J., 1978, *Astrophys. J.*, **221**, L73.

Hutchings, J., Massey, P., and Bianchi, L., 1987, *Astrophys. J.*, **322**, L79.

Ichikawa, T., Nakano, M., Tanaka, Y., Saito, M., Nakai, N., Sofue, Y., and Kaifu, N., 1985, *Publ. Astron. Soc. Japan*, **37**, 439.

Iijima, T., Ito, K., Matsumoto, T., and Uyama, K., 1976, *Publ. Astron. Soc. Japan*, **28**, 27.

Innanen, K., Kamper, K., Papp, K., and van den Bergh, S., 1981, *Bull. Amer. Astron. Soc.*, **13**, 507.

Ivanov, G., 1985, *Astrophys. Space Sci.*, **115**, 409.

Ivanov, G. and Golev, V., 1984, *Astronomical Circ.*, **1352**, 1.

Ivanov, G. and Sharov, A., 1986, *Astrophys. Space Sci.*, **124**, 329.

Iye, M. and Richter, O., 1985, *Astron. Astrophys.*, **144**, 471.

Jacoby, G., 1980, *Astrophys. J. Suppl.*, **42**, 1.

Jacoby, G., 1989, *Astrophys. J.*, **339**, 39.

Jacoby, G. and Ford, H., 1986, *Astrophys. J.*, **304**, 490.

Johnson, H. and Hanna, M., 1972, *Astrophys. J. Lett.*, **174**, L71.

Johnson, H. and Morgan, W., 1953, *Astrophys. J.*, **117**, 313.

Johnson, H., 1961, *Astrophys. J.*, **133**, 309.

Kalnajs, A., 1974, in *La Dynamique des Galaxies Spirales*, Colloq. Int. CNRS No. 241, ed. by L. Weliachew (CNRS: Paris).

Kennicutt, R., 1983, *Astrophys. J.*, **272**, 54.

Kennicutt, R. and Hodge, P., 1980, *Astrophys. J.*, **241**, 573.

Kennicutt, R., Edgar, B., and Hodge, P., 1989, *Astrophys. J.*, **337**, 761.

Kent, S., 1983, *Astrophys. J.*, **266**, 562.

Kent, S., 1987, *Astron. J.*, **93**, 816, and **94**, 306.

Kent, S., 1989, *Astron. J.*, **97**, 1614.

Kent, S., 1989, *Publ. Astron. Soc. Pac.*, **101**, 489.

Kent, S., Huchra, J., and Stauffer, J., 1989, *Astron. J.*, **98**, 2080.

King, I., 1962, *Astron. J.*, **67**, 471.

Kinman, T., 1965, *Astrophys. J.*, **142**, 1376.

Knapp, G. R., Kerr, F. J., and Rose, W. K., 1973, *Astrophys. Letters*, **14**, 187.

Koper, E., Dame, T., Israel, F., and Thaddeus, P., 1991, *Astrophys. J.*, **383**, L11.

Kormendy, J., 1987, in *Structure and Dynamics of Elliptical Galaxies*, edited by T. de Zeeuw (Reidel: Dordrecht).

Kormendy, J., 1988, *Astrophys. J.*, **325**, 128.

Kron, G. and Guetter, H., 1976, *Astron. J.*, **81**, 817.

Kron, G. and Mayall, N., 1960, *Astron. J.*, **65**, 581.

Kumar, C., 1976, *Pub. Astron. Soc. Pac.*, **88**, 323.

Kumar, C., 1979, *Astrophys. J.*, **230**, 386.

Kunitsch, P., 1987, *ESO Messenger*, **49**, 42.

Kutner, M., Verter, F., and Rickard, L., 1990, *Astrophys. J.*, **365**, 195.

Lada, C., Margulis, M., Sofue, Y., Nakai, N., and Handa, T., 1988, *Astrophys. J.*, **328**, 143.

Lallemand, A., Duchesne, M., and Walker, M., 1960, *Publ. Astron. Soc. Pac.*, **72**, 76.

Lawrie, D. and Ford, H., 1982, *Astrophys. J.*, **256**, 120.

Lee, M., 1988, in *Galactic and Extragalactic Star Formation*, edited by R. Pudritz and M. Fich (Kluwer: Dordrecht), p. 653.

Lee, M. and Hodge, P., 1987, *Bull. Amer. Astron. Soc.*, **19**, 683.

Light, E., Danielson, R., and Schwarzschild, M., 1974, *Astrophys. J.*, **194**, 257.

Lin, C., Yaun, C., and Shu, F., 1969, *Astrophys. J.*, **155**, 721.

Lindblad, B., 1956, *Stockholm Obs. Ann.*, **19**, No. 2.

Lohman, W., 1954, *Z. fur Astrophys.*, **35**, 159.

Long, K. and van Speybroeck, L., 1983, in *Accretion Driven Stellar X-ray Sources*, edited by W. H. G. Lewin and E. P. J. van den Heuvel (Cambridge Univ. Press: Cambridge).

Lyngå, G., 1959, *Medd. Lunds. Astron. Observ.*, **2**, No. 137.

Madore, B., McAlary, C., McLaren, R., Welch, D., Neugebauer, G., and Matthews, K., 1985, *Astrophys. J.*, **294**, 560.

Makishima, K. and Ohashi, T., 1989, *Publ. Astron. Soc. Japan*, **41**, 697.

Marsden, B., 1988, *Sky and Telescope*, **75**, 253.

Martin, P. and Shawl, S., 1979, *Astrophys. J.*, **231**, L57.

Massey, P., Armandroff, T., and Conti, P., 1986, *Astron. J.*, **92**, 1303.

Massey, P., Hutchings, J., and Bianchi, L., 1985, *Astron. J.*, **90**, 2239.

Mateo, M., 1987, Ph.D. Thesis, University of Washington.

Mateo, M., 1987, in *Globular Cluster Systems in Galaxies*, edited by J. Grindlay and A. Philip (Reidel: Dordrecht).

Matsumoto, T., Murakami, H., and Hamajima, K., 1977, *Publ. Astron. Soc. Japan*, **29**, 583.

Matteucci, F. and Tornambé, A., 1988, European Southern Obs. Preprint No. 580.

Mayall, N., 1951, *Publ. Obs. Univ. of Michigan*, **10**, 19.

McClure, R. and Racine, R., 1969, *Astron. J.*, **74**, 1000.

McClure, R. and van den Bergh, S., 1968, *Astron. J.*, **73**, 313.

McElroy, D., 1981, Ph.D. Thesis, Arizona State University.

McElroy, D. 1983, *Astrophys. J.*, **270**, 485.

Moffat, A. and Shara, M., 1983, *Astrophys. J.*, **273**, 544.

Moffat, A. and Shara, M., 1987, *Astrophys. J.*, **320**, 266.

Morton, D., Andereck, C., and Bernard, D, 1977, *Astrophys. J.*, **212**, 13.

Mould, J., Cohen, J., Graham, J., Hamilton, D., Matthews K., Picard, A., Reid, N., Schmidt, M., Soifer, T., and Wilson, C., 1990, *Astrophys. J.*, **353**, L53.

Münch, G., 1962, in *Problems of Extra-Galactic Research*, edited by G. C. McVittie (Macmillan: New York), p. 119.

Nakai, N. and Sofue, Y., 1982, *Publ. Astron. Soc. Japan*, **34**, 199.

Nakano, M., Ichikawa, Y., Tanaka, N., Nakai, N., and Sofue, Y., 1987, *Publ. Astron. Soc. Japan*, **39**, 57.

Nedialkov, P., Kouztev, R., and Ivanov, G., 1988, *Astrophys. Space Sci.*, **162**, 1.

Newton, K. and Emerson, D., 1977, *Mon. Not. R. Soc.*, **181**, 573.

Newton, K. and Emerson, D., 1978, *Mon. Not. R. Astron. Soc.*, **181**, 573.

Nikolov, N. and Tasheva, R., 1989, *Astrophys. Space Sci.*, **152**, 239.

Nolthenius, R. and Ford, H., 1987, *Astrophys. J.*, **317**, 62.

O'Connell, R., 1976, *Astrophys. J.*, **206**, 370.

Odewahn, S., 1987, *Astron. J.*, **93**, 310.

Payne-Gaposchkin, C., 1957, *The Galactic Novae* (North Holland: Amsterdam).

Peck, M., 1980, *Astrophys. J.*, **238**, 79.

Pellet, A., 1976, *Astron. Astrophys.*, **50**, 421.

Pellet, A., Astier, N., Viale, A., Courtes, G., Maucherat, A., Monnet, G., and Simien, F., 1978, *Astron. Astrophys. Suppl.*, **33**, 441.

Peterson, C. and King, I., 1975, *Astron. J.*, **80**, 427.

Peterson, C., 1978, *Astrophys. J.*, **221**, 80.

Peterson, C., Ford, W., and Rubin, V., 1977, *Astron. J.*, **82**, 32.

Pooley, G., 1969, *Mot. No. R. Soc. Astron.*, **144**, 101.

Pritchet, C. and van den Bergh, S., 1984, *Publ. Astron. Soc. Pac.*, **96**, 804.

Pritchet, C. and van den Bergh, S., 1987, *Astrophys. J.*, **316**, 517.

Pritchet, C. and van den Bergh, S., 1988, *Astrophys. J.*, **331**, 135.

Racine, R., 1967, *Astron. J.*, **72**, 65.

Racine, R., 1991, *Astron. J.*, **101**, 865.

Racine, R. and Shara, M., 1979, *Astron. J.*, **84**, 1694.

Reddish, V., 1962, *Z. für Astrophys.*, **56**, 194.

Renzini, A. and Buzzoni, A., 1986, in *Spectral Evolution of Galaxies*, edited by C. Chiosi and A. Renzini (Reidel: Dordrecht), p. 195.

Rich, R. and Mould, J., 1991, *Astron. J.*, **101**, 1286.

Rich, R., Mould, J., Picard, A., Frogel, J., and Davies, R., 1989, *Astrophys. J.*, **341**, L51.

Richer, H. and Crabtree, D., 1985, *Astrophys. J.*, **298**, L13.

Richer, H., Crabtree, D., and Pritchet, C., 1990, *Astrophys. J.*, **355**, 448.

Rickard, L., 1975, Ph.D. Thesis, University of Illinois.

Richstone, D., Bower, G., and Dressler, A., 1990, *Astrophys. J.*, **353**, 118.

Richter, N., 1971, *Astron. Nachr.*, **292**, 275.

Roberts, M., 1966, *Astrophys. J.*, **144**, 639.

Roberts, M. and Whitehurst, R., 1975, *Astrophys. J.*, **201**, 327.

Rood, H., 1979, *Astrophys. J.*, **232**, 699.

Rosino, L., 1973, *Astron. Astrophys. Suppl.*, **9**, 347.

Rosino, L., Cappacioli, M., D'Onofrio, M., and Della Valle, M., 1989, *Astron. J.*, **97**, 83.

Rubin, V. and Ford, W., 1970, *Astrophys. J.*, **159**, 379.

Rubin, V. and Ford, W., 1971, *Astrophys. J.*, **170**, 25.

Rubin, V., Ford, W., and Kumar, C., 1973, *Astrophys. J.*, **181**, 61.

Rubin, V., Kumar, C., and Ford, W., 1972, *Astrophys. J.*, **177**, 31.

Ruiz, M., 1976, *Astrophys. J.*, **207**, 382.

Ryden, B. and Stark, A., 1986, *Astrophys. J.*, **305**, 823.

Ryle, R., Smith, F., and Elsmore, B., 1950, *Mon. Not. R. Soc. Astron.*, **110**, 508.

Sandage, A., 1983, *Astron. J.*, **88**, 1108.

Sandage, A., Becklin, E., and Neugebauer, G., 1969, *Astrophys. J.*, **157**, 55.

Sandage, A. and Tammann, G., 1974, *Astrophys. J.*, **191**, 603.

Sargent, W., Kowal, S., Hartwick, F., and van den Bergh, S., 1977, *Astron. J.*, **82**, 947.

Sawa, T. and Sofue, Y., 1982, *Pub. Astron. Soc. Japan*, **34**, 189.

Schmidt, K., 1964, *Astron. Nachr.*, **288**, 19.

Schmidt, M., 1957, *Bull. Astron. Inst. Netherlands*, **14**, No. 480, 17.

Schmidt-Kaler, T., 1967, *Astron. J.*, **72**, 526.

Schmidt-Kaler, T. and Schlosser, W., 1973, *Astron. Astrophys.*, **29**, 409.

Schwarzschild, M., 1954, *Astron. J.*, **59**, 273.

Shane, W., 1975, in *La Dynamique des Galaxies Spirales*, edited by L. Weliachew (CNRS: Paris), p. 257.

Shane, W., 1978, in *Structure and Properties of Nearby Galaxies*, IAU Symp. 77, edited by E. Berkhuijsen and R. Wielebinski (Reidel: Dordrecht), p. 180.

Sharov, A., 1973, *Astron. Zh.*, **50**, 263.

Sharov, A., 1975, *Pis'ma Astron. Zh.*, **1**, 3.

Sharov, A., 1981, *Variable Stars.*, **21**, 485.

Sharov, A., 1983, *Soviet Astron. Lett.*, **9**, 208.

Sharov, A., 1984, *Soviet Astron. Lett.*, **10**, 243.

Sharov, A., 1990, *Astron. Zh.*, **67**, 723.

Sharov, A. and Alksnis, A., 1991, *Astrophys. Space Sci.*, **180**, 273.

Sharov, A. and Kholopov, P., 1979, *Variable Stars*, **21**, 219.

Sharov, A. and Lyutyi, V., 1980, *Astron. Zh.*, **57**, 449.

Sharov, A. and Lyutyi, V., 1983, *Astrophys. Space Sci.*, **90**, 371.

Sharov, A. and Lyutyi, V., 1985, *Soviet Astron. Lett.*, **11**, 248.

Sharov, A. and Lyutyi, V., 1989, *Astron. Zh.*, **66**, 241 and 462.

Simien, F., Athanassoula, E., Pellet, A., Monnet, G., Maucherat, A., and Courtes, G., 1978, *Astron. Astrophys.*, **67**, 73.

Sitko, M., 1984, *Astrophys. J.*, **286**, 209.

Sofue, Y. and Kato, T., 1981, *Publ. Astron. Soc. Japan.*, **33**, 449.

Soifer, B., Rice, W., Mould, J., Gillett, F., Robinson, M., and Habing, H., 1986, *Astrophys. J.*, **304**, 651.

Solomon, P. and Zafra, R., 1975, *Astrophys. J.*, **199**, L79.

Spasova, N. Staneva, A., 1984, *Pis'ma Astron. Zh.*, **10**, 275.

Spillar, E., Canterna, R., Carels, K., Benson, J., and Dyck, H., 1990, *Astrophys. J.*, **349**, L13.

Spinrad, H., 1962, *Astrophys. J.*, **135**, 715.

Spinrad, H. and Schweitzer, F., 1972, *Astrophys. J.*, **171**, 403.

Spinrad, H. and Taylor, B., 1971, *Astrophys. J. Suppl.*, **22**, 445.

Stark, A., 1977, *Astrophys. J.*, **213**, 368.

Stark, A., 1979, Ph.D. Thesis, Princeton University.

Stark, A., 1985, in *The Milky Way Galaxy*, edited by H. van Woerden, R. Allen, and B. Burton (Reidel: Dordrecht), p. 445.

Stark, A., Linke, R., and Frerking, M., 1981, *Bull. Amer. Astr. Soc.*, **13**, 535.

Stebbins, J., Whitford, A., and Johnson, H., 1950, *Astrophys. J.*, **112**, 469.

Sullivan, W., 1984, *The Early Years of Radio Astronomy: Reflections of Fifty Years after Jansky's Discovery*, (Cambridge Univ. Press: Cambridge).

Takase, B., 1967, *Publ. Astron. Soc. Japan.*, **19**, 427.

Takase, B. and Kinoshita, H., 1967, *Publ. Astron. Soc. Japan*, **19**, 409.

Taylor, B. and Kellman, S., 1978, *Astrophys. J. Suppl.*, **37**, 101.

Terzian, Y., 1972, in *External Galaxies and Quasi-stellar Objects*, IAU Symp. No. 44, edited by D. Evans (Reidel: Dordrecht), p. 75.

Teuben, P., Turner, E., and Schwarzschild, M., 1985, *Astrophys. J.*, **289**, 58.

Tonry, J., 1987, *Astrophys. J.*, **322**, 932.

Tremaine, S., 1976, *Astrophys. J.*, **203**, 72.

Trinchieri, G. and Fabbiano, G., 1991, *Astrophys. J.*, **382**, 82.

Tripicco, M., 1989, *Astron. J.*, **97**, 735.

Unwin, S., 1980, *Mon. Not. R. Astron. Soc.*, **190**, 551; **192**, 243.

Unwin, S., 1983, *Mon. Not. R. Astron. Soc.*, **205**, 773.

van de Hulst, H., Raimond, E., and van Woerden, H., 1957, *Bull. Astron. Inst. of Netherlands*, **14**, 480.

van den Bergh, S., 1958, *Publ. Astron. Soc. Pac.*, **70**, 109.

van den Bergh, S., 1963, *R.A.S.C. Jour.*, **57**, 63.

van den Bergh, S., 1964a, *Astrophys. J. Suppl.*, **9**, 65.

van den Bergh, S., 1964b, *Astron. J.*, **69**, 610.

van den Bergh, S., 1966, *Astron. J.*, **71**, 219.

van den Bergh, S., 1967, *Astron. J.*, **72**, 70.

van den Bergh, S., 1968, Comm. D. Dunlap Obs. No. 195.

van den Bergh, S., 1969, *Astrophys. J. Suppl.*, **19**, 145.

van den Bergh, S., 1972, *Astrophys. J.*, **171**, L31.

van den Bergh, S., 1975, *Astron. Astrophys.*, **41**, 53.

van den Bergh, S., 1977, in IAU Colloquium 37, *Redshifts and Expansion of the Universe* (CNEA: Paris), p. 13.

van den Bergh, S., 1981a, *Astron. Astrophys. Suppl.*, **46**, 79.

van den Bergh, S., 1981b, *Astron. J.*, **86**, 1464.

van den Bergh, S., 1981c, *Publ. Astron. Soc. Pac.*, **93**, 428.

van den Bergh, S., 1990, *Astrophys. Space Sci.*, **163**, 1.

van den Bergh, S., 1991, *Pub. Astron. Soc. Pac.*, **103**, 1053.

van der Kruit, P., 1972, *Astrophys. Lett.*, **11**, 173.

van Genderen, A., 1973, *Astron. Astrophys.*, **24**, 47.

van Speybroeck, L. and Bechtold, J., 1981, in *X-ray Astronomy with the Einstein Satellite*, edited by R. Giacconi (Reidel: Dordrecht), p. 153.

van Speybroeck, L., Epstein, A., Forman, W., Giacconi, R., Jones, C., Liller, W., and Smarr, L., 1979, *Astrophys. J.*, **234**, L45.

Vetesńik, M., 1962, *Bull. Astron. Inst. Czech.*, **13**, 180 and 218.

Vetesńik, M., 1962, *Bull. Astron. Inst. Czech.*, **13**, 180.

Vogel, S., Boulanger, F., and Ball, R., 1987, *Astrophys. J.*, **321**, L145.

Walker, M., 1974, *Pub. Astron. Soc. Pac.*, **86**, 861.

Walker, M., 1975, *Pub. Astron. Soc. Pac.*, **87**, 479.

Waller, B. and Hodge, P., 1990, in *Dynamics of Galaxies and their Molecular Cloud Distributions*, edited by F. Combes and F. Casoli (Kluwer: Dordrecht), p. 187.

Walterbos, R., 1986, Ph.D. Thesis, University of Leiden.

Walterbos, R. and Braun, R., 1991, *Astron. Astrophys. Suppl.*, in press.

Walterbos, R. and Schwering, P., 1987, *Astron. Astrophys.*, **180**, 27.

Walterbos, R., 1988, in *Galactic and Extragalactic Star Formation*, edited by R. E. Pudritz and M. Fich (Kluwer: Dordrecht), p. 361.

Walterbos, R., Brinks, E., and Shane, W., 1985, *Astron. Astrophys. Suppl.*, **61**, 451.

Walterbos, R. and Gräve, R., 1985, *Astron. Astrophys.*, **150**, L1.

Walterbos, R. and Kennicutt, R., 1987, *Astron. Astrophys. Suppl.*, **69**, 311.

Walterbos, R. and Kennicutt, R., 1988, *Astron. Astrophys.*, **198**, 61.

Welch, G., 1982, *Astrophys. J.*, **259**, 77.

Welch, D., McAlary, C., McLaren, R., and Madore, B., 1986, *Astrophys. J.*, **305**, 583.

Whitford, A., 1977, *Astrophys. J.*, **211**, 527.

Whitmore, B., 1980, *Astrophys. J.*, **242**, 53.

Whitney, C., 1971, *The Discovery of Our Galaxy* (Knopf: New York).
Wirth, A., Smarr, L., Bruno, T., 1985, *Astrophys. J.*, **290**, 140.

INDEX

Lightning Source UK Ltd.
Milton Keynes UK
UKOW031016010513

210028UK00001B/67/P

9 789048 141395